RECENT ADVANCES IN DEVELOPMENTAL BIOLOGY

Recent Advances in Developmental Biology

Vol. 4
Morphogenesis in Animals

A.P. DIWAN
N.K. DHAKAD

ANMOL PUBLICATIONS PVT LTD
New Delhi-110 002

ANMOL PUBLICATIONS PVT LTD
4374/4B, Ansari Road
Daryaganj, New Delhi-110 002

Recent Advances in Developmental Biology

First Edition 1995
ISBN 81-7488-039-9

PRINTED IN INDIA

Published by J.L. Kumar for Anmol Publications Pvt. Ltd. and
Printed at Mehra Offset Press, Delhi.

Preface

The fundamentals of Developmental Biology are presented within the framework of scientific discovery. Researches in Embryology have made almost incredible strides in the past few decades. Consequently, the existing concepts of the developmental biology have expanded. There has been a revolution indeed in this direction.

The text integrates the descriptive, experimental and biochemical approches into a conceptual framework for the analysis of development. All important points are illustrated diagrammatically. The title is not intended to be comprehensive, nor could it be at this length, but concentrates as putting across the basic principles of the subject as briefly and lucidly as possible. It does this with the aid of careful selected examples, some recent and other classic of the field and with numerous illustrations. The aim is to enthuse the reader with this active and exciting area of research and to lay a solid foundation on which further study of its various facts may be based.

Authors

Contents

1

Animal Cells

In all sexually reproducing organisms the development begins as soon as fertilization is completed. The zygote undergoes a series of mitotic divisions to form a number of identical cells. As the development proceeds further the individual cells progressively specialize as they form the functional tissues of the organism's body. Each cell of an embryo contains a nucleus, which is the chief repository of genetic information. The cell's cytoplasm also holds numerous other *organelles* — cellular components specialized for particular functions. Thus, as we shall discover here and in our study of tissue differentiation each functional cell type in an adult organism acquires its own particular complement of organelles arranged in a typical pattern. The precise, genetically determined architecture of a cell specializes it for its function: In other words, how successfully a cell fulfills its role depends on the genes it expresses and the structures it contains.

The Plasma Membrane

The *plasma membrane* is the outer limiting membrane of a cell, it is the barrier that separates the internal cell environment from the outside world. This membrane is highly selective, determining which substances can pass through it in either direction. In animals, cell-cell recognition occurs at the plasma membrane, and the outer membranes of both animal and plant cells contain receptor proteins that enable the cell to perceive hormones and other signals in its environment.

In electron micrographs the plasma membrane is 6 to 10

nm wide and can be seen to have three layers. This multilayer structure is called the *unit membrane*. Through the 1960s the unit membrane was thought of as a sandwich with a filling of lipid molecules covered on either side by a layer of protein.

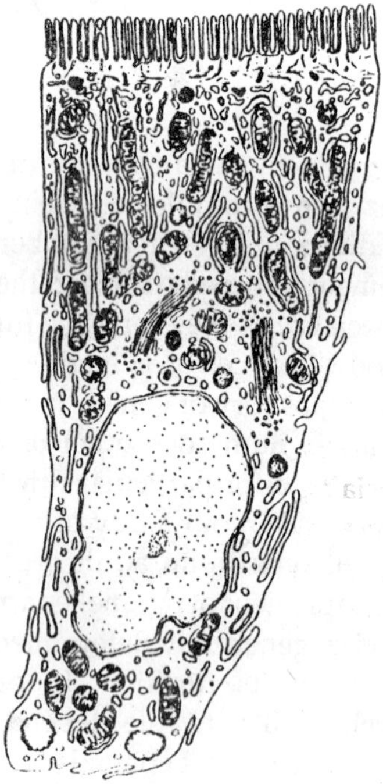

Fig. 1.1. An epithelial cell with microvilli.

However, more recent evidence from freeze-fracture microscopy shows that the lipid center also contains a mosaic of proteins. The proteins are likely buoys in a lipid sea, a conception of membrane organization known as the *fluid mosaic model*. This model stresses the fluid nature of the membrane, which allows proteins to move. Such protein movement can be illustrated by a simple experiment in which two different cells types are fused, each with different membrane protein components that are visible in the light microscope. At fusion, the different components, which serve as *cell surface markers* of the two cells, are separated, with each found on half of the fused cell. After about 30 minutes, however, the markers become mixed as the proteins move around within the now-common cell membrane.

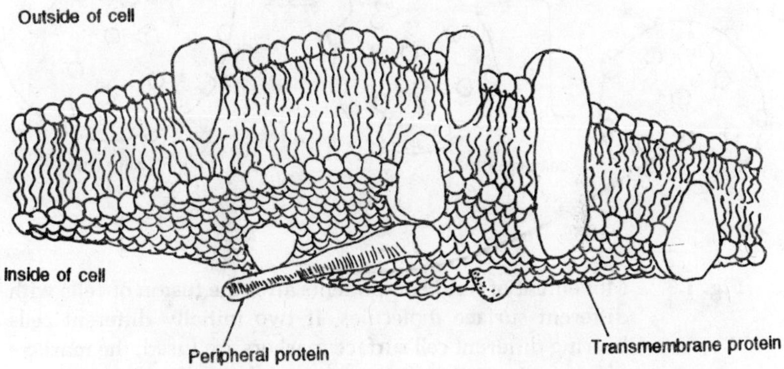

Fig. 1.2. Diagram of the fluid mosaic model of membrane organization. The hydrophobic regions of proteins are restricted to the hydrophobic interior of the membrane. Some proteins are loosely associated with the membrane, which others are deeply embedded in it. Transmembrane proteins and protein complexes span the membrane from the inner to the outer surface.

The plasma membrane of most cells is composed of about 40 per cent lipid, 55 per cent protein, and 5 per cent short chains of sugar molecules (oligosaccharide and carbohydrate). This composition can vary in specific cell types, however; for example, *Schwann cells,* which wrap their cell membranes around peripheral nerve fibers, have a plasma membrane composed of almost 80 per cent lipid and only 20 per cent protein. Accordingly, the freeze fracture images of Schwann cell membranes have few protein "bumps," a fact reflecting their low content of protein relative to lipid. This high lipid content makes the Schwann cell membrane an excellent insulator, helping to preserve the electrical energy of the nerve cell as it is propagated through the tissues of the body.

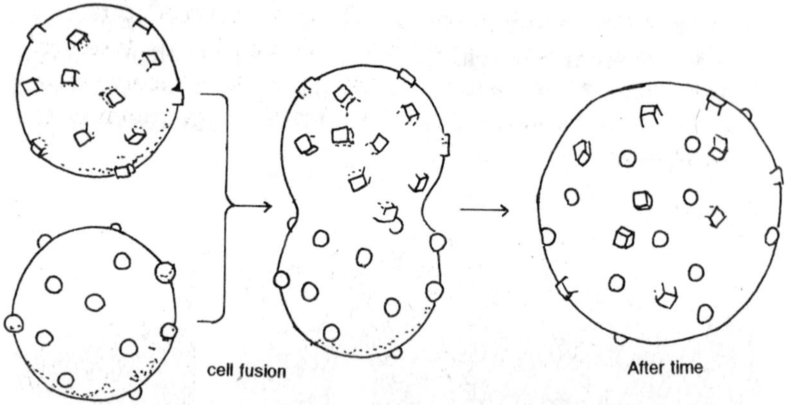

Fig. 1.3. Movement of fluid components after the fusion of cells with different surface molecules. If two initially different cells bearing different cell surface markers are fused, the markers intermingle on the surface of the fused cell within an hour.

The major lipid molecules found in cell membranes are *phospholipids, cholesterol* and *galactolipids*. The exact amount of any of these fatty constituents vary among cell types; plant cells have virtually no cholesterol. Protein constituents of cell membranes are divided into two classes, based on the ease with which they can be extracted from the membrane. Thus, *peripheral membrane proteins*, or *extrinsic membrane proteins*, can be easily washed off the membrane; *intrinsic membrane proteins* can only be removed by disrupting the membrane with detergents. The intrinsic proteins lie deep within the membrane and are often linked to oligosaccharides (forming *glycoproteins*) or to phospholipids (forming *lipoproteins*). In general, membrane proteins have been *hydrophobic* ("water loving") and *hydrophobic* ("water hating") domains, and the precise position of a protein molecule within a membrane depends on the shape and size of these domains and their interaction with the hydrophobic, lipid-rich interior of the membrane and the aqueous interface with the cytoplasm or cell exterior. Thus, if a protein or protein complex has a charged surface — that is, it is primarily hydrophilic — it will not sit deep within the hydrophobic lipid region of the membrane and is likely to be extrinsic. Conversely, primarily hydrophobic proteins are intrinsic, lying among the lipid molecules.

Movement of Materials

Substances can move through the plasma membrane by several methods, including passive diffusion, facilitated diffusion, active transport, endocytosis, and exocytosis. All of these are involved in the *selective transport* of molecules across the plasma membrane; these methods differ in how efficiently they move molecules and whether cellular energy is required.

Diffusion is the movement of molecules down a concentration gradient to equalize their concentration; diffusion is a passive process, requiring no energy from the cell, and is selective only in that some molecules move through the plasma membrane readily and others do not. Most molecules move through the membrane in tiny, protein-lined *pores* extending through the lipid bilayer. Small non polar substances such as oxygen and carbon dioxide can diffuse freely across the bilayer membrane structure, but most charged molecules cannot. The other selective aspect of diffusion is that, in general, the rate of diffusion depends on the size of the molecule: Smaller molecules move more quickly. However, many large lipophilic molecules can diffuse into the membrane easily.

Facilitated diffusion is a selective process of molecule movement through specialized membrane pores lined with proteins. Two types of pores participate in this process; some, lined with charged proteins, allow the movement of specific ions, such as Ca^{2+} or Na^{+}, other pores lined with uncharged proteins allow lipid-insoluble, uncharged molecules to pass, such as urea and glycerol. Pores exert additional specificity when *receptor proteins*, or *carrier proteins*, on the outside surface of the cell bind specific molecules and then transport them to the pores to facilitate the diffusion of these molecules through the membrane.

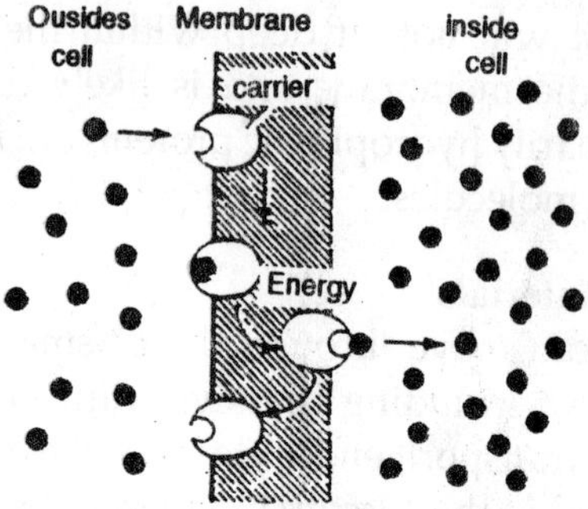

Fig. 1.4. Showing translocation across the Membrane.

Although facilitated diffusion allows cells to move molecules along a concentration gradient selectively, cells often expel or take up molecules, particularly salts, against a concentration gradient. Such transport of materials against a gradient requires *active transport,* because energy is required to drive the pumping mechanism that moves substances into and out of the cell. Active transport takes place on carrier proteins that traverse the plasma membrane; these proteins are *membrane-bound ATPases,* which use energy generated by the hydrolysis of ATP to transport ions and other small molecules. All carrier proteins are enzymes with specific substrates; for example, there are separate carriers for Na^+ and Ca^{2+} ions. In both facilitated diffusion and active transport, one important principle operates: Molecules transported into or out of cells move along a continuous protein pathway. Carrier proteins traverse the membrane to create a *specific channel* for each type of solute.

Specialization of the plasma membrane allow the cells to take up and release particles too large to diffuse or to be actively transported. For example, the membrane engulfs viruses, dust, and food particles in the process of *endocytosis.* This activity is common in macrophages, the scavenger cells of higher animals. Once a bubble of plasma membrane takes up a particle, it becomes a cytoplasmic vesicle by pinching off from the main membrane. As will be described later, these vesicles usually fuse to a lysosome, where the incoming material is enzymatically digested. The rate at which particles can be ingested by endocytosis can be very rapid; between 2 per cent and 20 per cent of the cell surface of an active macrophage can be internalized in one minute.

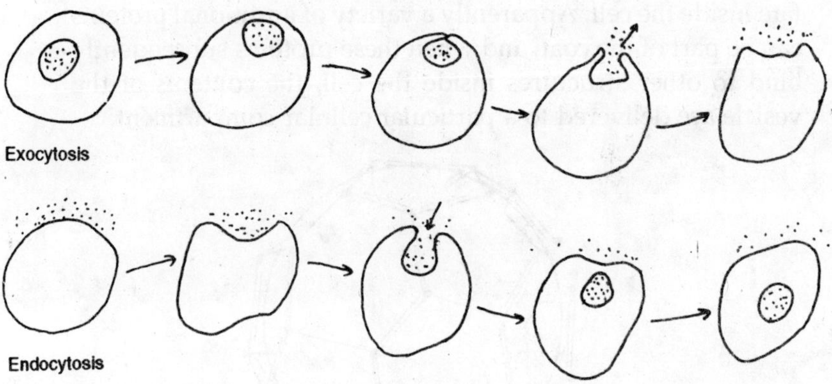

Fig. 1.5. Endocytosis and exocytosis. The uptake of materials involves the formation of single, membrane-bound vesicles in the cytoplasm after endocytosis. Exocytosis involves fusion of a cytoplasmic vesicle with the cell surface.

Specificity in endocytosis is made possible by specific cell receptors that allow cells to concentrate molecules from dilute surroundings without ingesting a large amount of fluid. When a receptor binds its target molecule, termed the *ligand*, the ligand-receptor complex migrates to a specialized region of the cell surface known as a *coated pit*. The pit is a depression in the cell surface; on the cytoplasmic side of the plasma membrane, a layer of protein coats the complex, and these coated complexes are rapidly budded off as vesicles into the cytoplasm. The coated vesicles have a remarkable structure composed of at least two proteins. One of them is a large, fibrous protein known as *clathrin.* This protein forms a quaternary complex called a *triskelion,* which contains three clathrin polypeptides and three smaller polypeptides; several dozen triskelions then link together to form a polyhedral, latticelike structure around each vesicle.

One of the most interesting areas of research in cell biology is discovering how individual coated vesicles acquire a specific fate inside the cell. Apparently a variety of additional proteins can be part of the coat, and when these proteins subsequently bind to other structures inside the cell, the contents of the vesicle are delivered to a particular cellular compartment.

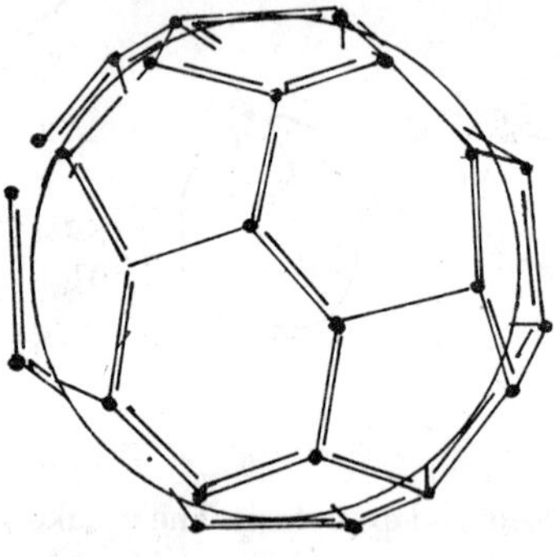

Fig. 1.6. Triskelion building blocks are composed of two protein molecules, one long clathrin chain and a shorter polypeptide. The clathrin ends hooks together to form a hexagonal lattice around vesicles, thus creating a coated surface.

Exocytosis, the reverse of endocytosis, is the process by which cells secrete many substances. Small membrane vesicles packed with the secretion are formed in the cytoplasm; these vesicles then fuse with the plasma membrane, invert, and release their contents outside the cell. Active exocytosis is a hallmark of secretory cells, such as gland cells producing a hormone or other product that is active outside the cell.

The Glycocalyx

In many cells, the outer surface of the plasma membrane is coated with a layer of molecules. This outer membrane coat is called the *glycocalyx,* which forms a fuzzy surface composed of the oligosaccharide side chains of the glycolipids and glycoproteins of the membrane and other proteins. The glycocalyx is hypothesized to perform an essential role in cell-cell recognition, because its fuzzy coat is the first site of contact when two cells meet; because this coat can be very sticky, cells may adhere tightly to other cells or to an artificial substrate such as laboratory petridish. Because different types of cells have different molecules in the glycocalyx, some cell types stick to each other while other cells glide by. Cells that migrate to new locations during development, such as nerve cells and cells derived from the neural crest, make and break thousands of *cell contacts* before reaching their final destination. The nature of the molecules that guide these migratory cells is only beginning to be understood, but it is likely that oligosaccharides and proteins in the glycocalyx are the signposts that make such activities possible.

Cell Organelles

The *cytoplasm* is the region of the cell between the plasma membrane and the nucleus. Within it a variety of membrane-bound organelles serve as discrete compartments that carry out specialized functions. In this section we shall consider the class of organelles with just one limiting membrane, leaving the organelles bounded by a double membrane for a later section.

The numerous organelles limited by a single membrane synthesize and transport materials around and out of the cell and isolate or store materials within it. Each type of cytoplasmic membrane has its own characteristic set of enzymes and likewise its own characteristic assortment of lipid components. We shall first consider one of these single-membrane organelles, the endoplasmic reticulum.

The Endoplasmic Reticulum is the Site Where Lipids and Some Proteins are Synthesized

The *endoplasmic reticulum (ER)* is a membrane system found in both plant and animal cells. In electron micrographs it appears either as a system of tubules or as flattened sacs surrounding cavities called *cisternae*. The ER is divided into two types, rough and smooth, according to appearance.

The *rough endoplasmic reticulum (RER)* is so named because the surface facing the cytoplasm is studded with numerous protein-synthesizing ribosomes. Both *replacement membrane proteins* and *proteins destined for export* from the cell are synthesized on the RER. These two classes of proteins are separated from each other as they are synthesized; sthose destined for export pass through the RER membrane into the cisternae, whereas replacement membrane proteins are folded and embedded into the RER membrane as they are generated. Not surprisingly, an extensive RER is a hallmark of secretory cells, because it is the site where the exported proteins are manufactured.

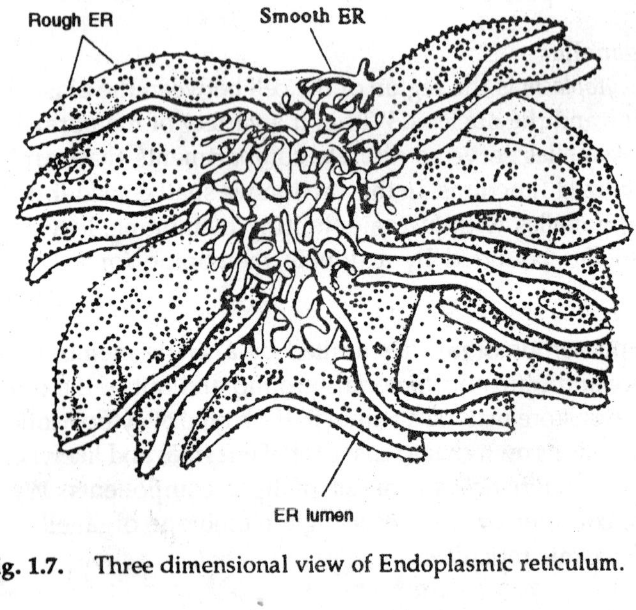

Fig. 1.7. Three dimensional view of Endoplasmic reticulum.

The *smooth endoplasmic reticulum (SER)* has no ribosomes but is otherwise similar in structure to the RER. Complex organic molecules such as cholesterol and the steroid hormones are synthesized here, along with membrane lipids constructed by SER enzymes from fatty acids and glycerol molecules that originate elsewhere in the cell. Newly synthesized proteins in the cisternae of the RER enter the SER through connections between the two membrane compartments. In the SER, lipids and some of the proteins are combined to make lipoprotein components of cell membranes. Proteins to be exported from the ER system are packaged in small membrane bags known as *transition vesicles*, which pinch off continuously from the rough and smooth ER and move to the Golgi apparatus.

The Golgi Apparatus

Transition vesicles from the ER system move through the cytoplasm and fuse with each other to form the *Golgi apparatus*, also known as the *dictyosome*. In electron micrographs the Golgi is seen as a series of flattened, convex membrane discs with large discs on one edge and vesicles at the opposite edge; the discs are connected by membrane channels. The Golgi apparatus is polarized, and the two edges perform different functions. Thus, on the one hand the large Golgi discs fuse with function vesicles; after the vesicle contents are processed within the Golgi, they are packaged for release down the line within smaller vesicles. These vesicles carry molecular tags that determine their destination.

The materials within the Golgi discs are destined for either export from the cell or insertion into cell membranes. Within the apparatus carbohydrates are synthesized, some of which are then attached to membrane proteins to form glycoproteins; many glycoproteins are destined for export from the cell to form the glycocalyx and other secretory products. Material destined for secretion leaves the Golgi packaged in secretory vesicles that move toward the cell surface, fuse with the plasma membrane, and release their contents by exocytosis. Other classes of small vesicles transport proteins from the Golgi to other cellular membranes where the vesicle contents enter the new membrane.

One of the most exciting areas of research in cell biology is the effort to determine how the traffic of membrane-bound vesicles is organized in the cell. How do vesicles know to fuse with a particular membrane? The proper sorting of membrane-bound organelles bobbing around in the cytoplasm requires that each organelle have specific surface structures so that it can interact and fuse with just one other membrane type. Just as molecular recognition mediates cell-cell interaction, the numerous collisions of organelles will result in fusion only if each organelle recognizes the appropriate tags on the other. Much current research is testing whether these tags are the proteins on the surface of each vesicle type.

Lysosomes

Lysosomes are small, round vesicles that contain battery of powerful *hydrolytic enzymes* that can break down cellular constituents Lysosomes destroy foreign matter, such as bacteria, brought into the cell by endocytosis. The lysosomal membrane fuses with the membrane of the endocytotic vesicle, dumping the hydrolytic enzymes into it; these enzymes then digest the vesicle's contents. Lysosomes also digest other cell organelles as necessary to make way for changes in the intracellular structure, but organelles are consumed in this way only when they are encased in an extra layer of membrane produced by the ER or Golgi. Under normal conditions, however, lysosomes do not fuse with either the ER or Golgi membrane, and the nature of the signal that controls lysosomal fusion with the extra membrane around organelles destined for destruction is not known.

Lysosomal enzymes also destroy debris from dead cells: When a cell dies, the lysosomes within it burst, digesting the cell contents in a process called *autolysis*. As we shall see throughout this book, such programmed cell destruction is an important aspect of development, for example, it plays a critical role in the metamorphosis of insect larvae to adult forms and of amphibian tadpoles to adult frogs and foads.

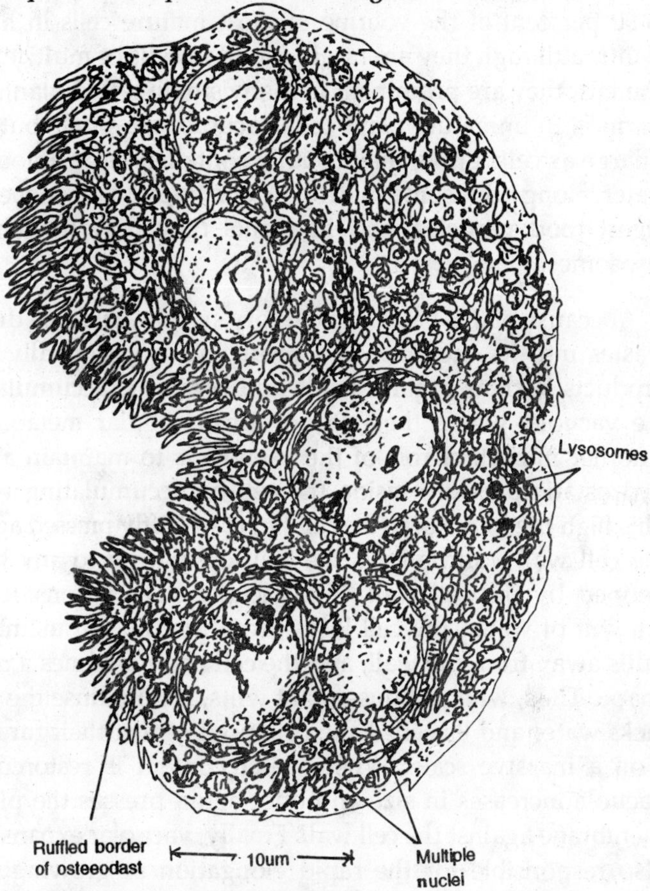

Fig. 1.8. Lysosomes in an Osteoclast.

Several important heritable diseases in which particular substances accumulate in cells arise from a deficiency in the lysosomal enzyme which normally degrades that metabolite. For instance, in Tay-Sachs disease, a complex brain lipid known as ganglioside GM2 accumulates as the result of a deficiency in the lysosomal enzyme N-acetyl-glucosaminidase A. The buildup of the ganglioside impairs brain function in afflicted infants and ultimately result in death.

Plant Lysosomes

Vacuoles are round, fluid-filled vesicles that constitute up to 90 per cent of the volume of most mature cells in higher plants although they are rarely found in cells of multicellular animals, they are prominent in many single-celled plants and animals. In immature plant cells vacuoles are small, but they enlarge as cells grow; they usually contain a large amount of water, along with some hydrolytic enzymes. In amoebae they digest food particles, and in plants they may function as lysosomes.

Because plants have no excretory system, each cell dumps wastes into its vacuole. As a result, many potentially toxic products such as organic acids and pigments accumulate in the vacuole, where they do not affect cellular metabolism. Another important role of the vacuole is to maintain a high hydrostatic pressure inside the cell by accumulating water. This high pressure keeps the cell *turgid* — stiffly pressed against the cell wall - and allows the cell to conform to any shape defined by the cell wall. If a vacuole loses a considerable amount of water, the cell loses turgidity, the cell membrane pulls away from the wall, and the cell body assumes a round shape. Thus, when a house plant wilts, it is because the plant lacks water and its cells have collapsed — lost their turgidity - on a massive scale. When water content is restored, the vacuole increases in size and once again presses the plasma membrane against the cell wall. Finally, vacuolar expansion is also responsible for the rapid elongation of growing plant cells. This expansion is evident in the elongation of roots into the soil during seed germination.

Mitochondria and Plastids

The double — membrane — bound *mitochondria* found in all eukaryotic cells and the similarly structured *plastids* found in all plant cells are the only cellular organelles other than the nucleus that contain their own DNA and machinery for synthesizing protein. However, these organelles are only semiautonomous; although some of the proteins within them are coded for by genes contained in the organelles themselves, most are coded for by genes in the nucleus. As a result, the genomes of the organelle and nucleus must co operate to build the organelle. A prime example involves ribulose bisphospate carboxylase, the enzyme responsible for fixating carbon in photosynthesis. This enzyme contains eight large and eight small polypeptide subunits. The gene for the large subunit is encoded in the plastid genome, and the protein is synthesized on plastid ribosomes; the small subunit is encoded by a nuclear gene, and its protein is synthesized on cytoplasmic ribosomes and then transported into the organelle.

In addition to possessing a semiautonomous genetic system, mitochondria and plastids are also *self-replicating cellular compartments*. This is, they multiply independently of mitosis and cell division, so their number can vary from just a few to hundreds or even thousands per cell. *Most important, the number and structure of these organelles is a characteristic of individual cell types.*

Mitochondria Supply Cellular Energy

Mitochondria, the cellular compartment in which the energy gained from the oxidation of sugars in converted into ATP, have a double membrane surrounding an internal *matrix*. The outer membrane is smooth, but the inner membrane folds and extends into the matrix; these extensions, the *cristae*, greatly increase the surface area of the inner membrane. Embedded in the cristae are the *electron transport components* and *ADP phosphorylation complexes* responsible for converting ADP to ATP in cellular respiration. Each of these complexes is composed of many protein subunits, and like the carboxylating enzyme of plastics, some of the protein subunits are mitochondrial gene products while others are coded for by nuclear genes. The mitochondrial DNA and the enzymes responsible for carbon oxidation are found in the matrix.

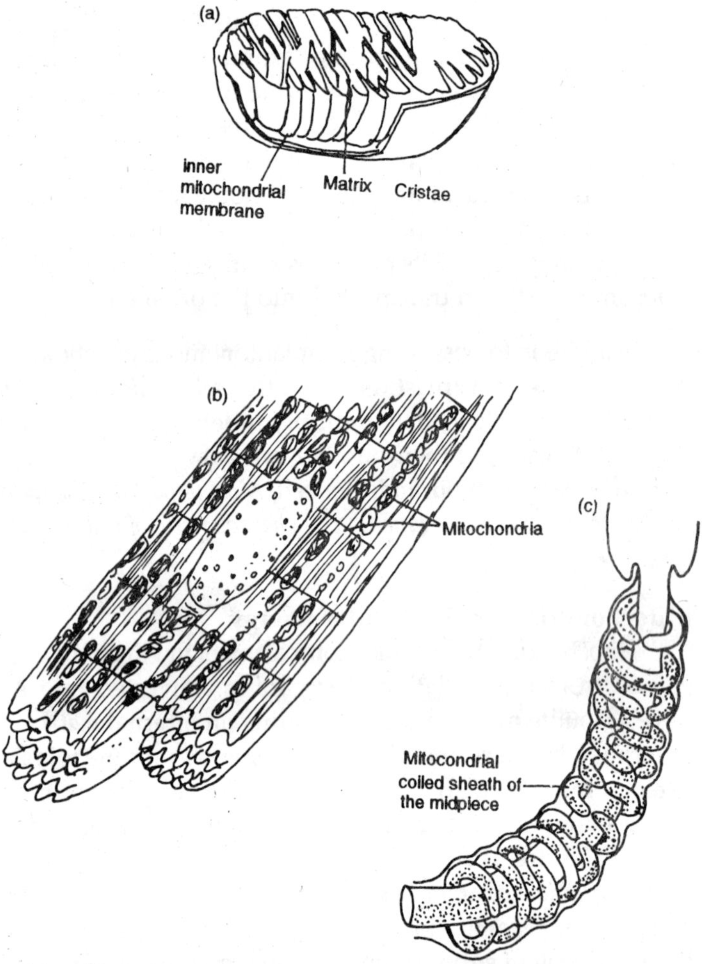

Fig. 1.9. Variation in mitochondrial morphology, (a) Cross-sectional view of bat pancreas cell mitochondrion, (b) Longitudinal section of bat cardiac muscle. The mitochondria are numerous and aligned in columns between the myofibrils. (c) A human sperm tail. The mitochondria are coiled to form a spiral starting just behind the sperm head.

The shape and number of mitochondria and their distribution in the cell are tailored for various cell types. For example, small, round, or sausage-shaped mitochondria dispersed throughout the cels are typical of many cell types in contrast, cardiac muscle, which has a very high energy requirement, has mitochondria lined up in columns between the muscle fibers. These numerous mitochondria supply the energy for a constantly beating heart and are spatially arrayed for the efficient production of energy to support continual muscle action. By contrast, the mitochondria in the midpiece and tail of the sperm cell in many animal species are long and coiled like the strands of a rope these mitochondria supply ATP to dynein, and when this ATP is hydrolyzed, the released energy drives the microtubule sliding that creates the whiplike propulsion of the sperm tail.

The Plastids of Plants

In addition to the mitochondria found in all eukaryotic cells, plant cells also contain plastids. The best-known type of plastid is the *chloroplast*, found in green leaves where photosynthesis occurs. The chloroplast has a double membrane surrounding an internal space called the *stroma*. Sheets of inner membrane called *lamellae* protrude into the stroma, and the lamellae in turn often stack up to form *grana*. Each layer of a grana stack is called a *thylakoid*. The lamellar membrane system surrounds the innermost compartment of the plastid, the cisternal space. The lamellar membrane contains the electron transport apparatus, ADP phosphorylation complexes and the proteins binding chlorophyll, the green pigment that captures light energy. Within the stroma are the plastid DNA and the enzymes of the *Calvin cycle*, which are required to fix carbon dioxide into sugar. The organization of the chloroplast is more complex than that of the mitochondrion, but in both structures electron transport and ATPase activity occur on a membrane, while sugar metabolism takes place in the "cytoplasm" of the organelle.

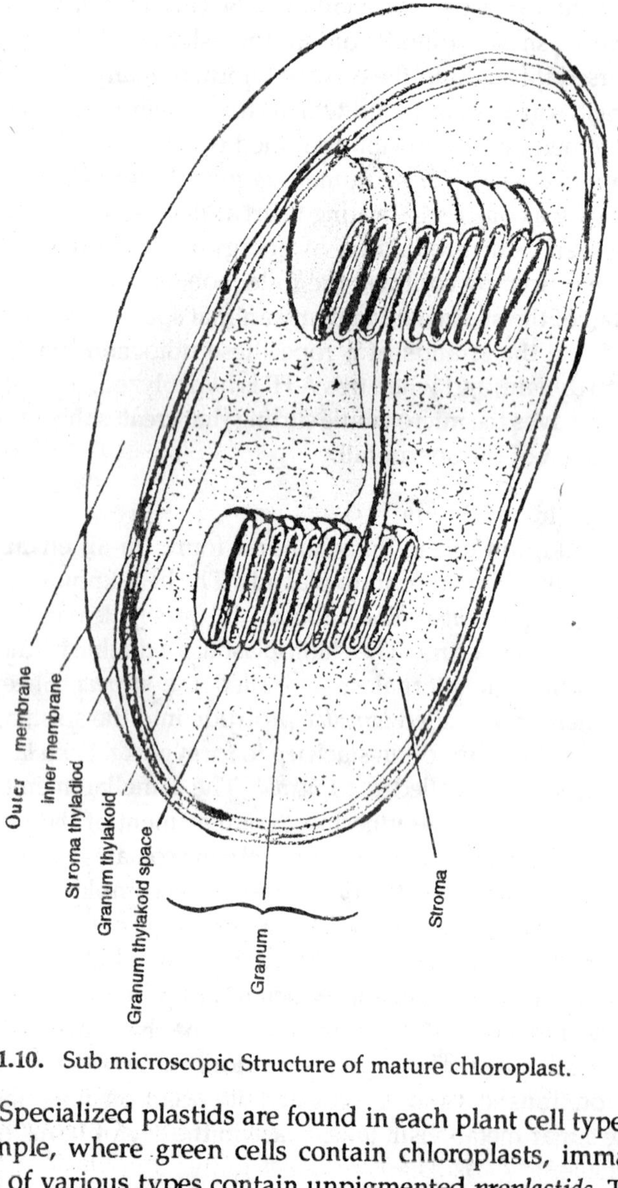

Fig. 1.10. Sub microscopic Structure of mature chloroplast.

Specialized plastids are found in each plant cell type. For example, where green cells contain chloroplasts, immature cells of various types contain unpigmented *proplastids*. These contain only a simple, unfolded inner membrane, but in cells destined to become leaf cells, proplastids enlarge, accumulate chlorophyll, and develop an extensive lamellar system typical of a chloroplast. In petals the proplastids usually do not become green and instead store pigments, such as yellow carotenoids or purple anthocyanins; ultimately they develop into *chromoplasts*, the plastids that give flower petals their colour. In the cells of the potato, the proplastids differentiate to starch-storing *amyloplasts* rather than chloroplasts.

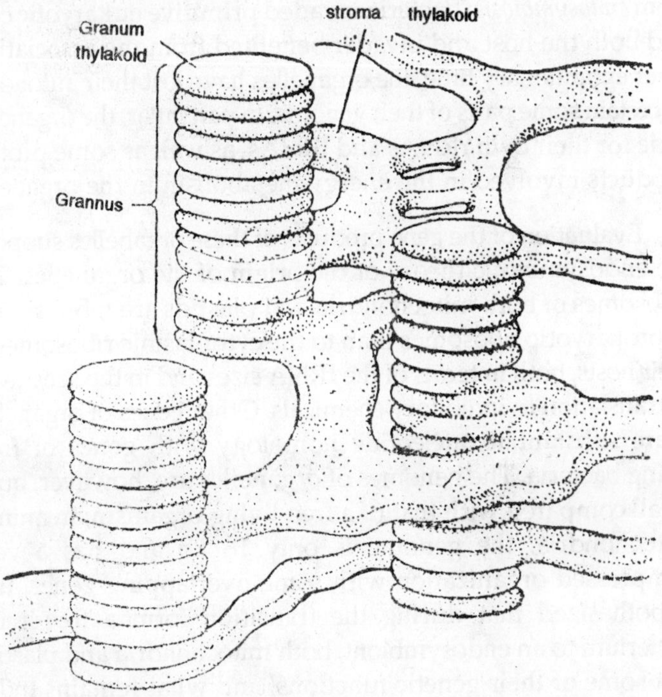

Fig. 1.11. Diagram showing the three dimensional arrangement and inner relationships of grana.

It has already been seen that many plant cells can readily dedifferentiate and generate new structures by re-forming a meristem. Thus plastids must also be able to dedifferentiate back to proplastids; for example, chloroplasts of a leaf explant must return to the proplastid state in the forming meristem and then be able to redifferentiate to all the other plastid types. Hence the state of differentiation of a particular plastid is reversible, rather than being the result of a terminal developmental process. Like the mitochondrion, plastid type depends on cell type and hence is ultimately controlled by the genes in the nucleus of a plant cell.

Symbiotic Nature of Mitochondria and Plastids

The likeliest origin of the DNA-containing organelles of is from *endosymbiosis*: Bacteria invaded primitive eukaryotic cells and both the host and invader benefited from the association. Over evolutionary time, the organelles have lost their autonomy but retain some parts of their genome; in particular, the organelles code for their own rRNAs and tRNAs, as well as some protein products involved in the energy metabolism in the organelle.

Evaluation of the gene products of these organelles supports the endosymbiotic theory of the origin of the organelles. The ribosomes of both mitochondria and plastids are more similar to prokaryotic ribosomes than to the cytoplasmic ribosomes of their hosts, both in terms of the rRNA sizes and in the sensitivity of translation to inhibitory chemicals. Other genes of organelles share substantial nucleotide homology with genes of free-living bacteria. The genomes of organelles are, however, quite small compared with that of a free-living organism; in animal mitochondria, the genome is only 16 kb and has a very compressed organization with some overlapping genes. It is hypothesized that during the transition form a free-living bacterium to an endosymbiont, both mitochondria and plastids lost some of their genetic functions, and what remains today is merely a remnant of the original genomes.

The Nucleus: Control Center of The Cell

The *nucleus*, the defining feature of eukaryotes, is the most conspicuous organelle of most cells. It is surrounded by a double membrane called the *nuclear envelope*. The outer membrane of the envelope is often continuous with the ER of the cytoplasm. In places the inner and outer nuclear membranes are interrupted to form *nuclear pores.* 70 to 75 nm in diameter, and these pores serve as channels of communication between the nucleus and cytoplasm Nuclear pores are selective about what moves through them, however; for instance, one *ribonucleoprotein particles* of newly synthesized mRNA and ribosomal subunits leave the nucleus, they cannot re-enter.

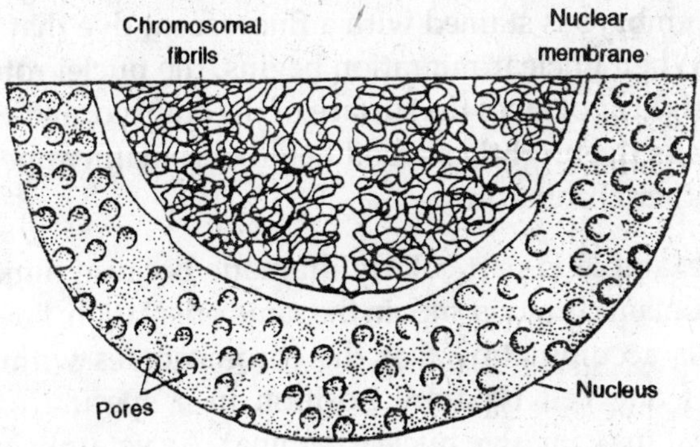

Fig. 1.12. Nucleus with the chromosomal fibrils and nuclear pores.

Major Components of the Nucleus

The interior of the nucleus has two major components: the *nucleoplasm* and the chromatin. Nucleoplasm is the matrix of the nucleus and contains enzymes important in the processing of RNA to make functional mRNAs A prominent feature of most nuclei is the *nucleolus,* the site where recently synthesized rRNA and protein is packaged into ribonucleoprotein bodies for transport to the cytoplasm. A third component of the nucleus is a protein network that lines the nucleoplasmic surface of the nuclear envelope, the *nuclear lamina,* which is composed of intermediate-like filaments built from polymers of proteins called *lamins.*

Variation in Nuclear Structure

Numerous variations in unclear morphology are possible. In addition to changes in unclear shape that distinguish one cell type from another, the individual chromosomes in cell nuclei appear to be held in different three-dimensional arrays specific to particular cell types. This positioning of the chromosomes is best illustrated by the precise orientation of the chromatin within the migrating nuclei of a *Drosophila* embryo. Early in development the embryo is a single, multinucleate cell with the nuclei near the center; these nuclei then migrate to the cell surface. Each nucleus has a single, large, heterochromatic area that is readily visualized as a bright spot when the embryo is stained with a fluorescent dye that binds to DNA. When nuclear migration begins, the nuclei rotate so that the bright spot faces the surface of the embryo, and within each nucleus the brightly stained heterochromatin appears to lead' the other chromatin.

This example is particularly dramatic because hundreds of nuclei contain chromosomes in the same orientation; however, evidence is accumulating that the chromosomes within the nucleus of any cell type are position in a specific way (a process that may involve nuclear lamina). As yet unknown is whether—and if so, how—such positioning of the chromosome influences gene expression.

Extracellular Matrices : The Cellular Landscape

Cells are characterized not only by their internal organelles but also by specific extracellular materials, collectively called the *matrix*. Matrix components are produced inside the cell and transported outside, where they are assembled into the matrix structure. This structure varies with the function of the associated cell: for instance, the cells of the connective tissues such as cartilage (discussed in the next chapter) support body tissues and produce a particularly abundant form oF extracellular matrix component called *collagen*. We shall briefly consider the collagen molecular complex as an example of an extracellular matrix constituent.

Collagen, making up as much as 6 per cent of human body weight, is the single most abundant protein in the human body. It provides tissue support in the adult and also molds tissues during development. There are numerous forms of the collagen molecule, and cells in different tissues or cells at different developmental stages in the same tissue can produce different molecular forms.

In general, collagen is an assemblage of polypeptides. The basic building blocks of this multimeric structure are the *alpha collagen polypeptides*, proteins coded by various members of a multigene family. All of these proteins share an unusual amino acid composition: One-third of the amino acid residues are glycine, one-third proline or hydroxyproline, and only one-third all the other amino acids. The primary amino acid sequence forces the molecule to assume a stiff, helical structure; three of these strands then intertwine to make a helical *tropocollagen* molecule, which is held together by hydrogen bonding among the alpha polypeptides. The *collagen fibril* visible in an electron micrograph is formed by the head-to-tail and side-by-side association of numerous tropocollagen molecules to make a *supramolecular array*. This molecular structure of long, woven strands (like a rope) gives collagen two qualities vital for tissues - flexibility for body movement and strength for support.

The composition of the alpha chains in the tropocollagen molecules varies in different forms of collagen. These assorted forms are coded by diverse genes and a subset of types accumulates in a range of cell types to serve different functions. Other proteins associate with collagen to tailor the extracellular matrix for specific functions; three of these proteins are *elastin, fibronectin,* and *laminin.* Elastin, as its name implies, allows tissues to stretch and then reassume their form. Fore example, the walls of the large arterial blood vessels, which expand and contract as the heart pumps blood through them, contain many elastin fibers. Fibronectin has attracted considerable attention in recent years because of its wide distribution in numerous body tissue types and in cells in tissue culture. Different embryonic tissues produce fibronectin at specific times during development, and it may provide a sticky substrate that cells use to crawl from one place to another.

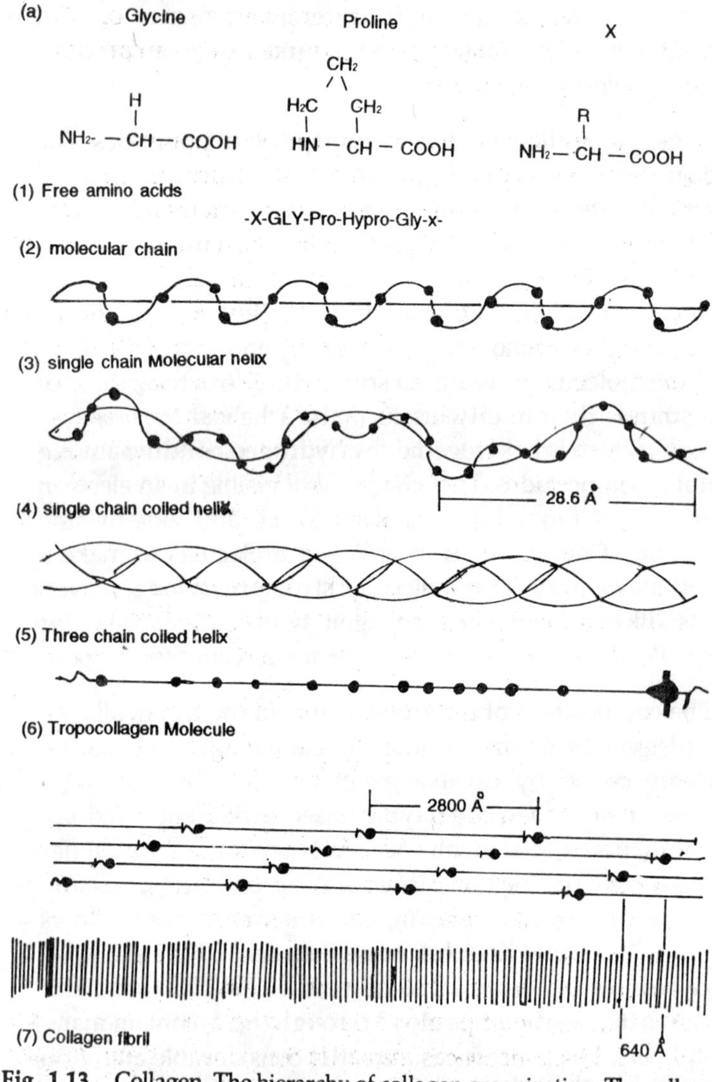

Fig. 1.13. Collagen. The hierarchy of collagen organization. The collagen fibril is built of numerous collagen polypeptide chains. These chains associate in groups of three of form the tropocollagen molecule. These molecules associate in a head-to-tail fashion to form long tropocollagen fibers. The fibers then associate side by side at a precise displacement of 0 nm, and this spacing creates the banded appearance of collagen fibers observed under the microscope.

2

Morphological Changes in Cell Culture

The history and evolution of cell populations *in vitro* stem initially from characteristics of the tissues or organs used for explantation. As developmental products, each of these carries a spectrum of identifying markers: distinctive patterns of enzyme activities or unique metabolic products, diagnostic cytoarchitecture or elements of fine structure, and other specificities recognizable microscopically or biochemically. Phenotypic changes in early cultures of animal cells thus commonly derive from forward or retrograde shifts in state of differentiation, rather than from alterations at the genetic level. Progressive differentiation may in some instances continue to occur *in vitro*, particularly when intact embryonic rudiments are isolated from the body as a whole. In a classic study. Fell and Robison (1929) showed that the precursor of the chick embryonic femur is capable of progressive development if explanted to the surface of a plasma clot. Histotypically normal bone arose from cells of the initial primordium; while in external configuration, the head, trochanter, and condyles of the normal femur appeared in clearly recognizable form. Conditions in culture rarely permit progressive morphogenesis to this degree, but it is noteworthy that even when cells from embryonic rudiments are physically dissociated by tryptic digestion the isolated cells can still reassociate and reconstitute normal tissue fabrics. The potential for re-establishment of tissue fabrics is type specific, as was shown by constructing aggregates form cell suspensions of two different tissue types, e.g., chick mesonephros and cartilage cells from wing bud mesenchyme. Within such aggregates, type-specific reassortment occurs, with epithelial tubules formed from kidney cells and cartilage masses from chondrogenic cells.

Phenotypic shifts are more often regressive when populations of fully differentiated cells are placed in culture. Disappearance or reduction in expression of cell-type specific markers is in fact commonly observed for most cells in conventional monolayer cultures. Distinctions between related cell types become difficult to make after one or more serial passages, and the cells in culture assume more generalized patterns. Thus, patches of a simple pavement epithelium appear in cultures of kidney or other epithelial organs, in lieu of the distinctly different subtypes apparent *in vivo*. Similarly, stellate or spindlelike fibroblasts arise from diverse sources, such as cartilage, muscle, linings of blood vessels, and nerve sheaths, as well as from connective tissue per se. To early investigators, this convergence to simpler growth forms *in vitro* suggested reversion to a more generalized embryonic state. But true dedifferentiation in the sense of regaining broad developmental potentials has never been demonstrated in cultures of animal cells, although limited conversion between cell types (transdifferentiation) does occur. In other cases, there is a firm commitment to differentiation, even though the phenotypic expression of cell type-specific features may decline. Despite the apparent loss of differentiated properties as a result of these regressive shifts, it is often potentially reversible, and re-expression of specialized features may occur under altered culture conditions. Reversible transitions of this sort in cellular phenotype, without change in basic commitment, come under the heading of modulation as defined by Weiss and may occur in cultures of both normal and tumor cells.

Phenotypic change in animal cells takes on a new dimension when populations are maintained for prolonged periods in serial culture. Early investigators hoped to isolate pure strains of each cell type, which could then be subcultivated indefinitely *in vitro* without change, like cultures of microorganisms. Initially, such efforts did meet with some success, e.g., pure cultures of chick iris epithelium chick monocytes, and thyroid epithelium. These primary isolates from fresh tissue could not, however, be maintained indefinitely in their original form. Eventually it became clear that primary populations of cells taken from the intact organism will eventually die out in serial culture, or else give rise to variants that deviate from the normal precursor cell types. These two patterns can be exemplified by the fate of cells in cultures isolated from human fetal tissues and mouse embryos respectively. The classic studies of Hayflick and Moorhead (1961) showed, for example, that fibroblasts from human tissues proliferate for only about 50 serial passages when divided at each transfer on a 1:1 basis. Limited viability in this sense becomes apparent as population growth in such cultures slows to a standstill, debris accumulates, and further cell propagation cannot be obtained. By contrast, cells derived from mouse tissues do not show limited viability but instead undergo transformation at the population level to form established or permanent cell lines. In this process, freshly explanted mouse cells proliferate luxuriantly for a few weeks and then go into a sharp growth decline. This static state, however, is temporary and can be reversed if the population is maintained with periodic fluid changes. Proliferation begins again and, after a few more weeks, the population of mouse cells is multiplying as vigorously as before. The cells can then be subcultivated indefinitely by serial transfer as an established or permanent line. The emergence of such lines can be viewed as a mass progression or transformation, in which the original diploid cells become aneuploid, marked changes in growth properties occur, and cells may also alter in morphology and other features,, including the ability to produce tumors in many cases if injected into the hosts of origin. Cells from a number of other species tend to follow the sequence described for mouse cells: cultures from chick embryos show limited viability, as do those from human tissues. combinations of these patterns may also be seen, but in no case do cells taken directly from the intact organism proliferate indefinitely in unmodified form.

Evolutionary and morphogenetic changes may thus condition the properties of animal cells *in vitro,* but heritable changes in phenotype can also stem from discrete events within single cells. A classic problem here, faced earlier with microorganisms, is whether a given change in cellular phenotype results from adaptation within the population as a whole, or by spontaneous variation in one or a few cells, which then spreads by selection to characterize the entire system. In the case of bacteria, the fluctuation test, a statistical approach devised by Luria and Delbruck (1943), provided an experimental answer. Acquired resistance in *Escherichia coli* to an infecting virus (bacteriophage, or simply phage) was used as a model system. From a mass culture of E. *coli* containing a low frequency of variant cells resistant to phage, a series of sublines were initiated. The inocula used were too small to contain any resistant cells initially. When each of the subline cultures had grown to a large size, samples from all were tested for the frequency of variants resistant to bacteriophage, along with a similar number of replicate samples obtained from the original undivided mass population. Statistically, variance among the replicate samples from the undivided culture was small and essentially identical to the mean frequency of resistant variants in these cultures, as would be expected for a Poisson distribution. However, variance among samples taken from different sublines was very large, suggesting that variation had occurred independently, at random, and at different times in the separate subline cultures. The simplest explanation for this disparity was that genetic mutation was the random event; and that such mutants were then selected for by subsequent exposure to phage. This inference for bacteria was later borne out by a varied array of direct genetic tests which show that most, if not all, bacterial variants do indeed arise from mutational changes in the DNA coding system. Fluctuation tests have also been adapted for use with cultures of animal cells, and have aided in the recognition of genetic mutants in these populations as well.

It is important to recognize, however, that random and spontaneous changes in cell populations do not necessarily mean as such that genetic mutation has occurred. A positive result in fluctuation tests does not indicate the nature of the event concerned; this information must be obtained by other means. Specifically, fluctuation tests do not reveal where in the cell spontaneous and random changes have occurred, or even if a given variation is genetic or non-genetic in nature. These caveats need to be emphasized because in recent years it has been found that in addition to genetic mutants, phenotypically stable variants, based on non-genetic changes, can arise in cultures of animal cells. These so-called epigenetic variants, like genetic mutants, give positive results with a fluctuation test, and their frequency of appearance can likewise be calculated by this means. As later discussion will make clear, the essential difference is that epigenetic variants result from a stable shift in gene expression rather than from a structural change (mutation) at the nucleotide level. The spontaneous change picked up in fluctuation tests with these variants appears to be a random but stable shift in DNA methylation pattern, which is then perpetuated in subsequent cell replications. Because DNA methylation is known to affect gene expression, a stable shift in methylation pattern may accordingly lead to an equally stable change in cellular phenotype, without altering the basic DNA coding system. Epigenetic as well as genetic variants thus exist in cultures of animal cells, and when present in low frequency may masquerade as gene mutations.

Modulative Shifts in Phenotype

One of the earliest examples of modulation to be studied *in vitro* was the formation of cartilage by chondrocytes in serial culture. The advantages of this experimental system include the ready availability of chondrocytes in nearly pure form, when obtained from such sources as the vertebral rudiments of 10-day-old chick embryos. The freshly isolated cells can be dissociated and cultivated as monolayers, reaggregated masses, or in gels, and numerous cell type-specific markers are available for study at morphological and biochemical levels. In their initial experiments. Holtzer and colleagues examined the properties of chondrocytes when grown serially in monolayer form. Although at explantation the cells retained the typical morphology of chondroblasts or chondrocytes, formed matrix, and incorporated ^{35}S in this process, these cell type-specific features disappeared in subsequent subcultures. The population as a whole continued to grow vigorously but became fibroblast-like in appearance, with cells no longer surrounded by matrix as detected with appropriate staining procedures. Synthesis of chondroitin sulfate in these cultures, as determined by incorporation of ^{35}S, also progressively declined. Attempts by Holtzer and co-workers to reverse the transition by altering culture protocols were unavailing, and the conclusion reached was that progressive and irreversible dedifferentiation had occurred.

Later work by Coon (1966) showed that the loss of specialized function of chondrocytes in monolayer culture was more apparent than real. With improved media and altered cultural conditions the differentiated state of freshly isolated chondrocytes could be stabilized in serial culture, without progressive decline in cartilage-making function. Similarly, if dedifferentiated fibroblast-like populations were produced by subcultivation or chondrocytes in mass populations, it was possible to restore cartilage formation in these cells by suitable shifts to more optimal conditions. In this process, non-expressing cells regained the polygonal morphology of chondrocytes, formed typical cartilage capsules, and showed increased uptake of ^{35}S as sulfate was regained. In addition to demonstrating the reality of modulation for chondrocytes *in vitro*, these findings demonstrate that specialized function of cells is not necessary for heritable retention of the differentiated state. Success in restoring cartilage-making function was due in part to the use of improved synthetic media, combined with fetal calf serum, in lieu of the undefined horse serum-chick embryo extract employed for earlier work. But a more important factor was the innovative cultivation by Coon of chondrocytes at low population densities instead of passaging the cells with high cell numbers in serial culture. Coon noted that cartilage-making function declined progressively if mass populations were used, even in optimal media, although the cells continued to grow vigorously. But if similar cultures were initiated at low cell densities, isolated colonies developed from single cells. In the improved media used for this work the cells of these colonies took on the typical morphology of chondrocytes, which they retained even if passed serially for four or more transfers in the same way. While such observations do not define the underlying mechanisms of modulation or stabilization of the differentiated state, they do indicate that regulation is based on both extrinsic and intrinsic factors.

In more recent experiments, the emphasis has shifted to molecular patterns of synthesis, as cartilage cells undergo modulative shifts between specialized and unspecialized states. The studies of Benya and Shaffer (1982) with rabbit auricular chondrocytes are representative of the data obtained. For this work, dedifferentiated chondrocytes were harvested after five passages at high density in monolayer culture and were placed in nutrient medium containing 0.5% agarose, a procedure that immobilizes individual cells and allows expression of cartilage-making function even when relatively high inocula are used. As others have found, Benya and Shaffer observed that differentiated chondrocytes are distinguished at the molecular level by the synthesis of type II collagen and cartilage-specific proteoglycans. This biochemical phenotype disappears in serial monolayer culture and is replaced by a complex collagen phenotype consisting primarily to type I collagen and a low level of proteoglycans synthesis. When dedifferentiated chondrocytes are transferred to agarose, type II collagen reappears and the level of proteoglycans synthesis rises to a normal level. These changes coincide with a shift in morphology of the cells from a flattened form to the spherical morphology of anchorage-independent cell in suspension culture. Such observations suggest that cell shape may play a primary role in the initiation or maintenance of specialized function in differentiated cells.

Stabilization of differentiation in other culture systems seems also to depend on an interplay between explanted populations and microenvironmental conditions. Mammary epithelial cells, with their distinctive structural and functional markers, offer a second illustration of these relationships. As with other types of freshly isolated epithelia, primary cultures from mammary gland on a solid substrate soon give rise to a population of flattened epithelial cells, in which specialized functional activities decline or are lost. The kinetics of this process were demonstrated with particular clarity in the experiments of Ebner *et al.* (1961) with bovine mammary cells. Following the relative activity of several functional components in these cultures over a 15-day period, they were able to shown that the rates of decrease are non-parallel and distinctively different for individual markers. Thus, lactose production was initially high in freshly isolated cells but fell rapidly and could not be detected after 24 hrs in culture, UDPGal-4-epimerase activity decreased more slowly, disappearing at 7-10 days after explanation. Synthesis of β-lactoglobulin declined even more gradually and stabilized at a plateau level after 15 days. These data are of particular interest because they indicate that declines in functional activity are not controlled by a single cell-wide mechanism at quantitative level, although the potential for modulative return to a more active state was directly demonstrated here.

Cultures prepared directly from mouse mammary gland have been used extensively in recent years for studies on morphogenesis and modulation *in vitro* (reviewed by Levay-Young *et al.*, 1987). Freshly isolated cells in monolayers were used initially to determine the requirements of epithelial cells for differentiation. In such experiments, Emerman and Pitelka (1977) showed that cells if grown directly on a plastic surface give rise to a thin sheet with the surface polarity of a transporting epithelium and a poorly developed internal secretory apparatus. On a surface coated with collagen there was modest morphological increase for secretion of rough endoplasmic reticulum and Golgi complexes. Striking differentiation of mammary cells occurred, however, if the collagen membrane was freed physically from the substrate, so that it contracted sharply and floated freely in the medium with the explanted cells on the upper surface. As Michalopoulos and Pitot (1975) had found previously with liver cells, this leads to a change in shape of the cultured cells and onset of a more differentiated state. Mammary cells on floating collagen gels become cuboidal to columnar in shape, with long microvilli on the apical surface, and formation of a basal lamina separating epithelial cells from the gel below. Internally, the ultrastructure of these cells shows rough endoplasmic reticulum filling most of the cytoplasm and stacks of cisternae typical of actively secreting cells.

Definition of conditions for stabilizing the differentiated state in cultures of mammary epithelium enabled Emerman and Pitelka (1977) to demonstrate that the flattened non-secretory epithelia grown directly on plastic had not been irreversibly altered. When these dedifferentiated cells were subcultured to floating collagen gels, the cells regained the normal appearance of secretory cells in 2 days, and were then indistinguishable from populations explanted directly by this procedure. Using floating collagen gels, Emerman *et al.* (1977) were further able to define hormonal requirements for maintenance of the differentiated state. In media containing insulin alone, the cells had few secretory organelles and only a small amount of casein as detected by radioimmunoassay. When the nutrient was supplanted with insulin, cortisol, and prolactin, prominent development of the internal secretory apparatus occurred, with large increases in the amount of casein intracellularly and in the culture medium. These experiments show that mouse mammary cells on floating collagen membranes differentiate in response to hormones as do the same cells *in vivo.*

Modulation in the production of secreted proteins by mouse mammary cells was examined in more detail in experiments by Lee *et al.* (1984). These studies showed that cells on floating collagen gels secrete y-casein and also secrete α_1-, α_2-, and β-caseins into the culture medium. None of these were detectable in media from cells grown on plastic, and only to a slight extent for populations maintained on attached collagen gels. Contraction of gels on detachment does seem significant, since cells on glutaraldehyde-treated gels, which do not contract on removal from plastic, did not secrete any of the caseins. Two other proteins in milk, probably transferrin and butyrophilin, were only partially modulated in secretion by changes in the surface substrate, while acidic protein was synthesized abundantly by freshly isolated mammary cells, but appeared in only very small amounts in cultured cells regardless of the substrate used. These observations again suggest that while maintenance of mammary epithelial cells on different substrates can profoundly alter secretory patterns, regulation is distinct for individual proteins and is not governed by any single mechanism.

Since casein secret on is a direct indicator of differentiation an mammary cells, the question arises whether sensitivity of this marker to the extracellular matrix in floating collagen gels is mediated by secretion, synthesis, or stability of the proteins produced. In their original experiments. Lee *et al.* (1984) measured the amounts of intracellular and media caseins for mouse mammary cells maintained on different substrates. For cells on floating gels, considerable quantities of casein accumulated in the media, and only a small amount remained in the cells. Although cells on plastic did not secrete casein into the medium, only the same low level was found within the cells. These data show that casein found in the culture medium of floating gels had not simply accumulated without release within cells on plastic. In a later study. Lee *at al.* (1985) showed that cells on plastic rapidly lose the ability to secrete caseins, mRNA is produced and synthesis of milk proteins continues at a low level in these cells, but the newly synthesized caseins are degraded intracellularly. In comparison, large increases in mRNA for casein are observed for cells on floating gels, and the casein synthesized are not degraded but secreted instead into the extracellular medium. These investigations illuminate factors at the molecular level that underlie differentiation of mammary cells as well as modulative shifts to alternative functional states.

Stable Changes in Gene Expression

Differentiation in embryos and stem cell systems implies a predetermined pattern of gene action for each of the tissues or organs concerned. While modulation may occur within these populations *in vitro,* the final morphogenetic state is unchanged. This leads to an interesting question: Can developmental changes continue within individual cells once the overall process of differentiation is complete? Answers to this question in experimental terms are not obvious. One might expect that if morphogenetic change continued, any further differentiation would take on a new dimension. In this context, gene expression might be visualized as random, unprogrammed events in single cells, or in response to an external inducer. However, these same phenomena are often described as characteristics of a mutational process. This ambiguity means that it is not possible at the present time to say definitively whether such complex processes as the emergence of tumors *in vivo,* or appearance of variants in cell cultures depend on genetic changes, epigenetic shifts, or more probable to a combination of the two. In the present climate of opinion, variation in animal cells, as in microorganisms, is thought to follow predominantly a genetic model. Much evidence for this view can be found in numerous reviews and monographs currently available as sources of information. However, less attention has been given to the possibility that epigenetic shifts may participate as nongenetic mechanisms for variation in animal cells, and the discussion that follows will emphasize this topic with data that have become available in recent years.

Spontaneous phenotypic changes in cultures of neoplastic and normal cells on occasion can serve as a basis for epigenetic studies, but alterations in gene expression induced by direct exposure to external inductors have led to more useful experimental systems. The early observations of Charlotte Friend that virus-transformed mouse erythroleukemic cells were blocked in differentiation provided the impetus for a broad spectrum of experiments on gene expression in these cells, which continue to the present time. Friend cells will grow indefinitely in culture as unspecialized paraproerythroblasts, and they can be induced at any time to differentiate to normoblasts, with the production of hemoglobin, by brief exposure to a disparate series of chemical inducers, e.g., sodium butyrate, dimethylsulfoxide (DMSO), hexamethylenebisacetamide (HMBA), hormones, and inhibitors of ion transport. Substances such as butyrate that are naturally occurring compounds without mutagenic potential are particularly interesting, because they indicate clearly that induction is based on a change in gene expression rather than genetic alteration in the reaction system. Moreover, butyrate-treated Friend cells respond by synthesizing many new species of proteins that are not detectable in control cells. These *de novo* changes in protein and RNA synthesis are not seen when Friend cells are induced with other agents such as DMSO and are rapidly reversible when butyrate is removed. The mode of action of butyrate as an inducer is unknown. The co-ordinate induction of histone hyperacetylation may be significant, but it is not clear how the latter could provide a mechanism or working model for the induction process. Similarly, much has been made of molecular configurations in polar-planar inducers such as DMSO or HMBA, but the role of these unique properties for the induction of gene expression remains to be determined.

Epigenetic changes can be examined in a broader format with another inducer of gene expression, the nucleoside analogue 5-azacytidine (5-aza-CR). Experimental analysis of mechanisms in this case is facilitated by the known effects of this agent on DNA methylation, which provide a molecular model for induced cell variation. The first indication that 5-aza-CR affects gene action came from a study on transformation of mouse C3H/10T1/2 cells by cancer chemotherapeutic agents. Fully functional striated muscle fibers appeared in large numbers after brief exposure of this fibroblast-like line to micromolar concentrations of 5-aza-CR. These multinucleated myotubes were found to have elevated levels of myosin ATPase and showed a twitching response to added acetylcholine (ACh). Myogenic activity had never been observed in previous studies of 10 T 1/2 cells exposed to many different agents, although the effect was highly reproducible after treatment with 5-aza-CR. Moreover, the potential for altered differentiation extended to other end products as well. Taylor and Jones (1982) showed that phenotypic conversion by 5-aza-CR included the appearance of biochemically differentiated chondrocytes and adipocytes, which appeared later and in somewhat smaller numbers than induced muscle fibers. Cloning experiments indicated that conversion proceeded by induction rather than by selection of previously existing variants. Five subclones of 10T1/2 cells all expressed both muscle and fat cell phenotypes after exposure to 5-aza-CR, and in later studies even single cells, plated at limiting dilution in isolated wells, show a high frequency of conversion if treated with 5-aza-CR. More surprising was the discovery that cells from the 10 T 1/2 line that had undergone neoplastic transformation were still capable of forming typical striated muscle fibers on exposure to 5-aza-CR. Thus, phenotypic conversion can occur in cells with tumorigenic properties as well as in normal cells The implication of these studies seems to be that 5-aza-CR induces a reversion in target cells to a more pluripotent state, from which redifferentiation can proceed via alternative pathways.

Although most of these experiments were performed with 10 T 1/2 mouse cells, other established murine lines and even a diploid Chinese hamster line (Sagar and Kovac, 1982) respond to 5-aza-CR with the formation of muscle, cartilage, and adipocytes. In searching for a common mechanism of action, early observations showed that 5-aza-CR is incorporated into both RNA and DNA (see review by Vesely and Cihak, 1978). However, since 5-aza-3-deoxycytidine (5-aza-CdR) is an even more potent inducer of myogenesis and is incorporated into DNA only, evidence pointed to a process associated with DNA synthesis. In a classic study, Jones and Taylor (1980) linked incorporation of 5-aza-CR and induction of muscle with a clear-cut inhibition of methylation in newly formed DNA. As background, it should be pointed out that many investigations have shown that DNA methylation patterns affect gene expression. Since 5-aza-CR also has little or no mutagenic activity a non-genetic mechanism mediated by methylation changes is an attractive explanation for the effects of 5-aza-CR. In the experiments mentioned, Jones and Taylor showed that newly formed strands of DNA containing 5-aza-CR remained unmethylated at CpG sequences, the primary site for methylation in mammalian DNA. The maximum effect was observed at 48 hrs after treatment with the inhibitor, a time point which coincided with the peak for induction of myotubes as well. This correlation of phenotypic conversion in 10 T 1/2 cells with systematic undermethylation of newly formed DNA agrees with many other investigations, in which hypomethylation of DNA can be linked directly with increased gene activity, e.g., synthesis of chicken β-globin (McGhee and Ginder, 1979).

A more direct indication that morphogenetic induction by 5-aza-CR stems from structural modifications in DNA has now been provided by studies with 10 T 1/2 cells at the molecular level. These investigators reasoned that myoblasts arising by normal differentiation, as well as those originating from 10 T 1/2 cells by exposure to 5-aza-CR, must contain a putative regulator locus, since activation triggers a developmental programme leading to the co-ordinated appearance of a series of enzymes and other muscle-specific markers. Such a regulative locus should be active (unmethylated) in myogenic cells, while methylated and inactive in nonmyogenic tissues. High-molecular weight DNA was accordingly isolated from unmodified 10 T 1/2 cells, 10 T 1/2 cells that had been induced by 5-aza-CR to form myoblasts and embryonic mouse myoblasts. Clear-cut conversion to muscle fibers was observed after transfection with genomic DNA from both types of myogenic cells, but not with DNA from untreated 10 T 1/2 cells. Specificity in these inductions was documented by assays using monoclonal antibody produced against myosin heavy chain, as well as with the sensitive immunoperoxidase technique used to detect this protein in single cells. In further experiments, Davis *et al.* (1987) showed that expression of a single transfected cDNA is sufficient to convert 10 T 1/2 cells to myoblasts. To obtain the specific cDNA sequences active in this process, cDNAs were prepared from a normal mouse myogenic line and from a line or 10 T 1/2 cells treated with 5-aza-CR. Subtracted probes were then obtained by hybridizing the cDNAs extensively with poly $(A)^+$RNA isolated from unmodified 10 T 1/2 cells. These probes were then used to screen a myoblast/myotube cDNA library. Of numerous sequences obtained, one designated MyoD1 induced efficient formation of muscle fibers when transfected into 10 T 1/2 cells.

It is not yet clear whether the MyoD1 sequence is identical to the azacytidine gene (i.e., the locus activated *in situ* by treatment of 10 T 1/2 cells with 5-aza-CR), and/or the active gene in myoblast DNA which induces muscle formation when transfected into 10 T 1/2 cells. That more than one gene can serve as activator is suggested by the studies of Pinney, Pearson-White, Konieczney, Latham, and Emerson (1988), who induced myogenesis in 10 T 1/2 cells with a genomic cosmid sequence obtained from a non-myogenic cell type (human leucocytes).

Southern blot analyses showed that the sequence in question, which has been designated *myd*, is not identical with the MyoD1 gene. One or more genes may therefore serve as regulatory loci in myogenesis but since MyoD1, at least, does not promote adipogenesis it seems unlikely that these are general differentiation agents. More recent studies have concentrated on the expression product of MyoD1, which has been shown to be a phosphoprotein present in the nuclei of proliferating myoblasts and differentiated myotubes, but not detectable in 10 T 1/2 or other non-muscle cell types. As a precursor, MyoD1 cDNA contains an open reading frame of 318 amino acid residues, with a sequence of 22 residues which resemble a region that is highly conserved in the family of c-Myc proteins. Site directed mutagenesis of MyoD1 cDNA was employed to analyze functional domains in MyoD1 protein. Comparison of deletion mutants showed that extensive regions could be removed at either and of the MyoD1 protein without interference with biological activity. Expression of a continuous sequence of only 68 amino acids from MyoD1 was sufficient to activate myogenesis in 10 T 1/2 cells. The active portion includes a very basic region (residues 102-135), without which both nuclear localization and myogenesis were blocked. The c-Myc related region (residues 143-162) was likewise essential for myogenic induction, but in the absence of this sequence nuclear localization of the rest of the probe continued to occur. The exact chromosomal location of MyoD1 has not yet been determined, but the genetic analyses performed map this gene to mouse chromosome 7 and human chromosome 11.

Much progress has thus been made in linking normal and 5-aza-CR-induced myogenesis with structural modifications in a regulatory locus at the DNA level. The MyoD1 model service to explain the co-ordinated programme of changes which appear on 5-aza-CR treatment, but the role of regulator and final effect or genes in this integrated pattern remain to be determined. Less complex systems for analysis of single gene changes can be found in the analysis of stable variations which arise spontaneously or by induction in cell cultures. At an earlier time it was believed that essentially all such variants were the result of gene mutation. It is now clear, however, that a number of these are in fact epigenetic in origin, particularly those obtained by treatment of cells with 5-aza-CR, DNA methylation changes are particularly useful as an explanation for such variants, which seem to be based on stable shifts in gene expression rather than on genetic alteration in the DNA coding system. The characteristics which permit this interpretation include the following pertinent features (reviewed in the general references cited above) (1) the presence or absence of methyl groups on specific CpG sequences throughout the genome which forms a heritable and reproducible pattern that persists through successive cell division. This is so because of the semiconservative character of methylase activity, in which a methyl group is added to cytosines in newly formed DNA only if a corresponding methyl group is present on the complementary CpG of the pre-existing partner strard. Second, if a methyl group at a particular site is gained or lost by spontaneous or induced change, the new pattern is perpetuated for the same reason through successive replications, although fidelity is not absolute. In combination, these characteristics provide a working model for at least one class of heritable epigenetic changes. DNA methylation patterns here play the role of a second information system, parallel to, but independent from the primary DNA coding sequence.

A good example of epigenetic variation in these terms is provided by the experiments of Compere and Palmiter (1981) on the metallothionein-1 (MT-1) gene in W7 mouse thymoma cells. Metallothioneins are specialized proteins which bind heavy metals, and can be induced in certain cell lines on exposure to cadmium or glucocorticoid hormones. Although this inductive response does not ordinarily occur in untreated W7 populations, it can be stimulated to take place if W7 cells are previously exposed to 5-aza-CR. The molecular basis for this shift was analyzed by Compere and Palmiter, using the restriction enzymes Hpall and MspI, which in combination show the degree of methylation at CpG sequences. In untreated W7 cells, all the available Hpall sites in the vicinity of the MT-1 gene were fully methylated. After exposure to 5-aza-CR, the same sites were unmethylated, along with associated flanking sequences. Thus, DNA methylation, which exists at control sites in the W7 cells, seems to block inducibility of the MT-1 gene, a locking mechanism which can be removed by demethylation of these sites with 5-aza-CR.

Cells deficient in thymidine kinase (TK) and other specific enzyme activities offer additional study systems for epigenetic variation induced by 5-aza-CR. Variant lines lacking TK activity have been produced by many investigators through prolonged exposure to 5-bromodeoxyuridine (BUdR). IN V79 Chinese hamster cells, TK—variations were found to emerge in this way by a two-stage process. In the first stage, there was an abrupt mutation-like event, later found to be stochastic drop in copy number at the TK locus. This conferred partial resistance to BUdR, along with decline of TK activity to a lower level. The second phase was distinctively different. During this interval, partially resistant first-stage cells were propagated serially for months in media containing BUdR. Cells in these populations continued to lose TK activity on a logarithmic basis for 28 weeks or more. The terminal populations were stably and completely resistant to BUdR, lacking entirely in TK activity and unable to grow in HAT medium. Studies by Wise (1985) and Wise and Harris (1988), subsequently showed that transition to complete BUdR resistance and the TK—state were associated with gradual hypermethylation in thes' region of the TK gene, without further change in copy number. Repression by methylation is again seen here as a mechanism for silencing gene activity. The interpretation given above is strongly supported by experiments designed to induce re-expression of activity in the V79 TK$^-$cells that had been obtained. Brief exposure of such populations to 5-aza-CR resulted in mass conversion to HAT$^+$ state, with one-step increases of 10^6

in frequency of TK^+ cells over background levels. Although the incidence of HAT^- cells arose slightly in cultures treated with the chemical mutagen ethyl methanesulfonate, the increase obtained was orders of magnitude less than effects from treatment with 5-aza-CR. Large increases in frequency of HAT^+ cells were also obtained with n-butyrate and L-ethionine, both of which affect gene expression but have no mutagenic potential. Re-expression of TK activity in this system seems clearly to be an epigenetic event, which reverses the gradual extinction in TK expression imposed by BUdR-associated methylation. In all cases, however, TK activity in these revertants was significantly lower than levels in wild-type V79 cells, indicating that there had been no reversal of the initial drop in copy number at the TK locus.

Other examples of epigenetic variation have been found in cell lines that require L-asparagine for growth and lack detectable levels of asparagine synthetase. In the earlier study, made with Jensen rat sarcoma cells, asparagine-independent variants were obtained in high frequency from the parental auxotrophs after treatment with 5-aza-CR. All revertants were phenotypically stable, whether maintained in the presence or absence of asparagine. Asparagine synthetase preparations isolated from revertants were indistinguishable in biochemical properties from those of the normal enzyme obtained from rat liver or other rat cell lines. These comparisons included measurements of thermolability and several kinetic parameters, as well as immunological properties. Other studies showed that the parental Jensen rat sarcoma cells contained no immunologically cross-reacting material. Taken as a whole, these data suggest that asparagine synthetase gene is repressed but intact in asparagine auxotrophs, and can be restored to a normal functional state by spontaneous or induced epigenetic change. One further instance of epigenetic change is worth mentioning because the marker in question, the well-known requirement of CHO-K1 Chinese hamster cells for proline, has long been regarded as a prototype for auxotrophic genetic mutations in mammalian cells. Such an interpretation was put forth by Kao and Puck (1967) when this naturally occurring line was first characterized, and is consistent with several observed properties: (1) the proline requirement is phenotypically stable, (2) spontaneous reversion to the pro^{+}state occur with a frequency of less than 10^{-7} as measured by fluctuation tests, (3) there is a small increase in incidence of revertants after treatment of CHO-K1 cells with mutagens, and (4) biochemical differences between pro^{+} and pro^{-} cells are readily apparent when proline synthesis is measured in cell extracts. However, no director proof of a mutational origin was forthcoming, and all the characteristics mentioned above are equally compatible with the hypothesis that lack of proline synthesis in CHO-K1 cells stems from suppression of proline synthesis, perhaps by hypermethylation, rather than from gene mutation. Recent experiments suggest that in fact the epigenetic explanation is the correct one. Thus, exposure of CHO-K1 cells to 5-aza-CR induces a 10^{5} to 10^{6} increase over background conversion to the proline-independent state. The revertants are stable phenotypically, with enzymatic activities reexpressed for pyrroline 5-carboxylate synthase and ornithine aminotransferase, as in other proline-synthesizing lines. These observations deserve careful consideration, since the view often expressed has been that stable variants in culture are in general genetic rather than epigenetic in origin. A more salutary approach may be to require direct evidence in each system for the mechanisms involved, rather than simple analogy to a preconceived model.

3

Rearrangements of DNA in Eukaryotes

Various stable changes in structural and functional properties have been reported in cells in a variety of situations. Several chapters in this volume discuss such changes and the circumstances and mechanisms associated with their occurrence. Stable changes in the structural and functional properties of cells are generally based on modifications at the level of DNA. Five kinds of modifications are known: (1) elimination of DNA sequences; (2) amplification of selected sequences; (3) rearrangement of DNA segments, with no loss or gain of sequence; (4) methylation of DNA; and (5) changes in transcriptional activity of DNA mediated by transcription factors. Most stable changes in cells come about by the turning on and off of the expression of particular genes by trans-acting transcriptional factors. Some modulations are reversible and some are normally permanent, as in the terminal differentiation of cells in multicellular organisms. These kinds of changes, as well as changes in transcription achieved through methylation of DNA, are discussed in other chapters. This chapter is concerned with change in cells brought about by the gain, loss, or rearrangement of DNA sequences in the eukaryotes of animal cells.

Gain of DNA Sequences

It is a well known fact that gene is a specific part of DNA. Its expression is done under two steps: transcription and translation. Many primary transcripts are produced from a single gene; each messenger RNA (mRNA) molecule formed from a primary transcript is, in turn, amplified into many copies of an encoded protein. By these two amplifications, sufficient copies of a polypeptide can be produced from a single pair of alleles in most kind of diploid cells, even in cases in which an enormous amount of a single protein type is required in a short time, e.g., hemoglobin in differentiating red blood cells and silk protein in silk gland cells.

A few exceptions to this generality are known. In *Drosophila*, ovarian follicle cells synthesize large amounts of proteins that form the chorion of the egg in just a few hours during egg formation. The genes encoding these proteins are expressed at high levels in follicle cells at the appropriate time. Before synthesis of the chorion proteins, genes encoding them increase in copy number by differential replication. The genes occur in two clusters, which, between them, contain single copies of genes for the six major chorion proteins. One cluster, located in the X chromosome, undergoes a 15-fold amplification, and the second cluster, in chromosome III, is amplified. The amplified copies in both clusters are produced by extra rounds of bidirectional replication beginning in one or more origins of replication and extending for variable distances (in successive rounds) up to 40-50 kilobase pairs (kbp). The amplified DNA remains in the chromosome in an onion skin arrangement. This is in contrast to amplification of genes encoding ribosomal RNA (rRNA) in amphibian and certain insect oocytes, in which the extra gene copies are extra-chromosomal and form multiple nucleoli.

The *trans*-acting molecular signals that trigger and regulate the number of extra replication and the signals themselves have not been identified. Such signals must recognize specific *cis* elements in the two sets of genes, presumably located in the origins of replication. Production of the *trans* signals must be developmentally regulated within the ovarian follicle cells. Two genes have been identified by mutation that are involved in generating the trans signal to both chorion gene clusters.

This work is an important step toward deciphering the developmental mechanism that regulates the specific amplification of the chorion genes.

Specific amplification of the chorion genes does not occur in the silk moth. Instead, in the moth, formation of the chorion takes place more slowly and the moth germ-line genome also contains a multiplicity of genes for chorion proteins. These two factors presumably obviate the need for differential replication of genes to meet the protein requirement of chorio—genesis.

Amplification of specific DNA sequences was described many years ago in polytene chromosomes of the midge *Rynchosciara* by Pavan and Da Cunha (1969). Polytene chromosomes in insects, plants, and protozoa provide individual cells with multiple copies of all genes. In insects and presumably in plants (but probably not in protozoa), polytene chromosomes represent a strategy for quickly accommodating demands for large amounts of gene products. In polytene chromosomes of *Rynchosciara*, however, further amplification of a few specific genes is superimposed on the generalized amplification of polytenization. The selective amplifications are recognized as puff formation in certain chromosome band in which DNA replication precedes RNA synthesis in the puff. A protein produced in a large amount after these amplifications has been identified as a candidate for the coding function of a DNA puff.

The first example of specific gene amplification to be discovered is the differential replication of DNA encoding rRNA in amphibian oocytes, a phenomenon subsequently found to be widespread, but not universal, among animals, including insects. Replication occurs just before meiosis or in meiotic prophase. At least one copy of the DNA segment encoding 28S, 18S, and 5.8S rRNA plus the DNA spacers between coding regions is produced from the chromosome. The extra DNA separates from the chromosome and is somehow circularized.

It then replicates many times by the rolling circle mechanism, producing many additional DNA circles of various contour lengths. Each circle contains integral multiples of the basic unit encoding one each of the three rRNAs and DNA spacers. The extrachromosomal circles form separate nucleoli, in which rRNA is synthesized and assembled into ribosomes. In this way, a large store of ribosomes (about 10^{12} ribosomes in a frog oocyte) is built up, to be used in embryonic development. In some animal species lacking amplification of rRNA genes, large amounts of rRNA are synthesized in nurse cells surrounding an oocyte, and the RNA is transferred in the oocyte. This apparently accomplishes the same end as amplification of rRNA genes.

In frog oocytes, amplified rDNA accounts for 75% of the total DNA in the oocyte. After fertilization of the egg, the chromosomal DNA replicates with each cell cycle but the extrachromosomal rDNA does not. Therefore, the extra copies of rDNA become rapidly diluted to an undetectable level early in embryonic development.

Differential amplification of rDNA and other genes occurs in the macronucleus in ciliated protozoa, as discussed in Section 5.15. These several examples (chorion genes in *Drosophila*, rDNA in oocytes, and rDNA and a few other genes in ciliates) demonstrate that gene amplification has evolved several times as a possible way to meet a demand for large amounts of a gene product. However, differential amplification of genes is clearly not a general strategy used by somatic nuclei to achieve *transient* increases in gene expression. Permanent amplifications of a few genes in the germ lines of most eukaryotes, i.e., genes encoding rRNA, transfer RNA (tRNA), and histones, are well-known.

At least two cases of experimentally forced amplification of specific genes are known. When mammalian cells are grown in culture in the presence of methotrexate, an occasional cell (perhaps 1 in 10^3) undergoes; amplification of the single copy of the gene encoding dihydrofolate reductase. Dihydrofolate reductase catalyzes synthesis of methylene tetrahydrofolate, the methyl-contributing coenzyme required for synthesis of thymidine monophosphate (TMP) from deoxyuridine monophosphate. Methotrexate inhibits dihydrofolate reductase, blocking DNA replication and cell reproduction. In the occasional cell that undergoes amplification of the gene encoding dihydrofolate reductase, the inhibition of methotrexate is overcome by expression of the amplified copies of the gene. Up to 1000 extra copies of the gene may be built up. The amplified copies of the gene can occur as segments of DNA integrated into a chromosomal DNA molecule at the site of the original gene. In such case, they form a homogeneously staining region in the chromosome. The amplified genes may also be present in an unstably maintained form as small chromosome fragments called double minutes. Since double minutes lack centromeres, they cannot be evenly distributed during mitosis and are rapidly lost from a cell clone when the selective pressure (methotrexate) is removed.

Selective amplification of other genes in cultured cells has also been induced by treatment with other toxic substances. For example, amplification of genes encoding the protein metallothionein occurs in cells chronically exposed to a toxic concentration of Cd^{2+} ions. However, amplification of genes in response to such drugs as methotrexate (mTx) or poisons such as Cd^{2+} ions has not been reported for the germ line of animals.

Rearrangement of DNA Sequences

Change in arrangements of DNA sequences occurs widely among eukaryotes but, with few exceptions, these are not achieved by physical shifting of DNA segments. Most rearrangements are mediated by transposable elements, most or all of which are transcribed into RNA, reverse transcribed into DNA, and inserted in chromosomal DNa at new locations. Genes at the point of insertion may be mutated or otherwise inactivated by the insertion. Removal of a transposable element by deletion (excision) may result in reactivation of gene expression at the site of the excision. Gene conversion can also provide a change in DNA sequence arrangement without actual physical shifting of DNA segments. Examples are switching of mating type in yeast and switching of genes encoding surface antigens in trypanosomes. Since neither transposable elements nor gene conversion involves rearrangement of DNA segments directly, they are not considered further here.

Two phenomena that involve cutting and splicing of DNA have crucial genetic consequences. During the development of an animal, the diversity of genes encoding heavy and light chains of antibodies is generated by combining segments of DNA. A given V segment is spliced to a given J segment following deletion of all intervening V and J segments. In this situation, the process is regular in that deletions occur as specific blocks of multiple V and J segments, rather than as deletion of completely random-sized stretches of DNA.

The formation of antibody genes by deletion is a normal developmental process. The second instance of cutting and splicing of DNA is the more or less random breakage and rejoining of chromosome segments known as translocations. Translocation is probably always the effect of agents that break DNA directly or that cause breakage by interference with DNA replication. Therefore, translocations are generally viewed as abnormal events. In no case do they represent a normal or regulated rearrangement of DNA sequences.

Translocations can produce various effects including mutation of genes at break points, fusion of genes, and changes in gene expression by placing genes under the influence of a different promoter. The latter effect contributes to the transformation of normal cells into transformed cells through enhancement of transcriptional activity or proto-oncogenes. Such enhancement of activity is one of several ways in which proto-oncogenes are converted to oncogenes.

Most kinds of transformed cells have elevated rates of spontaneous translocation producing a variety of aneuploid phenotypes. In addition to translocations, many types of transformed cells lose and gain chromosomes through a propensity for mitotic non-disjunction, which further contributes to karyotype variation. The reasons for the elevated rates of translocation and non-disjunction are not known.

Loss of DNA Sequences

Loss of entire chromosomes, parts of chromosomes, or DNA sequences dispersed within chromosomes occurs in a few animal species during their development, but it is a relatively rare phenomenon. The losses occur in very early development as cell lineages for future somatic nuclei and future germline nuclei are established; losses are restricted to the somatic cell lineage. Ammermann (1985) distinguishes three kinds of DNA in these situations: G DNA is present in nuclei destined to serve as the germline; S DNA is the part of G DNA present in somatic nuclei; and E DNA is the part of G DNA that is eliminated in the formation of somatic nuclei.

Elimination of DNA is a normal part of development in a few species in four taxonomic groups: arthropods, nematodes, insects and ciliated protozoa. Chromatin diminution was described in the nematode *Ascaris* by Boveri (1887) a century ago. The eliminated DNA is in large heterochromatic blocks; elimination occurs during the first cleavage divisions. Measurements of how much DNA is eliminated are not in agreement. The estimates range from 22% of G DNA (Mortiz and Roth, 1976), to 27%, to 34% (Pasternak and Barrell, 1976) to 56%. The estimates of remaining DNA (S DNA) range from 0.15 to 0.46 pg. In *Parascaris equorum* (a horse parasite), 85% of the DNA is eliminated.

In *Ascaris*, the eliminated DNA (E DNA) is made up at least in part of repetitive sequences that form a satellite in a CsCl isopyknic gradient (Roth and Moritz, 1981; Mueller et al., 1982). As much as 50% of the E DNA could be composed of unique sequences according to Tobler *et al.* (1972) and to Goldstein and Straus (1978).

According to Roth and Moritz (1981), about 90% of the S DNA is made up of unique sequences and 10% of repetitive sequences, but only a tiny fraction of the germ-line repetitive sequences correspond to the E sequences that undergo elimination. Mueller *et al.* (1982) conclude from hybridization experiments that E DNA (before elimination) is not transcribed. Whether E DNA in *Ascaris* has any function is not known.

Chromatin elimination also occurs in several species of copepod crustaceans. In the fourth through seventh cleavages in the early blastula of three species of Cyclops, about 50% of the G DNA is eliminated. In one species, the E DNA is present in large heterochromatic clumps, which in another species are at the ends of chromosomes and in another are present both at ends and at kinetochores: in a third species, the heterochromatic E DNA blocks are scattered throughout the chromosomes. Thus, the normal elimination of E DNA requires regulated breakage and rejoining of chromosomes. During the stages of elimination, circular strings of nucleosomes of <1μm to many microns in length appear in the cells. This suggests that elimination involves an intrachromosomal recombination event in which E DNA forms circles during its excision from chromosomes.

Among some insects of *Diptera*, entire chromosomes are lost from somatic cells during early cleavage but are retained in germ-line cells (White, 1954). In gall midges, these chromosomes, composed entirely of DNA, are necessary for development of spermatocytes and oocytes. E chromosomes are apparently active in RNA synthesis during oogenesis, in keeping with the idea that E chromosomes in insects contain genes essential for gametogenesis.

The gains and losses of DNA among multicellular organisms (animal) discovered to data are listed in Table 3.1. There are no proven cases of direct rearrangement of DNA sequences; in general, DNA-sequence rearrangement involves transcription of DNA into RNA and reverse transcription into

DNA or gene conversion. By far the most extensive changes in DNA occur in ciliated protozoa.

The Ciliate Genome

Ciliates contain two kinds of nuclei, a micronucleus and macronucleus, which are quite different from each other. Some species contain two or more of each kind. For example, *Urostyla grandis* (a hypotrich) contains more than 100 macronuclei and many micronuclei. The multiple micronuclei in a cell are, in principle, all identical to each other in structure and function, and the multiple macronuclei are all structurally and functionally the same.

TABLE 3.1

Gains and Losses of DNA

Organism	*DNA changes*
	Normal changes
Some nematodes	Elimination of DNA from somatic cells during early cleavage
Some *Diptera*	Eliminating of DNA from somatic cells during early cleavage
Copepods	Eliminating of DNA from somatic cells during early cleavage
Drosophila	Amplification of chorion genes in follicle cells
Many animals	Amplification of rDNA sequences in oocytes
Lymphocytes	Deletion of DNA sequences in formation of immunoglobulin genes
Ciliated protozoa	Elimination of DNA sequences, amplification of DNA sequences, and rearrangement of DNA sequences during macronuclear development Abnormal changes
All eukaryotes	Chromosomal translocations and deletions
Vertebrates	Amplification of various genes (e.g., gene-encoding dihydrofolate reductase, metallothionein, protooncogenes), usually under some cytotoxic or genotoxic environmental condition

The genome of micronucleus is made up of a set of chromosomes (usually diploid) and divides by mitosis. The genes in a micronucleus are contained in high-molecular-weight chromosomal DNA. However, micronuclear genes are not expressed (e.g., no RNA synthesis) in vegetative cells. The macronucleus (or macronuclei) provides all the RNA (except RNA made in mitochondria) needed for cell metabolism and reproduction. In sharp contrast to micronuclear DNA, all the DNA in a macronucleus is in subchromosomal molecules; i.e., DNA molecules have lower molecular weights than do micronuclear (chromosomal) DNA. Also, the macronucleus divides amitotically. Amitosis is imprecise; one daughter macronucleus usually receives a little more DNA than the other. This random mode of nuclear division is successful because every DNA molecule is present in many identical copies (sometimes thousands), and each daughter macronucleus receives some copies of every molecule, hence receives some copies of every DNA sequence to be found in macronuclei of the species.

The micronucleus is a germ-line nucleus. It undergoes meiosis during call mating (conjugation). In cell mating, haploid micronuclei are exchanged through a cytoplasmic bridge between the two cells of a mating pair. After it is exchanged, a haploid micronucleus fuses with a stationary haploid micronucleus in the recipient cell, forming a diploid micronucleus. About this time, the mating cells separate. The new diploid micronucleus in each exconjugant cell divides mitotically without accompanying cell division, and one of the daughter micronuclei develops into a new macronucleus. While this is occurring, the old macronucleus (or macronuclei) and all unused haploid micronuclei are destroyed. Variations of this pattern occur in different species, but the overall principles are the same.

DNA Changes in Ciliates

Extensive changes in genomic DNA occur in the development of the macronucleus from a micronucleus in ciliates. Most of the studies have been done on a few species among the holotrichs, chiefly in the genus *Tetrahymena* and on a few species of hypotrichs in the genera *Euplotes, Stylonychia* and *Oxytricha*. A consistent picture has emerged, with some major differences between the holotrichs and the hypotrichs. The main features of macronuclear development are outlined here to illustrate how these organisms manipulate their genomes. Details of the processes are described in reviews and the original research reports.

Changes in the ciliate genome during formation of a macronucleus include the following:

1. Elimination of chromosomes
2. Conversion of chromosomal DNA to subchromosomal molecules
3. Elimination of repetitive and unique sequences between genes
4. Splicing of DNA segments
5. Elimination of intron-like sequences
6. Rearrangement of sequences into different orders
7. Addition of repeat sequences (telomeres) at the ends of the subchromosomal molecules

These changes are discussed briefly in the next several sections.

Elimination of Chromosomes

The micronucleus of the hypotrich *Stylonychia lemnae* contains a variable number of chromosomes estimated to range between 100 and 180, with an average of 140, (haploid numbers). Only 35 or 36 of these chromosomes participate in formation of a macronucleus; most chromosomes (containing exclusively E DNA) are eliminated from the micronucleus just as it embarks on its development into a macronucleus. The eliminated chromosomes are condensed (heterochromatic) at the time of elimination. This elimination is quite similar to the elimination seen in *Ascaris* and copepods, in which germ-line DNA (G DNA) is composed partly of DNA that is eliminated (E DNA) and partly of DNA that is retained (S DNA) in formation of somatic cells. The situation in these ciliates is more complex, however, since most of the DNA sequences in the retained 35 or 36 chromosomes (mutative S DNA) are eliminated later in macronuclear development when individual genes are excised from the chromosomes.

Chromosome elimination has not been documented in other hypotrichs, although it might occur but not be prominent enough to be readily seen cytologically. Some sequences are known not to participate in the subsequent DNA replication that characterizes macronuclear development (Spear and Lauth, 1976), and these may have been eliminated. They may in fact be eliminated in a manner similar to elimination in *Stylonychia*.

In the holotrich *Tetrahymena*, DNA representing about 15% of the DNA sequence complexity in the micronucleus is eliminated. This diminution does not occur as a visible cytological event in the form of chromatin elimination as in *Stylonychia*, but it does occur very early during the rounds of DNA replication that are part of macronuclear development. The sequences in question are not simply diluted out by underreplication but are actively and specifically destroyed.

Thus, both hypotrichs and holotrichs undergo DNA diminution at the start of macronuclear development, an event that is extensive and visible as chromosome elimination in *Stylonychia* but much less extensive and not cytologically discerned in *Oxytricha* or *Tetrahymena*. The molecular mechanism by which the E DNA is recognized and then destroyed is completely unknown.

THE SECOND ELIMINATION OF DNA SEQUENCES IN MACRONUCLEAR

Development in Hypotrichs

After the first elimination of DNA in hypotrichs, the remaining DNA undergoes repeated replication to form polytene chromosomes. By contrast, the early elimination of sequences in the holotrich *Tetrahymena* apparently leaves the DNA in subchromosomal fragments. These replicate repeatedly (after telomere addition; see Section 5.1.3.) but do not form polytene structures. These events in holotrichs are summarized in the diagram. A more detailed account of sequence elimination in holotrichs (and hypotrichs) is given in Section 5.1.4.

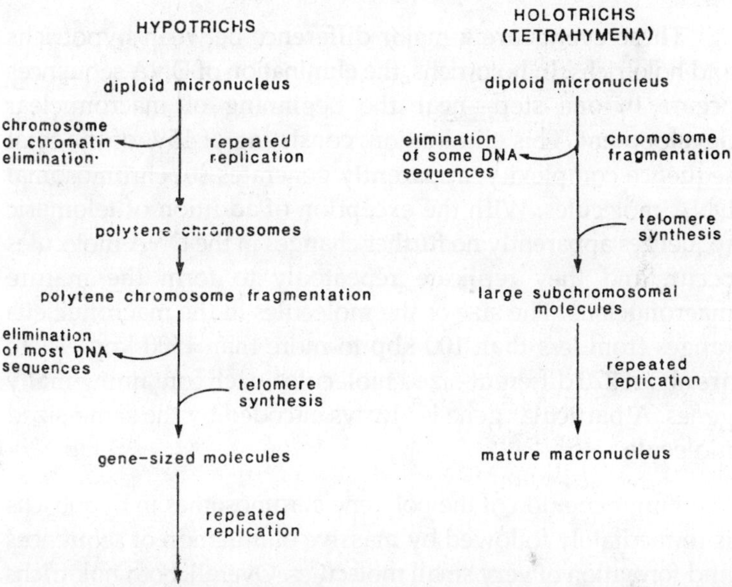

Fig. 3.1. Comparison of the processing of DNA in hypotrichs and holotrichs during macronuclear development.

The formation of polytene chromosomes in hypotrichs is complete about 2 days after cell mating. The chromosomes then become fragmented. Fragmentation in its earliest stage is detectable by electron microscopy (Kloetzel, 1970). A septum, believed to be composed of protein, forms through every interband of every polytene chromosome. The septa subsequently extend to form separate envelopes around each individual chromosomal band. In effect, vesicles are formed, each containing the material present in a chromosome band plus one half of each adjacent interband. Most of the DNA in each vesicle is then degraded, leaving small molecules ranging in size from a few hundred base pairs to 15 kbp. The number average size is about 2200 bp in *Oxytricha*. Each molecule contains a single coding sequence (gene), and a particular gene is usually encoded by one but sometimes several size of DNA molecules.

These events are a major difference between hypotrichs and holotrichs. In holotrichs, the elimination of DNA sequences occurs in one step—near the beginning of macronuclear development. This elimination, consisting of 15% of the total sequence complexity, apparently generates subchromosomal DNA molecules. With the exception of addition of telomeric sequences apparently no further changes in the DNA molecules occur, and they replicate repeatedly to form the mature macronucleus. The size of the molecules in the macronucleus ranges from less than 100 kbp to more than 1500 kbp. There are about 270 different-sized molecules, each containing many genes. A particular gene is always encoded by the same-sized molecule.

Fragmentation of the polytene chromosomes in hypotrichs is immediately followed by massive elimination of sequences and formation of very small molecules. Overall, both holotrichs and hypotrichs break up the DNA in their chromosomes and eliminate sequences. In holotrichs, large subchromosomal molecules, each containing many genes, are formed. The hypotrichs carry the process to an extreme, eliminating about 95% of all sequences and reducing all DNA to small molecules, each containing a single gene.

Addition of Telomeres

Fragmentation of chromosomal DNA and elimination of sequences generates large subchromosomal, linear DNA molecules in holotrichs and short, linear DNA molecules in hypotrichs. Telomeric sequences consisting of many repeats of 5' CCCCAA 3' in holotrichs and of a fixed number of 5' CCCCAAAA 3' repeats in hypotrichs are added to both ends of every molecule. The telomeric sequences differ somewhat in arrangement in the two groups. The telomeres are synthesized on the ends of the DNA molecules by an enzyme called telomere terminal transferase. The telomeres are important for the replication of the ends of the DNA molecules, hence for the maintenance (stability) of the molecules in vegetative macronuclei. They may also have functions in organization of the DNA into chromatin.

Intricacies of Sequence Elimination

The simplest scheme for the fragmentation of chromosomes and the elimination of sequences in holotrichs would be the recognition and destruction of specific short segments of DNA between longer sequences. The longer sequences would then be the subchromosomal molecules of the macronucleus. In fact, this probably happens, but additional events occur that make the process more complex, for example events in formation of the DNA molecule in the macronucleus encoding rRNA. The molecule is 11 kb long and contains two copies of the transcription unit for rRNA arranged in inverted order with a spacer between them. In addition, non-coding stretches of DNA are present just inside the telomeres at the two ends. This molecule is derived from a single copy of the rRNA transcription unit in micronuclear DNA. Thus, excision of the rDNA is accompanied by replication in a way that creates a palindromic molecule of 11 kb containing two copies of the transcription unit. Other subchromosomal molecules are formed by deletion of sequences between segments, the ends of which are then spliced together to make longer subchromosomal molecules. Although such cutting, deletion, and splicing occurs in a regular way, so that the samesized molecule always contains the same genes, alternative cutting and splicing can occur. In addition, some sequences that are eliminated in one stain of *Tetrahymena* may be retained in the macronucleus of another strain. Altogether, it is estimated that there are about 5000 sites in the micronuclear genome at which breakage, deletion, and splicing take place to yield the 270 different subchromosomal molecules.

The amount of sequence eliminated in connection with chromosomal fragmentation in hypotrichs is much greater than that which occurs in holotrichs. The final molecular products are much smaller, and more kinds are formed. In *Oxytricha*, about 24,000 different gene-sized molecules are present in the macronucleus. These molecules are not simply excised from chromosomes during macronuclear development but are formed by cutting, deletion, and splicing of DNA segments. Most, if not all, macronuclear molecules, as they are represented within micronuclear chromosomes, contain sequences of a few dozen base pairs up to 1000 bp that are deleted when the macronuclear version of the gene is generated. These sequences, called internal eliminated sequences (IES), interrupt regions that are destined to become a part of the transcribed segments in macronuclear DNA molecules or to become part of the non-transcribed segments (leaders and trailers) in the macronuclear molecules.

Some micronuclear DNA segments in *Oxytricha* undergo alternative cutting, deletion, and splicing, giving rise to several macronuclear DNA molecules of different-sizes that share a high degree of homology.

Finally, in the formation of the macronuclear molecule that encodes actin, IES are deleted and fragments spliced, but the fragments are spliced in a different order and polarity than that present in the micronuclear chromosome. This is the only case of a genetic rearrangement in DNA sequences in a ciliate that occurs as a normal part of ciliate development. All other changes in ciliates described as rearrangements are in fact deletions followed by splicing.

Enzymes are presumed to catalyze the various cutting—deletion—splicing—reordering events that transform transform micronuclear chromosomes into macronuclear genes, but none has yet been identified, except the telomere terminal transferase.

If remains a possibility that all sequence processing is accomplished through RNa intermediates (transcribed from micronuclear DNA) in a manner analogous to the cutting, deletion, and splicing of RNA in the removal of introns from primary transcripts, followed by reverse transcription into DNA. Such a process is responsible for the generation of pseudogenes in higher eukaryotes. Lipps (1985) detected reverse transcriptase activity in the developing macronuclei of *Stylonychia*. Models designed to accommodate all the events of macronuclear DNA formation in hypotrichs via RNA intermediates requires a set of complex assumptions. In holotrichs such as *Tetrahymena*, it is virtually certain that formation of macronuclear DNA does not involve RNA intermediates.

Differential Amplification of rDNA

During the formation of the macronucleus in *Tetrahymena*, the various subchromosomal fragments are replicated equivalently to form the DNA-rich macronucleus. The final copy number is about 45 for each molecule. However, the gene encoding rRNA is derived from a single copy of the gene in the micronucleus but is differentially amplified to yield about 9000 copies in the mature macronucleus. Likewise, in *Oxytricha* each of the 24,000 or so different molecules is present in about 1000 copies per macronucleus in the G1 period, but the gene for rRNA is present in 100,000 copies. The 100,000 copies are derived from probably two, but no more than a few, copies of the gene (per haploid genome) in the micronucleus of *Oxytricha*. Therefore, during macronuclear development, replication of the single gene for rRNA in *Tetrahymena* and of the two to several genes in *Oxytricha* is differentially regulated.

A few other DNA molecules undergo differential amplification during the clonal aging of *Stylonychia* Different clones show amplification of different molecules. Which molecules amplify is a genetic trait of the clone. The coding functions of these molecules are not known. Similar amplification of a few kinds of molecules has been observed in *Oxytricha* including amplification of molecules encoding rRNa to the increased level of 1.5×10^6 copies per macronucleus. Attempts to produce cell clones with differential amplification of metallothionein genes by chronic exposure to heavy metal ions have so far not succeeded.

Overall Organization of Genes in Ciliate Chromosomes

In holotrichs, the micronuclear DNA is cut into about 270 have sub-chromosomal segments, each containing many genes. The only known exception is the gene encoding rRNA; it is present by itself in a separate, sensitively small DNA molecule of 11 kb in two palindromically arranged copies. By contrast, in hypotrichs all micronuclear DNa is reduced to small molecules with a number average of 2200 bp, with a single gene per molecule. For the few molecules that have been examined in detail, roughly one half of each molecule consists of a coding sequence; the remainder form non-transcribed leaders and trailers. The hypotrichs, in effect, define the maximum length of a gene, including its transcription and replication accessory sequences [each molecule is also a separate replication unit]. This circumstance can be used to show how all the genes are arranged in chromosomes relative to one another and without needing to known the coding function of the genes. Hybridization mapping of large segments of cloned, micronuclear DNA has shown that most genes are grouped along the chromosome. Each group contains two to several genes separated by short spacers. One group is separated from the next by spacers many kilobases in length, consisting of both unique and repetitious sequences. Both rRNA genes that have been identified occur singly and are separated from adjacent genes to both sides by many kilobases. The destruction of spacers between genes accounts for a major part of the loss of DNA sequence complexity during macronuclear development. The presence of spacers between genes is a plausible explanation for the C-value (S-value) paradox in eukaryotes.

What advantages accrue to ciliates by the extensive processing of DNA is not completely clear. Some possible explanations are the following. Amplification of the total genome in the formation of macronucleus presumably allows for much larger cell size, which is important in the life-style of the organisms. Differentially higher amplification of genes encoding rRNA must occur to provide for adequate ribosome production. The differentially amplified rRAN genes form a band in a gel in which total macronuclear DNA has been electrophoretically distributed by size. Several more bands of varying intensities are present, suggesting that several additional genes may undergo differential amplification. The coding functions of these amplified DNA molecules are not known. The destruction of spacer DNA, eliminating all repetitious sequences and more than 90% of unique sequences in the genome of hypotrichs, eliminates the cost of carrying the large amount of non-essential DNA that would be present in the highly polyploid macronucleus required to support a very large cell size. However, *Tetrahymena* (and presumably other holotrichs) eliminates only 15% of its genome. The creation of thousands of subchromosomal DNA molecules (holotrichs) or millions of gene-sized molecules (hypotrichs) in the macronucleus may preclude division by mitosis for some reason. On the other hand, amitotic division of the macronucleus is only successful because of the high multiplicity of copies of each DNA molecule in ciliates. Finally, the DNA processing that accompanies macronuclear development may not be advantageous but may simply be necessary for transcriptional activation of genes, since micronuclear genes in general are never expressed.

Whatever the ciliates have gained by the unusual manipulations of their genomes, they have come about after the early separation of ciliates from other eukaryotes. Some events held in common with other eukaryotes, such as differential amplification of rDNA in oocytes, and cutting and splicing of DNA to generate Ig genes, probably arose independently in evolution. Further attestation to the singularity of the ciliates in evolution is the alteration of their genetic code. The codons UAA and UAG, which serve as stop codons in eukaryotes generally, are read as glutamine in ciliates. UGA is the only stop codon in ciliates, that does not code for any amino acid and thus stops the synthesis of protein.

4

Nucleic Acids in Differentiation

It is a well-known fact that all cells of the embryo are derived from the fertilized egg and the differences between cells are mainly in the specific proteins required for specialized function. Finally, we know that the structure of proteins is coded by the sequence of nucleotides in the DNA of the cell's chromosomes. If cells with identical genes become different, surely only certain genes in certain cells have been activated. This would lead to the synthesis of specific proteins in the specific cell types.

It has already pointed out that chemical factors are generally responsible for activating or repressing genes. For example, the factors that tell cells to active their genes for nerve-specific proteins are apparently passed from an inducing to a responding tissue. In this chapter we shall examine differentiation at the nucleic acid level, as this level plays a major role in controlling protein synthesis. We shall examine evidence that specific genes are indeed activated at specific time during the differentiation of specific cells. In addition, we shall look at possible means by which these genes are repressed or activated. First let us briefly consider some of the techniques used to reveal differences in cellular nucleic acids, which reflect activation of specific genes.

Methodology

Several techniques have been developed to determine whether the nucleotide sequences of two batches of DNA or of a batch of DNA and a batch of RNA are similar. This sort of information is needed to reveal the differences and similarities in nucleic acids during the differentiation process. These techniques are based on the pairing specificity that occurs between the bases of nucleotides in nucleic acids. We know that adenine (A) on one DNA chain pairs specifically with thymine (T) on the other chain, and that guanine (G) pairs with cytosine (C). We also know that RNA nucleotides contain uracil (U) instead of thymine. During RNA synthesis, uracil pairs with adenine on a DNA chain. These pairings occur by means of hydrogen bonds that form between the bases of the nucleotides.

The strands of the DNA double helix can be separated by mild heating, which breaks the hydrogen bonds between the nucleotides. The strands will reassociate (anneal) if the mixture is slowly cooled, with A binding T and C binding G. This process is the basis for the techniques we shall consider. Because two strands of DNA will associate only if the sequences of bases are complementary, we can compare DNAs from different organisms or tissues for similarities and differences in base sequence by separating them through heating and then promoting reassociation under carefully controlled conditions. If reassociation is rapid and complete, the DNAs have similar or identical base sequences. If reassociation does not occur, the base sequences are not similar. Such experiments are called hybridization experiments. Many variations of the basic hybridization experiment have been developed. Some involve binding RNA to DNA in the presence of other types of RNA. These experiments help determine whether the two types of RNA have similar nucleotide sequences. If we know whether two types of cells have similar or different RNA sequences, we can determine whether the same or different genes have been activated in these cell types. These experiments will be described shortly.

The experimental protocols used in nucleic acid hybridization experiments are complex, and the conditions under which the nucleic acid strands reassociate must be carefully controlled. No attempt will be made to describe the details of these experiments. For these, the student is referred

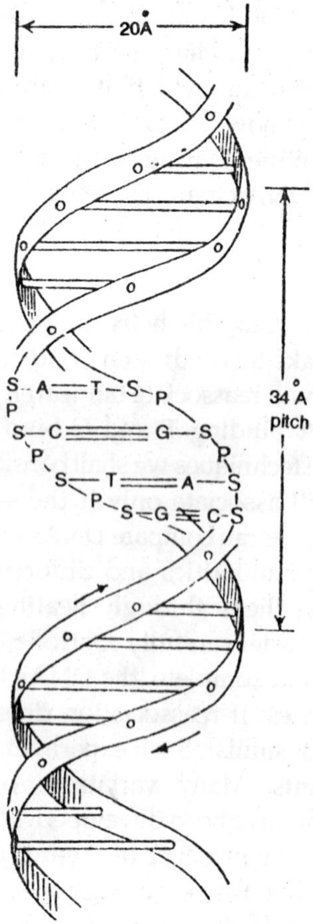

Fig. 4.1. DNA and its paired bases.

to the references at the end of the chapter. We shall, however, provide a general description of how these experiments are performed and what they can tell us about differentiation at the nucleic acid level. In the basic DNA-DNA hybridization experiment, radioactive DNA is produced by injecting a radioactive label, such as "hot" phosphate, into the organism.

Nuclei with radioactive DNA are isolated by centrifuging the homogenized tissue, and the nucleic acids are chemically extracted from the nuclei, RNA is then removed enzymatically from the DNA preparation, isolating the radioactively labeled DNA. Non-radioactive DNA is also used in these experiments; it is extracted from other specimens in the same way. The non-radioactive DNA is heated to break the hydrogen bonds between the nucleotides. The DNA helices thus dissociate into single strands. These non-radioactive single pieces of DNA are attached to a filter or imbedded in a column of gel. Single strands of radioactive DNA segments are then passed through the filter or column. If the base sequence of the DNA on the filter is similar to that of the radioactive DNA, base pairing will occur and the radioactive DNA will anneal with the DNA on the filter. The amount of radioactive DNA bound to the DNA on the filter can easily be measured with a scintillation counter. If the two DNA's are not similar, the radioactive DNA will pass through the filter without binding. In this way one can determine whether two DNAs have similar or different base sequences. The results of such DNA-DNA hybridization experiments suggest that there is species-specific differentiation of DNA. That is, DNA base sequences differ between species. Evolutionary relationships between different species have been confirmed using the DNA-DNA hybridization technique. In general, closely related species have more DNA base sequences in common than do distantly related species. In addition, such experiments suggest that there is no tissue-specific differentiation of the DNA base sequence in cells of the same organism. This makes sense because all cells in an organism are derived from the fertilized egg and should possess identical genes.

Major advances in nucleic acid technology include development of the technique of gene cloning. In this technique, purified messenger RNA is used t make many copies of the gene (DNA) that codes for the messenger RNA. This is accomplished with the enzyme reverse transcriptase (RNA-dependent DNA polymerase). This enzyme copies RNA templates to make complementary DNA strands (cDNA). Many copies of a given gene can be produced in this manner. Such gene clones are extremely useful in determining whether a given species of messenger RNA is present in a specific cell at a specific time. This is done by using DNA-RNA hybridization techniques to measure how much of the RNA under study is complementary to the specific cDNA. In this way cDNA is used directly to determine the presence or absence of a particular species of RNA. The study of differentiation at the messenger RNA level, therefore, has become far more direct through the use of cDNA probes.

Tissue-Specific Differentiation of RNA

There are three types of RNAs in Eukayotic organisms—mRNA, tRNA and rRNA. Are different RNA molecules present in different tissues? In other words, is there tissue-specific differentiation of RNA? Intuition says yes, since different tissues possess different tissue-specific proteins, and these proteins must be coded for by specific RNA molecules. We would guess further that although all cells in an organism possess identical genes, only some genes are active in given tissues. Thus tissue cells can differentiate by synthesizing the specific proteins required for their specialized functions. Let us briefly consider an experiment that shows that different tissues do indeed contain different RNAs. Note again that it is the base sequence in messenger RNa (and DNA) that determines the amino acid composition of specific proteins. A modification of the hybridization technique is used to determine whether tissue-specific differentiation of RNA exists. Segments of DNA are heated or chemically treated to separate them into single strands. These strands are adsorbed onto a nitrocellulose membrane filter. Radioactive RNA is then produced by injecting radioactive RNA precursors into an organism and extracting the RNA from different tissues, such as the spleen, liver and kidney. Radioactively labeled RNA from a specific tissue is then incubated with the DNA on the filter in the presence of increasing concentrations of unlabeled RNa from different tissues.

If unlabeled RNA from a given tissue significantly inhibits the labeled RNA from binding to the DNA, the base sequence in this unlabeled RNA must be similar to that of the radioactive RNA. Fewer labeled RNA molecules can bind to the DNA because the hybridization sites on the DNA template are being taken by the similar unlabeled RNA molecules. Likewise, if unlabeled RNA from a given tissue does not inhibit the radioactive RNA from binding to the DNA on the filter, the base sequence of the unlabeled RNA must not be similar to the base sequence of the radioactive RNA. The unlabeled RNA is binding to DNA strands different from those to which the labeled RNA binds. The student is referred to the references at the end of this chapter for a more in-depth analysis of the many parameters and problems associated with-these experiments.

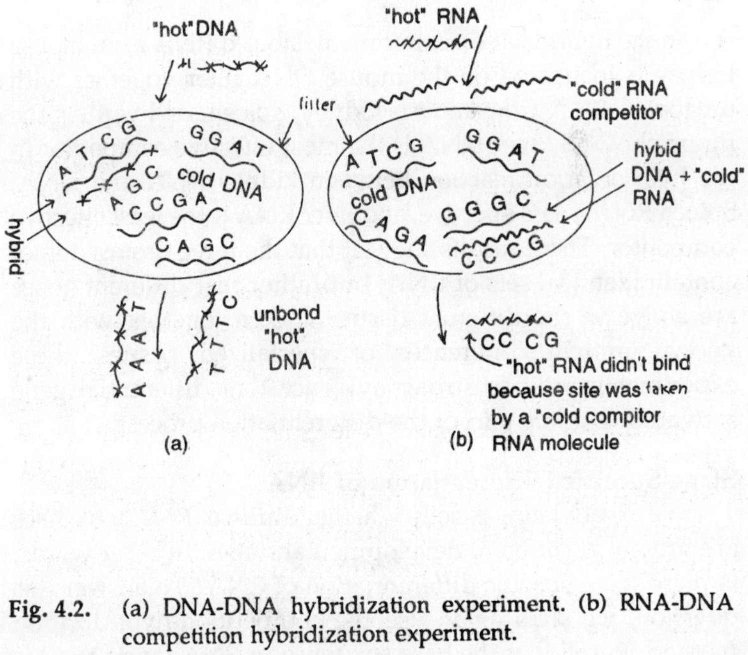

Fig. 4.2. (a) DNA-DNA hybridization experiment. (b) RNA-DNA competition hybridization experiment.

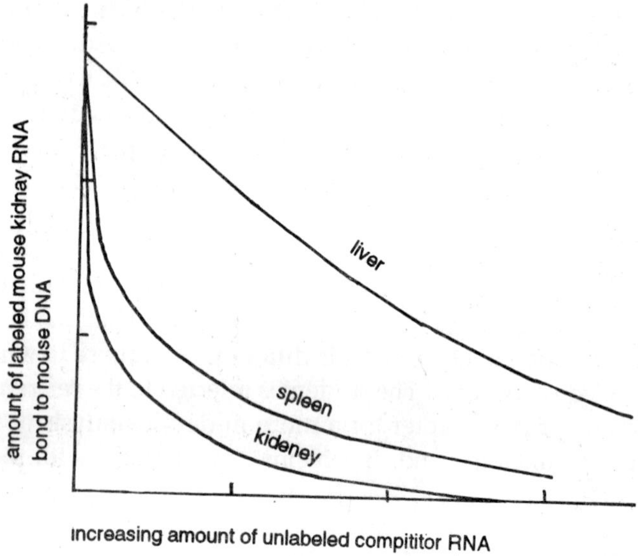

Fig. 4.3. Tissue-specific differentiation of RNA.

In the hybridization experiment, labeled RNA from mouse kidney is incubated on the mouse DNA filter, together with unlabeled RNA from mouse kidney, spleen, or liver. As the figure shows, kidney RNA is the most effective competitor in the hybridization reaction between kidney RNA and DNA, Spleen RNA is less effective, and liver RNA is the least effective competitor. These results suggest that the three tissues do not contain identical sets of RNA, but rather that different genes are active in the different tissues. Taken together with the known protein differences of specialized tissues, these experiments provide strong evidence that differential gene activation is a key part of the differentiation process.

Stage-Specific Differentiation of RNA

We do not know exactly whether different RNA molecules are present at different developmental stages. In other words, is there stage-specific differentiation of RNA? To answer this question, we shall again use the competition-hybridization method. Recall that the base sequence in RNA (and DNA) is the genetic code that controls the synthesis of each specific protein. So in order to identify similarities and differences in RNA during different developmental stages, we must examine base sequences. Similarities and differences between base sequences of nucleic acids can be determined using hybridization techniques. If segments of DNA or a segment of DNA and a segment of RNA hybridize with each other, the base sequences are complementary. If they do not, the base sequences are not similar.

To determine whether there is stage-specific differentiation of RNA during early embryonic development, segments of DNA double helix are heated or chemically treated so that the DNA separates into single strands. The strands are adsorbed on a filter. Radioactively labeled RNA from a specific developmental stage of an embryo of the same species is then incubated with the DNA on the filter. These incubations are carried out in the presence of increasing concentrations of unlabeled RNA from the same or a different developmental stage of the organism. If a batch of unlabeled RNa from a give developmental stage significantly inhibits the labeled RNA from binding to the DNA, the base sequence of this unlabeled RNA is similar to that of the labeled RNA. If, on the other hand, the unlabeled RNA from a given developmental stage does not inhibit the binding of the labeled RNA, the base sequence of the unlabeled RNA is not similar to the base sequence of the radioactive RNA.

In another hybridization experiment, labeled RNA from the prism (postgastrula) stage of the sea urchin embryo is incubated on the sea urchin DNA filter together with unlabeled RNA from different sea urchin stages. The graph shows that unlabeled prism RNA is the best competitor of the labeled prism RNA in binding to the DNA on the filter. The next best competitor is unlabeled gastrula RNA; unlabeled blastula and unfertilized egg RNA are significantly poorer competitor. Unlabeled *P. morganii* (starfish) RNA does not compete at all. From this experiment, we can conclude the following : (1) The

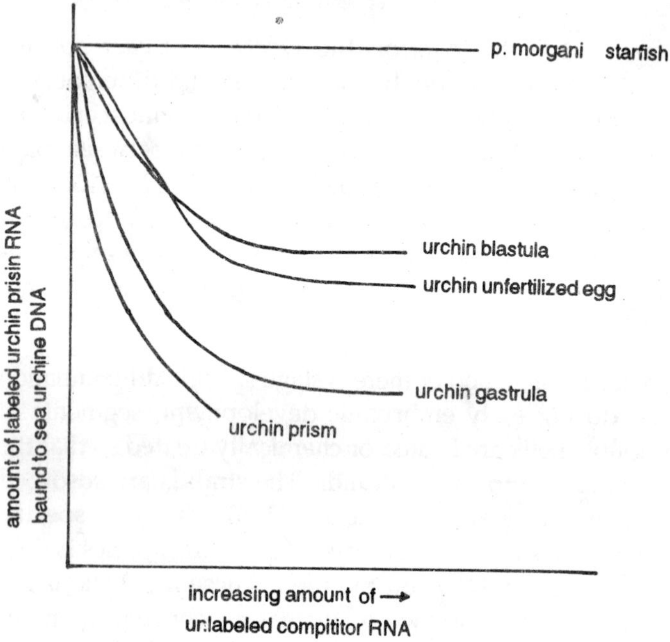

Fig. 4.4. Stage-specific differentiation of RNA.

sea urchin gastrula has many RNA sequences in common with the sea urchin prism stage. (2) The sea urchin blastula and unfertilized egg have fewer RNA sequences in common with the prism stage. (3) Starfish RNA is not similar to sea urchin prism RNA in terms of base sequence. This work suggests that there is indeed stage-specific differentiation of RNA. This makes sense, because suiely different genes are activated at different developmental stages to form the RNA needed to code for the specific and sometimes different proteins required for development to proceed at each stage. Before leaving the discussion of nucleic acid hybridization, we should mention some important information that this technique has revealed about the genome of the cells of higher organisms (eukaryotic cells). Eukaryotic cells—all cells above the bacteria and blue-green algae—have a well-defined nucleus. Prokaryotic cells (bacteria and blue-green algae) generally contain a single naked DNA double helix. Most eukaryotic cells contain thousands of times the amount of DNA found in prokaryotic cells. Much but not all of this DNA is required to code for and regulate the tremendous genetic diversity expressed during the differentiation process in higher organisms.

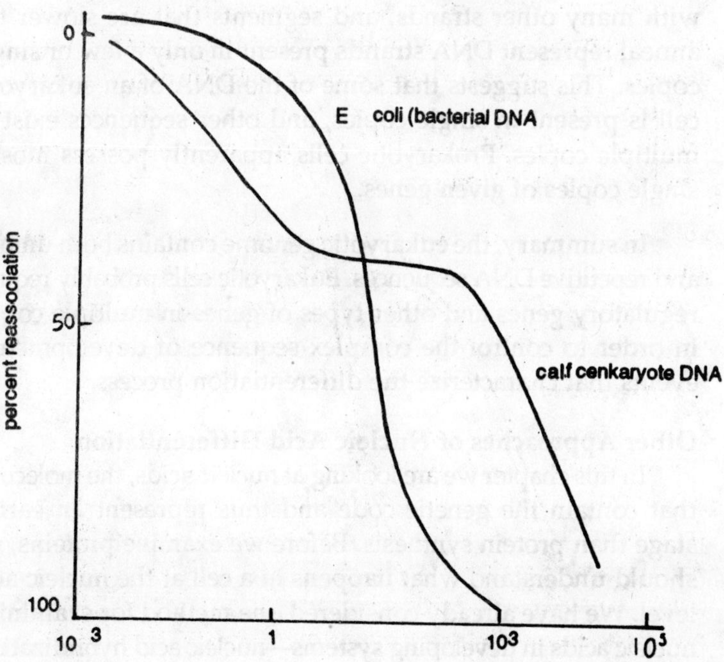

Fig. 4.5. **Reassociation of prokaryotic (bacterial) DNA and of eukaryotic (calf) DNA.**

Investigators using the DNA-DNA hybridization technique have shown that some single strands of eukaryotic DNA reanneal with each other rather quickly at low DNA concentrations. Other pieces, however, do not reanneal until the DNA concentration and incubation time increase. The more quickly DNA strands hybridize at low DNA concentrations, the greater the number of identical DNA copies present. Obviously, it is not difficult for a strand to find and bind with a complementary strand if many copies of the sequence are present. The DNA strands that are present in fewer copies, however, have less chance of colliding with a complementary sequence. Thus, the strands that reanneal quickly at low concentrations have base sequences in common with many other strands, and segments that are slower the anneal represent DNA strands present in only a few or single copies. This suggests that some of the DNA of an eukaryotic cell is present in single copies, and other sequences exist in multiple copies. Prokaryotic cells apparently possess mostly single copies of given genes.

In summary, the eukaryotic genome contains both unique and repetitive DNA sequences. Eukaryotic cells probably require regulatory genes and other types of genes in multiple copies in order to control the complex sequence of developmental events that characterize the differentiation process.

Other Approaches of Nucleic Acid Differentiation

In this chapter we are looking at nucleic acids, the molecules that contain the genetic code and thus represent an earlier stage than protein synthesis. Before we examine proteins, we should understand what happens in a cell at the nucleic acid level. We have already considered one method for examining nucleic acids in developing systems—nucleic acid hybridization. We have seen how this method can be used to reveal tissue-specific differentiation of RNA during development. Heretofore, we have examined only total cellular RNA, but this method also makes it possible to identify specific messenger RNA molecules. This is done by using the enzyme reverse transcriptase to make DNA that is complementary to purified molecules of messenger RNA. This complementary DNA (cDNA) is a very potent tool that can be used to determine whether its complementary messenger RNA is present in cells. Single strands of the DNA are adsorbed to a filter, and batches of RNA are passed through the filter. Any messenger RNA that is complementary to the cDNA will hybridize with the cDNA. These hybrids can be detected by any of the variety of methods previously described. Thus we can determine whether a specific message for a specific protein is present in a population of cells. We can also determine how many times this message was duplicated and from that learn a lot about specific gene activation at the nucleic acid level.

If only cytoplasmic messenger RNA molecules are used in hybridization experiments, complex results are obtained. In this experiment, mRNA populations from different developmental stages of the sea urchin were hybridized with DNA made using gastrula stage mRNA as a template. The experimental results indicate that the sea urchin oocyte possesses not only the same mRNA sequences as the gastrula, but also many mRNA sequences not present in the gastrula. Adult sea urchin tissues (tube-foot, intestine, and coelomocyte) possess only a small number of messages, some shared with the gastrula and others not. The unshared messages presumably code for specific proteins required in these specialized differentiated tissues. Like those presented previously, these result indicate that the characteristics of cellular RNA vary during embryonic development and differentiation. These results have been extended to other systems including plant embryos. Using the hybridization technology, Goldberg and colleagues found that there is a large overlap in the mRNA species of different developmental stages of the soybean emt ryo. However, they also identified a unique set of superabundant mRNA that occur in different amounts at different developmental stages. This set of mRNAs probably codes for a spectrum of seed proteins that are synthesized during maturation. Some of these proteins are probably metabolized by the rapidly growing seedling. The issue is made more complex by results obtained by Rosenthal, Hunt, Ruderman, and others, which indicate that different mRNA molecules need not be present at different developmental stages. For example, oocytes and embryos of the surf clam *Spisula solidissima* contain identical sets of mRNA.

Different groups of these message sets, however, are associated with ribosomes at the different stages. Thus in the surf clam embryo, stage-specific subsets of mRNA are taken from a set present at both stages (oocyte and embryo). Stage-specific differences in protein types at the different developmental stages probably result from such stage-specific use of specific messages from the large population of messages contained in oocytes and later embryos. Some of the mRNA molecules are probable masked at certain stages, which prevents them from associating with ribosomes and, hence, being translated. Future work should tell us more about mRNA masking and about how a message can be unmasked to become functional at a specific developmental stage.

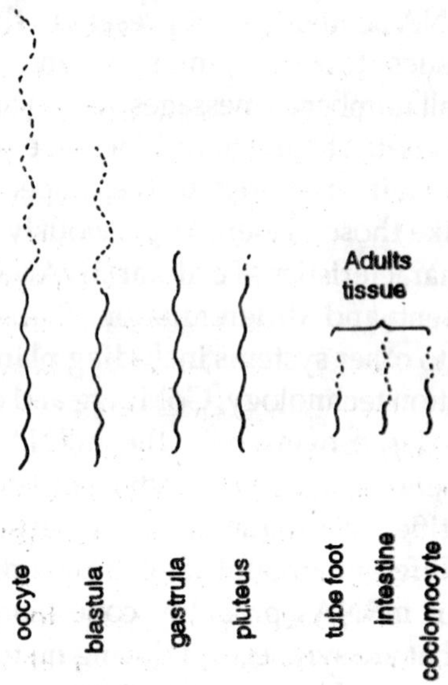

Fig. 4.6. Comparison of messenger RNA from sea urchin gastrula with messenger RNA from other sea urchin stages and tissues.

A second method for examining nucleic acid changes during differentiation uses the RNA synthesis inhibitor actinomycin D. Differentiating cells are incubated with this drug at various times. If actinomycin D prevents the synthesis of a specific protein, this suggests that the drug blocked the transcription of the messenger RNA for that protein. If, on the other hand, actinomycin D treatment does not block synthesis, the message for that protein was probably already present in the cell. In this way we can ascertain whether (1) a specific gene becomes activated shortly before protein synthesis occurs or (2) stable messenger RNA molecules were produced at sometime prior to incubation with actinomycin D. Determining when a specific gene is activated is an important part of understanding differentiation at the nucleic acid level.

A third approach to investigating the changes in nucleic acids during differentiation is to identify specific messenger RNA molecules by their ability to direct the synthesis of specific proteins. This can be done by incubating the RNA with the necessary components of the protein synthesis system. These components include ribosomes, amino acids, transfer RNAs, and a variety of factors and enzymes. They can be prepared from bacterial cells. We can determine whether messenger RNA molecules specific for a given protein are present by synthesizing that protein with such an *in vitro* protein synthesizing system.

These techniques have been used in many developing systems. For example, at a given point in the development of red blood cells, muscle cells, and pancreas cells, actinomycin D no longer prevents specific protein synthesis. This observation suggests that stable messenger RNA molecules accumulate in these cells in preparation for the extensive synthesis of cell-specific protein characteristic of the differentiated state. In the mouse, actinomycin D blocks the synthesis of hemoglobin in prospective red blood cells before day ten to gestation. After day ten it does not do so. Thus any hemoglobin message made before day ten is stable and can synthesize the protein after day ten.

We might now ask, how does messenger RNA become

stabilized? One possibility is that the rate rate of message synthesis is increased before day ten so that there are too many copies for the message to be totally degraded. Also, additional nucleotides, such as polyadenylic acid, may be attached to the messages as they are synthesized, and these nucleotides may combine with some protein to protect the messages from degradation. Along the same lines it has been proposed that most or all genes in the genome are transcribed in eukaryotic cell, but that only some are protected from degradation This sort of regulation might play a major role in determining which genes form messages that are active in synthesizing protein—many messages are transcribed but few reach the cytoplasm to synthesize protein. This concept (proposed by Wilt and others) may be one important piece of the differentiation puzzle. The chromosomes of higher cells are complex. Before we discuss differentiation at the protein level, let us look at another interesting proposal that may account for the observation that during differentiation certain genes are turned on and other are turned off.

Gene Transcription During Differentiation

Chromosomes in the cells of higher organisms contain a great deal of protein associated with the DNA. This protein may play an important role in differentiation by repressing certain genes while allowing others to become active. There are two major types of chromosomal proteins: histone proteins and non-histone proteins. Histones are low-molecular-weight proteins rich in basic amino acids such as lysine and arginine. Histones from different organisms and different cell types are structurally similar. Non-histon proteins associated with DNA are larger and much more diverse than the histones. Most are phosphorylated and are acidic rather than basic in nature. It has already been seen that non-histone proteins may prevent histone binding or remove histones from regions of DNA, thus allowing RNA synthesis. Experiments using DNA-RNA hybridization techniques indicated that estrogen-stimulated oviduct chromatin produces eight times as much oviduct-specific ovalbumin messenger RNA as does chromatin from oviducts in which estrogen treatment was halted. If the non-histone protein fraction from estrogen-treated oviduct chromatin is added back to DNA with histones from oviducts in which estrogen treatment was halted, this chromatin then transcribes specific ovalbumin messenger RNA in amounts characteristic of estrogen-treated oviduct tissue. In the reverse experiment, non-histone proteins from estrogen-halted chromatin lowered the amount of ovalbumin messenger RNA produced by DNA with histones from estrogen-treated oviducts.

To summarize, inducer complexes, hormone complexes, and other molecules that control differentiation may activate genes by binding to non-histone proteins on chromosomes. Another major step forward in understanding gene activation during differentiation has come from work by Keith Yamamoto and his colleagues at the University of California at San Francisco. Early work in this area showed that when glucocorticoid steroid hormones attach to a liver cell, for example, they specifically bind to glucocorticoid receptor molecules that are present in the cell membrane. The hormone-receptor complexes then make their way to the nucleus of the cell and attach to regions of DNA called *enhancers*. These enhancers are stretches of nucleotides that are located adjacent to or sometimes within genes. They serve to increase the rate at which genes are transcribed, which leads to a greater yield of gene products.

Yamamoto and other hypothesized that enhancers may serve as triggers for differentiation by producing a one-time, irreversible activation of certain genes, causing the cell to become fixed in its production of differentiated cell product. Such an irreversible activation may result from the permanent binding of a specific protein called a *transcription factor* to an enhancer region. Enhancers may also serve roles other than that of trigger. They may control the periodic increases and decreases in activity of turned-on genes, thus acting as modulators. For example, when hormone complexes reach one type of enhancer, they may either dramatically boost transcription of the gene or not affect it at all, depending on whether the gene was triggered or untriggered. Untriggered genes cannot be influenced by the regulator hormone-protein, whereas triggered genes can.

Two different types of enhancers may be involved in this complex regulation: the *trigger enhancer* and the *modulator enhancer*. One way of looking at this concept is to imagine the trigger enhancer as functioning as an on-off light switch. The modulator enhancer can then be compared to a dimmer control that can work only if the on-off trigger is in the on position. Clarification of the details of gene control in higher organisms is helping us understand the complexities of the differentiation process. This absorbing are of research is progressing rapidly. The exact means by which gene action is regulated during the differentiation of eukaryotic cells may soon be completely understood.

Role of Ribosomal RNA in the Amphibian Development

We have mainly considered the differences and similarities in messenger RNA during development because mRNA directly codes for the tissue-specific proteins that are important in differentiation. Before moving on to these proteins, we shall briefly consider a type of RNA that is required for assembling ribosomes, the seats of protein synthesis. The synthesis of ribosomal RNA is controlled by specific genes that become activated at specific times in development. This is nicely shown by the anucleolate mutant of the amphibian *Xenopus*. When a one-nu individual (an individual with only one nucleolus in its cells) mates with another one-nu individual, they produce offspring of the following types: homozygous wild type (two-nu—two nucleoli in their cells), heterozygous (like parents—one-nu), and homozygous mutant (zero-nu—no nucleoli). Although the homozygous mutants have no nucleoli and do not make ribosomes, they do possess ribosomes that were made by the heterozygous mother frog. Using these maternal ribosomes, the mutants can synthesize proteins for sometime during early development. New ribosomes, however, are apparently required by the time the embryo hatches. Since the mutants cannot make new ribosomes, they die by the early swimming tadpole stage.

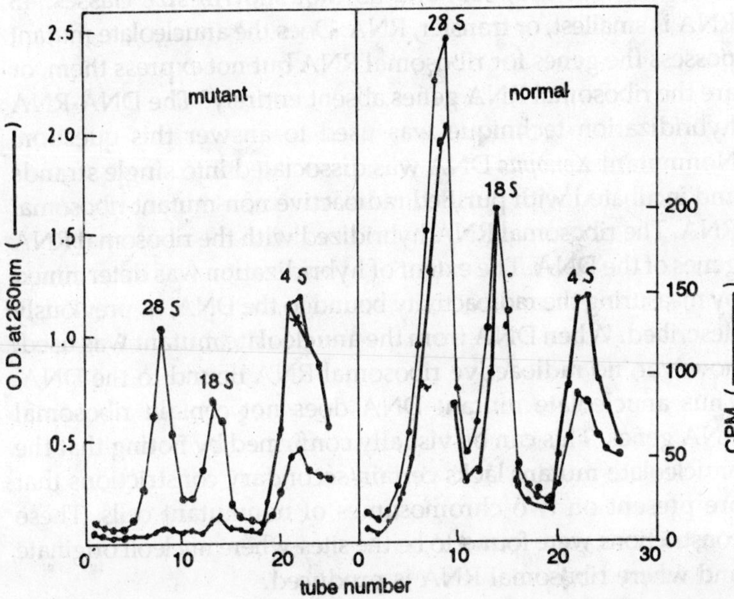

Fig. 4.7. RNA synthesis in normal and mutant *Xenopus* embryos at neurula-tailbud stage. Light lines indicate radioactivity incorporated into newly synthesized RNA. Dark lines show where the specific type of RNA appear in fractions from a density-gradient centrifugation experiment.

Let us look now at some specific evidence indicating that ribosomal RNA synthesis is controlled by specific genes active at specific developmental stages. Figure shows that ribosomal RNA is synthesized in normal *Xenopus* from the neurula stage to the tail bud stage. The figure also shows that ribosomal RNA is not synthesized in the mutant, though other RNA is still made. These results were obtained by incubating neurula-stage embryos with a radioactive label incorporated into RNA. Some of the different types of RNA synthesized were separated on the basis of their sedimentation rates (S, based on molecular size) in a gradient of substance such as sucrose. Ribosomal RNA includes 28S, 18S, and 5S (not shown) size classes. 4S RNA is smallest, or transfer, RNA. Does the anucleolate mutant possess the genes for ribosomal RNA but not express them, or are the ribosomal RNA genes absent entirely? The DNA-RNA hybridization technique was used to answer this question. Nonmutant *Xenopus* DNA was dissociated into single strands and incubated with purified radioactive non-mutant ribosomal RNA. The ribosomal RNA hybridized with the ribosomal RNA genes of the DNA. The extent of hybridization was determined by measuring the radioactivity bound to the DNA, as previously described. When DNA from the anucleolate mutant was used, however, no radioactive ribosomal RNA bound to the DNA. Thus anucleolate mutant DNA does not contain ribosomal RNA genes. This can be visually confirmed by noting that the anucleolate mutant lacks certain secondary constrictions that are present on two chromosomes of nonmutant cells. These constrictions were found to be the sites-where nucleoli originate and where ribosomal RNA is produced.

Studies by Donald Brown and his colleagues at the Carnegie Institution of Washington in Baltimore have provided insight into the mechanisms that turn genes on and off. In the maturing oocyte of *Xenopus*, the South African clawed frog, there are nearly 80,000 clustered copies of the gene that codes for 5S RNA, one of the three kinds of RNA found in ribosomes. These 80,000 5S genes are active only in the oocyte; 1600 other copies of 5S genes, called somatic 5S genes, are active not only in oocytes but also throughout the life of the frog. After fertilization, the oocyte 5S genes become inactive; only the somatic 5S genes remain active. How does this differential activation and inactivation of similar genes occur? The answer to this question is becoming clear. Apparently, individual 5S genes are transcribed when a group of three specific interlocking proteins attaches to a 5S gene in an area that is about 50 nucleotides in length. This region on DNA has been termed the *internal control region.* Once the proteins have attached to a 5S gene, the resulting DNA-protein structure is recognized by the enzyme RNA polymerase III, which is present in the nucleus. The RNA polymerase precisely binds to the internal control region with its attached proteins and then is able to transcribe 5S RNA genes, forming 5S RNA molecules.

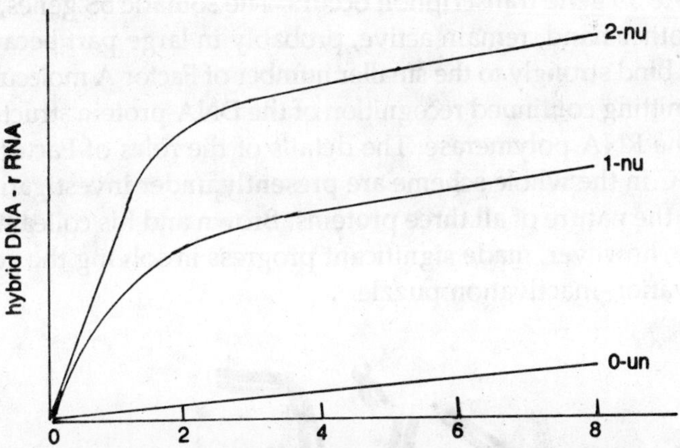

Fig. 4.8. Hybridization of ribosomal RNA with DNA. 2-nu DNA from normal *Xenopus* + ribosomal RNA. 1-nu DNA from heterozygous *Xenopus* + ribosomal RNA. 0-nu DNA from anucleolate mutant *Xenopus* + ribosomal RNA. From C.L. Markert and H. Ursprung, *Developmental Genetics*.

The three proteins involved in activation of the 5S RNA gene have been termed *Factor A, Factor B* and *Factor* C. Donald Brown and his colleagues have shown that, in the *Xenopus* unfertilized egg, there are about ten million molecules of Factor A for each 5S gene. Brown's group has also shown that the control region of the oocyte 5S genes differs by only three nucleotides from that of the somatic 5S genes. This difference, however, appear to give somatic 5S genes a far better ability to bind Factor A tightly. As there are many millions of Factor A molecules in the egg, there are enough to activate the 80,000 oocyte 5S genes, as well as the 1600 somatic 5S genes. After fertilization, however, the number of Factor A molecules drops dramatically so that there is only on Factor A molecule for every four to five 5S genes. Because the somatic 5S genes bind Factor A better than do the oocyte 5S genes, the somatic genes competitively bind most of the Factor A molecules. This leaves the oocyte 5S genes without Factor A. Then RNA polymerase cannot recognize and bind to the oocyte 5S genes, and no oocyte 5S gene transcription occurs. The somatic 5S genes, on the other hand, remain active, probably in large part because they bind strongly to the smaller number of Factor A molecules, permitting continued recognition of the DNA-protein structure by the RNA polymerase. The details of the roles of Factors B and C in the whole scheme are presently under investigation, as is the nature of all three proteins. Brown and his colleagues have, however, made significant progress in solving the gene activation-inactivation puzzle.

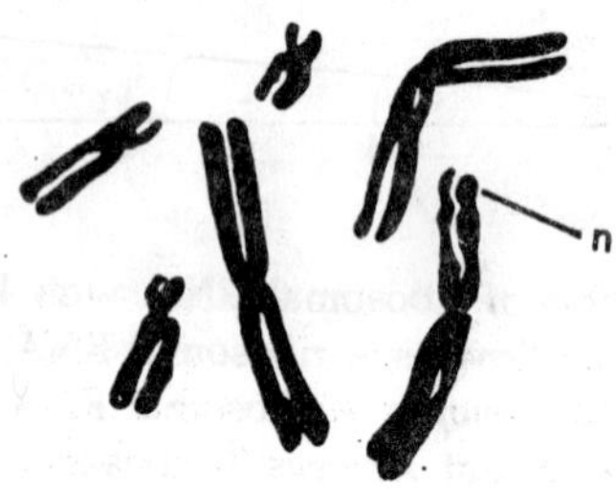

Fig. 4.9. Drawing of mitotic chromosomes of normal *Xenopus* larva.

The ability to transplant nuclei from a variety of embryonic cells into enucleated amphibian eggs has allowed scientists to address two questions: can a nucleus from a later developmental stage still support the development of the early embryo, and what factor turn genes on and off at various developmental stages? Gurdon and others transplanted nuclei from differentiated intestinal epithelium cells, tail-fin epithelial cells

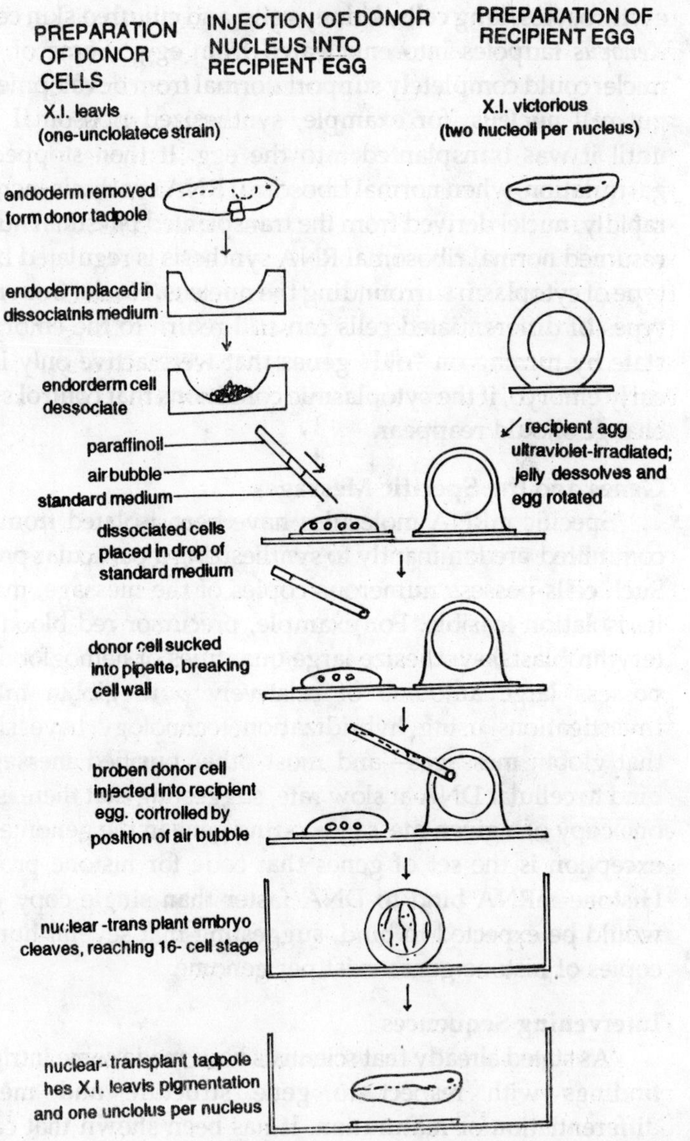

Fig. 4.10. Nuclear transplantation in *Xenopus*. Donor nuclei from one strain are transplanted to host eggs another strain. The tadpoles resulting from such transplants have characteristics of the nuclear parent, showing that the nuclei of the tadpoles were derived from the transplanted nucleus.

erythrocytes, lung cells, kidney cells, and cultured skin cells of *Xenopus* tadpoles into enucleated from eggs. Some of these nuclei could completely support normal from development. A gut cell nucleus, for example, synthesized ribosomal RNA until it was transplanted into the egg. It then stopped. At gastrulation, when normal ribosomal RNA synthesis increases rapidly, nuclei derived from the transplanted intestinal nucleus resumed normal ribosomal RNA synthesis is regulated by the type of cytoplasm surrounding the nucleus. Nuclei from many types of differentiated cells can still return to the embryonic state by turning on "old" genes that were active only in the early embryo, if the cytoplasmic conditions that control such a change should reappear.

Genes and the Specific Messages

Specific mRNA molecules have been isolated from cells committed predominantly to synthesizing a particular protein. Such cells possess numerous copies of the message, making its isolation feasible. For example, precursor red blood cells (erythroblasts) synthesize large quantities of hemoglobin and possess large amounts of relatively pure globin mRNA. Investigations using hybridization technology have shown that globin messages—and most other purified messages—bind to cellular DNA at slow rate, suggesting that there is only one copy of a given message-coding gene in the genome. One exception is the set of genes that code for histone proteins. Histone mRNA bind to DNA faster than single copy genes would be expected to bind, suggesting that several hundred copies of histone genes exist per genome.

Intervening Sequences

As stated already that scientists have made some intriguing findings with respect to gene structure and message differentiation or maturation. It has been shown that certain genes that code for mRNA molecules contain *intervening DNA base sequences*, or introns, that are not present in the messenger RNA codes by such genes. Introns have been found in many genes, including the mouse and rabbit β-globin gene, the mouse α-globin gene, the mouse immunoglobulin light chain gene, and the chick ovalbumin and chick lysozyme genes. Figure 4.11 shows the intervening sequences in the mouse β-globin gene. As the figure shows, the sequence coding for the β-globin polypeptide is interrupted by an intervening sequence between the bases coding for amino acids 30 and 31 and between those coding for amino acids 104 and 105. Intervening sequences are actually transcribed into RNA, but they are enzymatically excised from the newly transcribed RNA. The cut ends are then spliced to form the active mRNA.

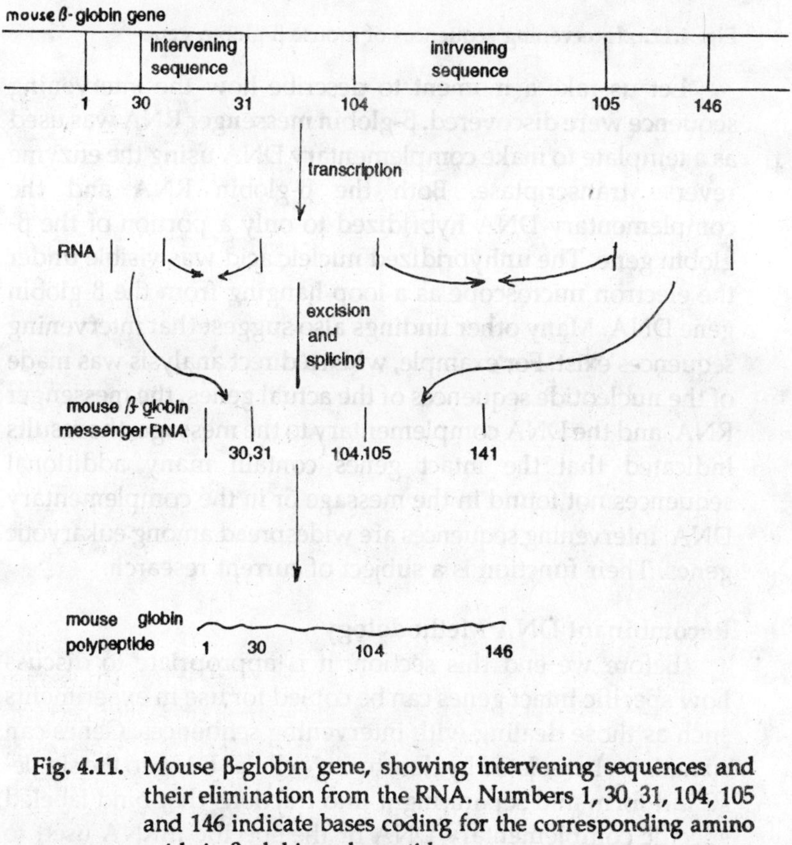

Fig. 4.11. Mouse β-globin gene, showing intervening sequences and their elimination from the RNA. Numbers 1, 30, 31, 104, 105 and 146 indicate bases coding for the corresponding amino acids in β-globin polypeptide.

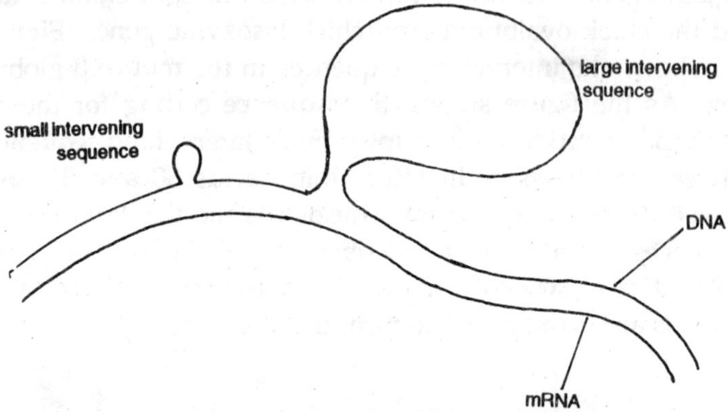

Fig. 4.12. Intervening sequences of mouse β-globin gene.

Let us take a moment to describe how the intervening sequence were discovered. β-globin messenger RNA was used as a template to make complementary DNA using the enzyme reverse transcriptase. Both the β-globin RNA and the complementary DNA hybridized to only a portion of the β-globin gene. The unhybridized nucleic acid was visible under the electron microscope as a loop hanging from the β-globin gene DNA. Many other findings also suggest that intervening sequences exist. For example, when a direct analysis was made of the nucleotide sequences of the actual genes, the messenger RNA, and the DNA complementary to the message, the results indicated that the intact genes contain many additional sequences not found in the message or in the complementary DNA. Intervening sequences are widespread among eukaryotic genes. Their function is a subject of current research.

Recombinant DNA Methodology

Before we end this section, it is appropriate to discuss how specific intact genes can be copied for use in experiments such as those dealing with intervening sequences. Genes can sometimes be isolated by denaturing cellular DNA to the single-strand form and separating it into fractions that bind labeled specific complementary DNA or the specific mRNA used to produce complementary DNA. Genes can also be chemically synthesized from nucleotides. To do this, one must know the exact amino acid sequence in a given protein or the actual nuleotide sequence of the gene. One of the genes that has been chemically synthesized in the human insulin gene.

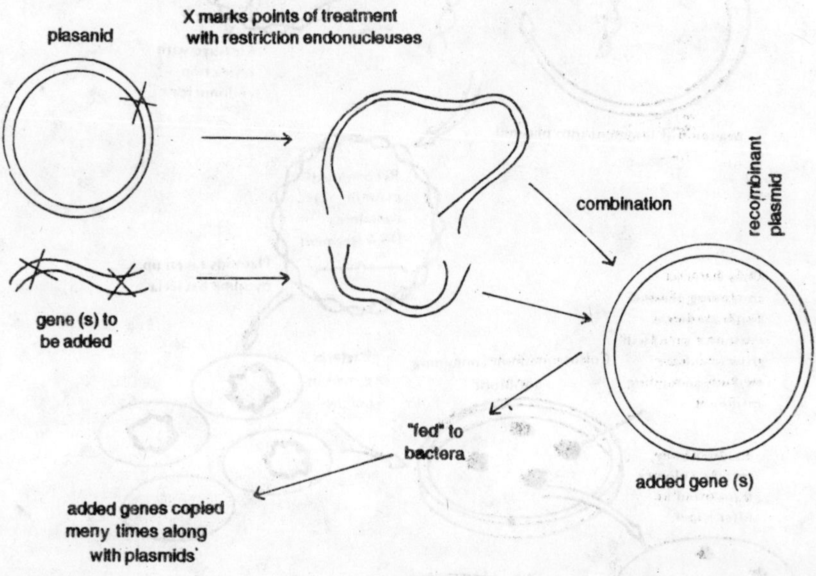

Fig. 4.13. Recombinant DNA technology, simplified drawing.

Once the gene is available, many copies of it must be made in order to provide enough for further experimentation. Enzymes called *restriction endonucleases* are used to insert the genes into small circular DNA molecules (*plasmids*) that replicate in bacteria such as *E. coli*. Restriction enzymes bread DNA strands at specific position determined by short nucleotide sequences. The broken ends are left "sticky". To be inserted into a plasmid, a gene must first be deleted from its DNA molecule; then the circular plasmid must be opened, using the same restriction endonuclease in both operations. The four

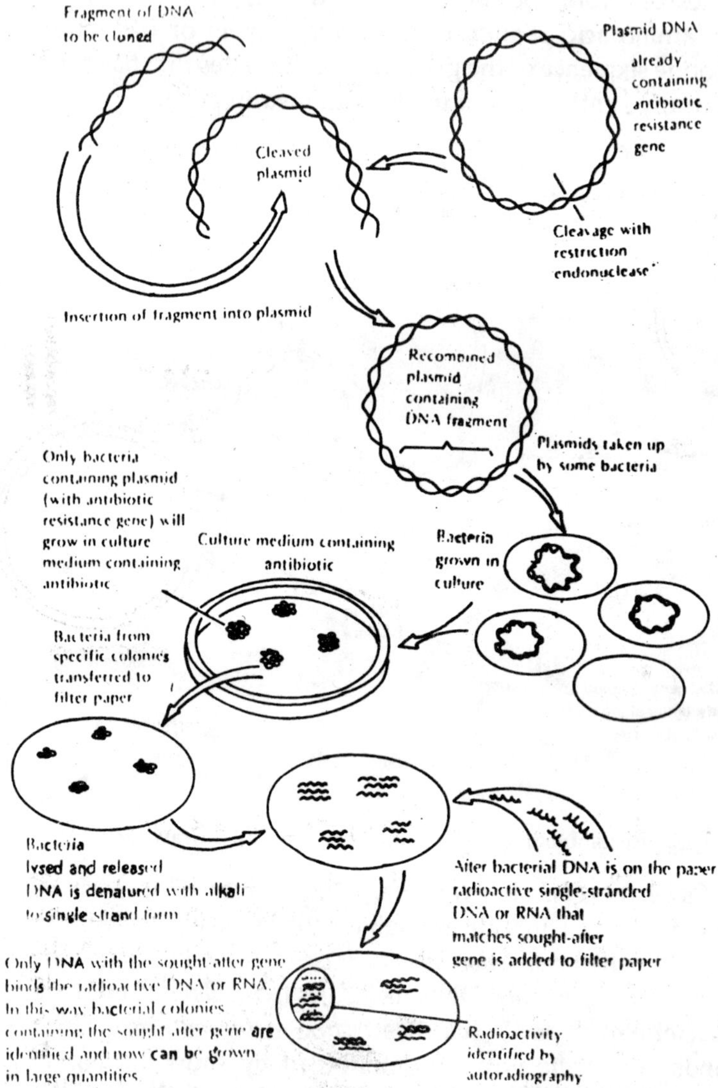

Fig. 4.14. Recombinant DNA technology and gene cloning.

broken ends then must rejoin so as to integrate the gene's DNA into the plasmid, reforming a circle. Because plasmid DNA, like that of viruses, can infect cells, the gene can now be inserted into any cell that the plasmid can attack. Procedures of this sort constitute *recombinant DNA technology,* which by now has revolutionized the study of genetics. Once the recombinant plasmid has been constructed, it can be "fed" to cultures of bacteria such as *E. coli.* The plasmids are constructed so as to give the bacteria that take them up a selective advantage. For example, a gene for antibiotic resistance may be included in the plasmid. If an antibiotic is added to the bacterial culture, only bacteria with the plasmid will grow. Thus cultures of bacteria containing plasmids can be developed. The plasmid DNA replicates in the bacteria, producing clones of specific genes. These genes can then be enzymatically recovered from the plasmids.

The potential of recombinant DNA technology is mind-boggling. Already, the human insulin gene has been cloned. *E. coli* possessing this gene can produce large quantities of life-saving human insulin. This technology is also revolutionizing the study of eukaryotic genes and probably will lead to a better understanding of differentiation in developing systems.

Differential Gene Expression

It is beyond the scope of this introductory text to go deeply into the exciting new work dealing with gene expression during differentiation in higher organisms. It is, however, important to understand some of the key concepts in this field. Gene expression in eukaryotic cells is apparently influenced by a number of mechanisms, including gene loss, gene amplification, and gene rearrangement. Gene are differentially transcribed, and the RNA transcripts are treated in different manners and used in a variety of ways. Families of neighbouring genes (multigene families) regulate the timing of gene expression as well as its amount and diversity. We shall briefly examine a few of these mechanisms in the context of cell differentiation.

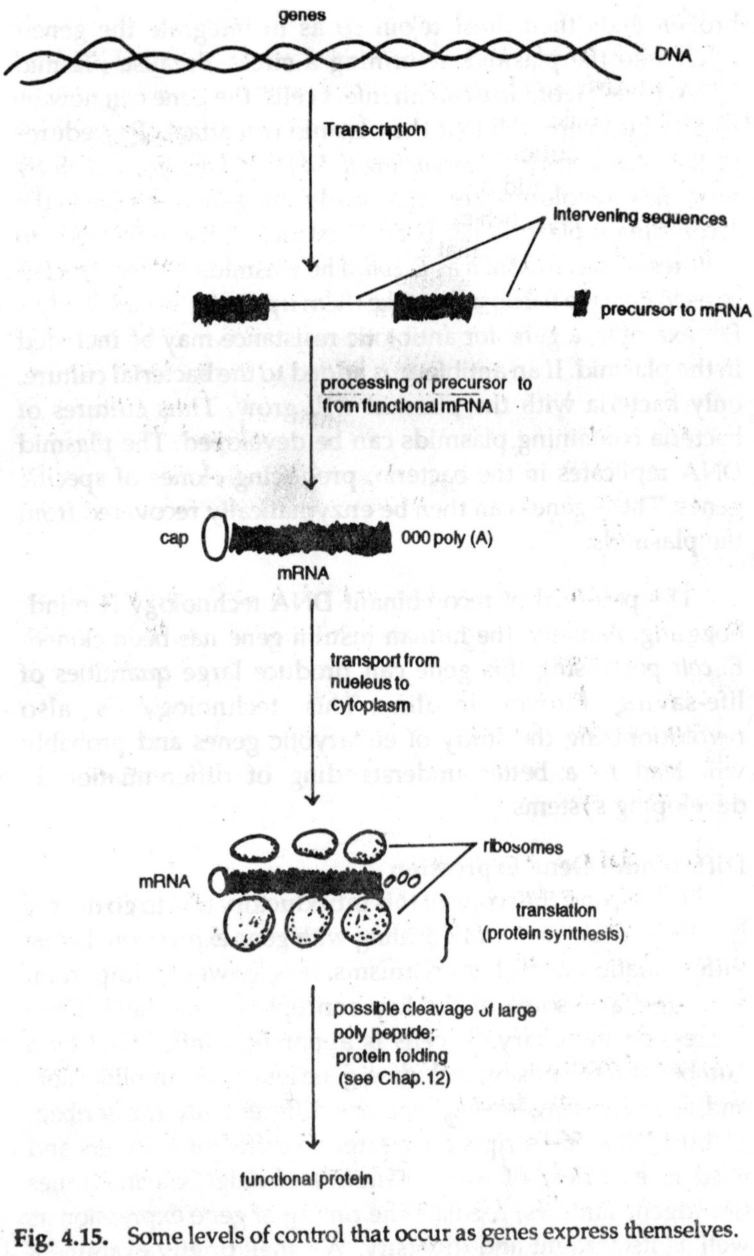

Fig. 4.15. Some levels of control that occur as genes express themselves.

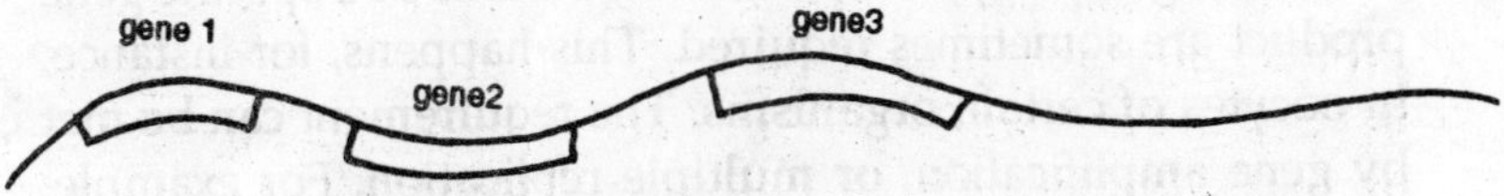

Fig. 4.16. Multigene families in development. Genes 1, 2 and 3 in a family, for example, may all be expressed at one time at a certain developmental stage, In other multigene families, gene 1 may be expressed at one stage, followed by genes 2 and at other stages.

Multigene Families

As development proceeds from stage to stage, some genes that are neighbours on chromosomes are expressed in sequence. The clustering of such genes into families may permit sequential activation to occur. In other cases, a cell turns groups of genes on or off at one time. Whole families of genes may be needed at a given developmental stage, and gene clustering could facilitate their activation or repression in a stage-specific manner. Although clustering of related genes does occur in prokaryotes, little evidence for such clustering is found in eukaryotes.

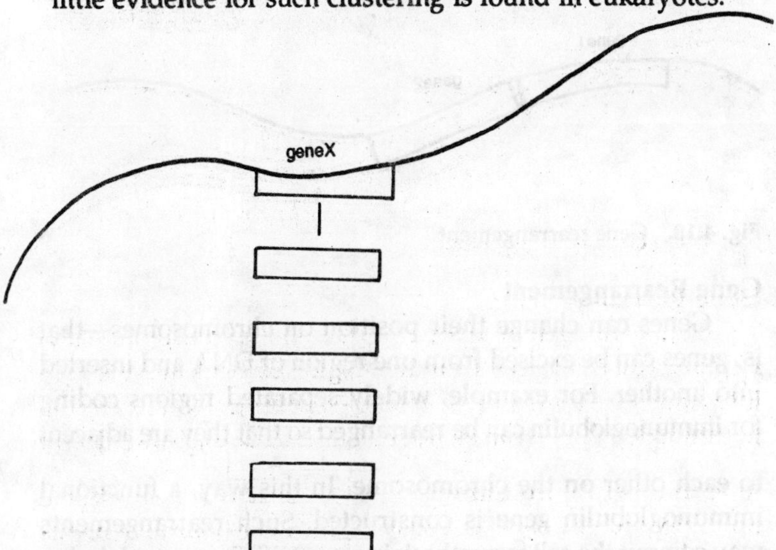

Fig. 4.17. Gene amplification.

Gene Amplification

During development, massive amounts of a specific gene product are sometimes required. This happens, for instance, in oocytes of certain organisms. The requirement can be met by gene amplification, or multiple replication. For example, ribosomal RNA genes are specifically amplified in oocytes of *Xenopus* so that thousands of times more ribosomes can be synthesized than are usually produced by a single body cell. In *Drosophila*, genes for chorion protein are amplified in ovarian follicle cells, which require large quantities of these proteins.

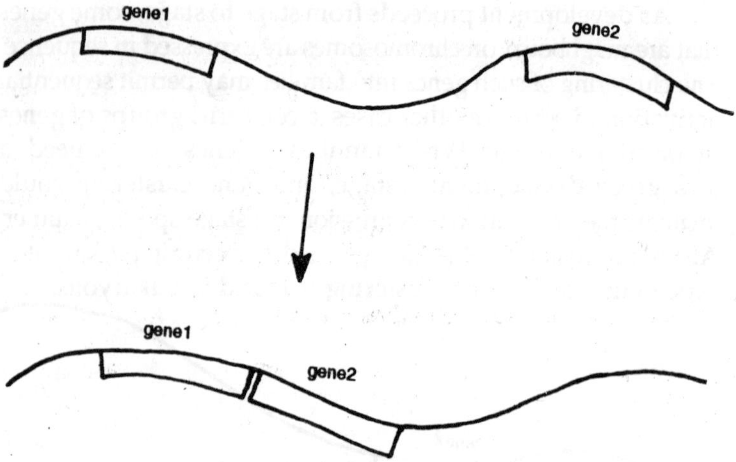

Fig. 4.18. Gene rearrangement.

Gene Rearrangement

Genes can change their position on chromosomes—that is, genes can be excised from one region of DNA and inserted into another. For example, widely separated regions coding for immunoglobulin can be rearranged so that they are adjacent to each other on the chromosome. In this way, a functional immunoglobulin gene is constructed. Such rearrangements may commit the cell to synthesizing a specific immunoglobulin.

It is not yet known whether gene rearrangements are of general importance in the differentiation of other cell types.

Other Mechanisms

Many other mechanisms play a role in the expression of specific genes in differentiating cell. For example, inverted GTP residues are often added to the 5' ends of an mRNA molecule. Similarly, poly (A) (polyadenylate) residues are often added to the 3' ends of mRNA. GTP caps may help bind the mRNA to ribosomes and enhance its stability. Poly(A) "tails" also stabilize mRNA, allowing it to stay active longer. Such alterations of mRNA may be important in controlling gene expression. The activity of a given mRNA may, in part, depend on such post-transcriptional events. Gene expression during differentiation and during the entire life of a cell may also be controlled at the level of translating mRNA transcripts. Factors may mask stored mRNA until it is needed. Such a mechanism may explain how some cells can suddenly begin to synthesize specific proteins, even though they are not actively synthesizing mRNA at that time. How messages are masked and unmasked is not well-understood. Although this chapter deals primarily with differentiation at the nucleic acid level, it is appropriate to mention that proteins are sometimes first synthesized as part of much larger proteins that are cleaved to form the functional protein. ACTH (adrenocorticotropic hormone) is formed from a larger precursor that contains other biologically active polypeptides as well. Insulin A and B chains are cleaved from a single polypeptide precursor. Thus, the cleavage of large polypeptides into more than one active protein may be a mechanism for controlling gene expression in certain systems.

Levels of Gene Control

Clearly, differential gene expression is controlled at many levels and by many mechanisms that are just beginning to be revealed. Future work will clear up mysteries that cloud the picture of gene expression in eukaryotic cells. Progress will probably be very rapid in this area because of recent technological advances. Several recent results have shown that the control of mRNA synthesis (transcriptional control) is essential to cellular differentiation. One experiment used the fact that estrogen stimulates the production of egg-white proteins in the chicken oviduct. Complementary DNA (cDNA) was formed from the mRNA molecules coding for the egg-white proteins ovalbumin, conalbumin and ovomucoid. Labeled nuclear RNA was then isolated from chick oviduct epithelial cells that had been either treated with hormone (estrogen) or left untreated. RNA from the treated cells hybridized 10-15 times as much cDNA as did that from untreated cells. This suggests that the production of egg-white protein mRNA in chick oviduct cells is controlled by hormone and the mRNA synthesis is a key step in the synthesis of these proteins. Similar result have been obtained for the estrogen-induced synthesis of mRNA for the proteins vitellogenin and very-low-density lipoprotein in the liver cells of roosters.

Additional evidence has come from studies on globin mRNA in the early chick. For the first few days after fertilization, chick embryos synthesize embryonic forms of α- and β-globin. Later, two adult forms are produced. It was shown that erythrocyte stem cells form one set of cells that synthesize only the embryonic globins and another set of cells that synthesize only the adult globins. Specific mRNA complementary to the cDNA for the embryonic globin gene was found in 5-day-old chick embryos, whereas the message for adult globin was found in 12-day-old embryos. These results support the notion that, during differentiation, specific cells produce specific proteins because specific mRNA molecules have been synthesized. The level of transcription, therefore, is an important control point for differential gene expression during differentiation. In another study it was shown that only adult mouse liver cells contain specific mRNA for a major protein excreted in the urine of adult mice (major urinary protein). Newborn mice do not secrete this protein and do not synthesize much mRNA for major urinary protein. Again, the presence of a specific differentiation product—in this case, major urinary protein—depends on the synthesis of the specific mRNA coding for that protein.

Specific destruction or degradation of messenger RNA is another control mechanisms in gene expression. During the differentiation of mammalian erythroblasts, the early developing cells synthesize many different mRNAs, including a small amount of globin mRNA. During the final few cell divisions, however, over 90% of the protein synthesized is globin. This suggests that globin mRNA are preserved, whereas many other mRNAs are destroyed. Destruction of specific mRNA has also been observed during the development of the slime mold *Dictyostelium discoideum*. If the developmental cycle is interrupted at a specific point, the slime mold cells destroy the new mRNAs that would have been used in synthesizing the specific proteins required for later developmental stages. Differentiation is also controlled at the translation level. For example, differential translation of mRNA molecules occurs in the cell cytoplasm of sea urchin eggs. Fertilized eggs synthesize protein much more actively than unfertilized eggs do, yet the same amount of mRNA is present in the cytoplasm before and after fertilization. Two-dimensional gel electrophoresis reveals that in the clam, the proteins formed before fertilization are different from those formed after, although the same mRNAs are present at both times. What about post-translation? After proteins are synthesized, several specific control mechanisms may come into play to influence their final concentration in the cell. For example, proteases present in the cytoplasm can cleave specific polypeptides into smaller polypeptides. This apparently occurs in the posterior pituitary gland, where cells can produce either adrenocorticotropin or beta lipotropin from the same primary translation product.

The Commitment to Differentiate

How do cells "decide" to differentiate? We have raised this question often because it is central to an understanding of embryonic development. Although the work cited below has brought us closer to answering this question, we are still far from a total understanding of this puzzle. In certain systems, DNA replication is not required for a cell to begin differentiating. This was shown by blocking DNA synthesis in certain systems with either are C or FUdR, two inhibitors of DNA replication. Rat myoblasts are converted to differentiated myotubes whether or not DNA synthesis is blocked by these inhibitors. In another study, erythroid differentiation was found to be unaffected by the inhibition of DNA synthesis.

Messenger RNA synthesis, followed by the synthesis of a specific protein, initiates the commitment to differentiate in certain systems. Inhibitors of mRNA (such as cordycepin or alpha amanitin) or inhibitors of protein synthesis (anisomycin or cycloheximide) block the differentiation of concanavalin A stimulated splenic lymphocytes and Friend cells into erythroid cells. An influx of calcium ions appears to be necessary for committing some cells to differentiate. For example, exposing lymphocytes to calcium ionophore A23187 commits them to differentiate. When concanavalin A is used to initiate so-called lymphocyte blast transformation, an influx of Ca^{2+} is observed prior to the transformation. When EGTA is used to remove Ca^{2+} from the medium, lymphocytes no longer respond to concanavalin A by differentiating. Similar results have been obtained in the Friend cell erythroid differentiation system. The puzzle of differentiation is becoming a little better understood. Intracellular calcium levels and messenger RNA synthesis are important in moving a cell toward the differentiated state.

Genetic Engineering : The Transfer of Genes to Mammals

Genes were injected into fertilized mouse eggs that were then implanted in the uteri of foster mother mice and allowed to develop. Some of the mice born to the foster mothers possessed the foreign genes. Three big questions remained. (1) Can such new genes integrate into the recipient's genome? (2) Can they be transmitted to future generations? (3) Can they function in the recipient? The last question is critical to developmental biologists because answering it could produce a far better understanding of how specific genes are turned on and off a various stages and in specific tissues. The answers to some of these questions are beginning to emerge. Work from many laboratories, including those of Mintz, Graham, Costantini, and Brinster, shows that 20-30% of the mice that develop from injected eggs carry the injected genes in their genomes and transmit them to their offspring. Thus the first and second questions have been answered in the affirmative.

The answer to the third question also appears to be affirmative, at least in some experiments. Brinster and Palmiter used a hybrid gene consisting of the viral thymidine kinase gene (*tk* gene) and control sequences from a mouse gene coding for the protein metallothionein. About 50% of the mice carrying the hybrid gene produced the viral thymidine kinase enzyme. Results have been mixed in studies by other investigators using different genes. In some cases, the genes were expressed; in others they were not. The key point, however, is that it is *possible* for cells to express transferred genes. The importance of such work cannot be overemphasized. Not only will it lead to a better understanding of the mechanisms of differential gene expression in developing systems, but it may lead to cures for many diseases that are known to result from genetic defects.

Giant Mice from Eggs Injected with Growth Hormone Genes

The work of Palmiter, Brinster, Hammer, Trumbauer, Rosenfeld, Birnberg, and Evans has exciting implications for accelerating animal growth and correcting genetic disease. A DNA fragment containing the promoter of the mouse metallothionein-I gene was fused to the structural gene of rat growth hormone. The resulting fusion gene was microinjected into the pronuclei of fertilized mouse eggs. Of 21 mice that developed from these eggs, 7 carried the fusion gene. Six of these grew significantly larger than their littermates. Several of these gene-injected mice had an extraordinarily high of fusion mRNA in their livers and a similar level of growth hormone in their serum. This is the first time that genetic engineering has altered the phenotype of an animal in such a profound manner. This technology may eventually lead to genetic stimulation of rapid growth in commercially important animals and to many other important practical applications. The age of helping humankind by introducing useful genes into the genomes has finally arrived.

5

Differentiation

Differentiation is the formation of identical cell from the properly differentiated cells to make the regeneration of cell possible. On the basis of molecular approval it is extremely difficult to interpret these changes in nucleic acids because it is uncertain which species of RNA are functional. Since proteins are the important products of nucleic acid messages and since their synthesis or lack of synthesis can be definitively determined, many investigators study proteins as markers of development and differentiation. In this chapter the main emphasis has been given to proteins in early development, methods of analysis in the protein field, and some of the specific proteins that differentiate cells from one another.

Protein in Early Development

As mentioned earlier, the sea urchin embryo is ideal for studying development because a vast number of embryos can easily been obtained and because normal development occurs in sea water. We can add a radioactive amino acid (such as radioactive methionine) to the sea water and measure the amount of this amino acid that is incorporated into newly synthesized protein. This is done by incubating the embryos with the "hot" amino acid for a short time, then placing the embryos in sea water without the radioactive amino acid. If we do this at various development stages, we can determine how much protein was synthesized at each specific stage by measuring how much radioactive protein was formed after we administered each pulse of "hot" amino acid. The amount of new protein that is synthesized can be measured with reasonable accuracy if we make sure that the radioactive amino acid is being taken up by the cells at all stages. We must also make sure that the intracellular supply (pool) of available amino acid remains constant at all stages. An example of the results obtained by investigators examining protein synthesis in the sea urchin embryo. Protein synthesis rapidly increases for the first three or four hour after fertilization and then declines, only to increase again at about the gastrula stage. In this experiment radioactive methionine was used and was shown to be taken up by unfertilized eggs and other stage. Amino acid uptake thus was not a problem in these experiments. Also, the supply (pool) of intracellular methionine was shown not 'to change appreciably in the different stage examined. Thus the pool of methionine was not a problem. These experiment suggest that different amounts of total protein are synthesized at various stages of embryonic development.

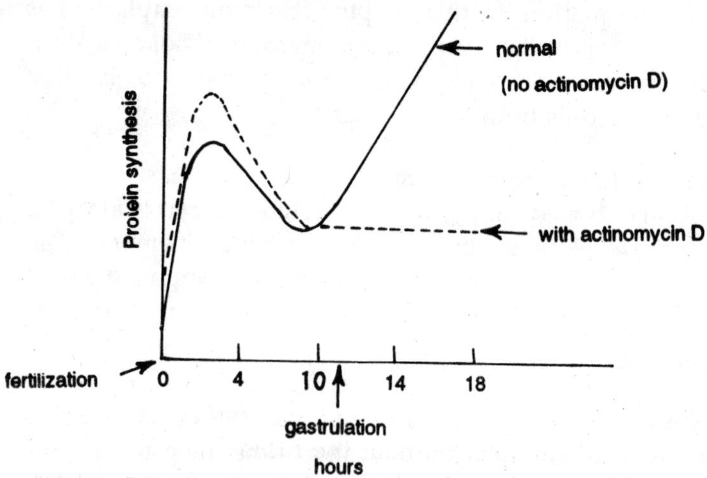

Fig . 5.1. Protein synthesis at various stages in sea urchin embryo development.

If embryos are treated with the RNA synthesis inhibitor actinomycin D, protein synthesis increases after fertilization but not at gastrulation. These results suggest that the proteins synthesized during early development use messenger RNA formed in the egg before fertilization. At gastrulation, however, new species of mRNA may be transcribed, a process that would be blocked by actinomycin D. More recent work has shown, however, that some specific proteins, such as histones, are synthesized in the early hours of sea urchin embryonic development using newly formed mRNA, as well as that stored in the egg before fertilization.

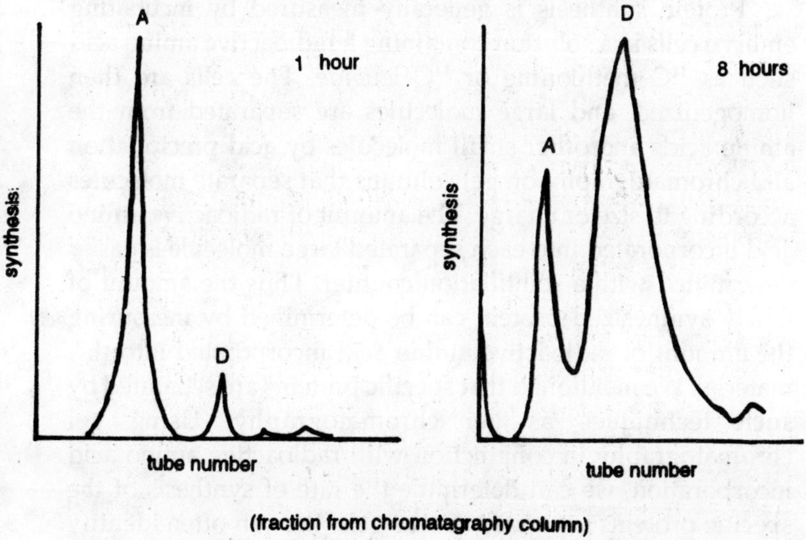

Fig . 5.2. Specific proteins synthesized at different stages of sea urchin embryo development.

We have described the synthesis of total protein at various developmental stages. What about specific proteins? Do all proteins in the cell follow the same pattern? Figure 5.2 shows that if individual marked proteins are separated chromatographically at different developmental stages, different amounts of particular proteins are found to be synthesized at different developmental stages. A lot of protein A, for instance, is synthesized one hour after fertilization in the sea urchin, but only a little of protein D is synthesized at this time. Eight hours after fertilization, however, less of protein A is synthesized and a lot of protein D is synthesized. The experiment shown in Figure 5.2 does not identify proteins A and D. In fact, these proteins may be similar and may represent different numbers of identical subunits or an association of the protein with cofactors. Later in the chapter we shall examine specific proteins and reaffirm that given proteins are often synthesized in different amounts at different developmental stages.

Measurement of Synthesis of Specific Protein

Protein synthesis is generally measured by incubating embryo cells in a solution containing a radioactive amino acid such as ^{14}C methionine or ^{14}C leucine. The cells are then homogenized, and large molecules are separated from the amino acids and other small molecules by acid precipitation and chromatography on gel columns that separate molecules according to size or charge. The amount of radioactive amino acid incorporated into each separated large molecule is easily determined with a scintillation counter. Thus the amount of newly synthesized protein can be determined by measuring the amount of radioactive amino acid incorporated into this material. We mentioned that specific proteins are separated by such techniques as gel chromatography. Using gel chromatography in conjunction with radioactive amino acid incorporation, we can determine the rate of synthesis of the specific proteins separated on the gels. We can often identify specific proteins by comparing their profiles on the gels with those of characterized protein standards. Electrophoresis (separation of proteins in electric fields) is also used to compare a protein sample with a known protein standard.

Immunological methods are also widely used to identify specific proteins. The specific protein is injected into rabbits or goats, which produce antibodies that can then be isolated from the blood of the animal. When the antibody is added to a specific protein in a mixture of many proteins, it complexes with the specific protein, forming a precipitate. Newly synthesized specific protein can be identified using the immunological technique coupled with the radioactive amino acid method. Cells are incubated with radioactive amino acid for a short time and then homogenized. The homogenate (or an extract thereof) is then incubated with antibodies against a single specific protein, causing the protein to precipitate from solution. The amount of radioactivity in the precipitate is then measured. This tells how much radioactive amino acid was incorporated into the specific protein precipitated by the antibody. In this way we can determine how much of the specific protein was synthesized while the cells were being incubated with radioactive amino acid.

This technique is very useful in identifying the pattern of synthesis of specific proteins during differentiation. Remember that just measuring the activity of a given enzyme or the presence or absence of a specific protein at a given developmental stage does not determine the time or rate of synthesis per se. Synthesis must be measured by a method such as that described above, in which the amount of *newly formed* protein is determined. The simple presence or absence of a specific protein or specific enzyme activity could result from the activation or inactivation of an enzyme by specific cofactors or from the differential degradation of the protein by degradative enzymes present in the cells. Before considering some specific proteins formed during differentiation, let us see how the immunological technique can be used to determine specific protein synthesis in one model system. These experiments are interesting because they will show not only how the technique is used, but also how protein synthesis and degradation are regulated. The system involves the synthesis of the enzyme tryptophan pyrrolase by the rat liver. This enzyme opens the indole ring of tryptophan, a precursor of the vitamin nicotinamide. The activity of this enzyme rises rapidly at birth. Its activity is also increased by the hormone hydrocortisone and by the amino acid tryptophan.

Purified tryptophan pyrrolase was injected into rabbits, which produced specific antibody against this enzyme. Rats were inoculated with hydrocortisone, tryptophan, or saline

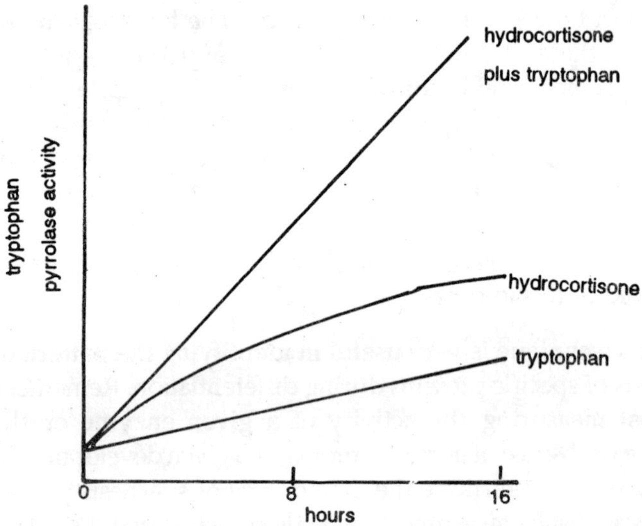

Fig . 5.3. Stimulation of tryptophan pyrrolase activity by hydrocortisone, tryptophan, or both.

and, after a few hours, radioactive amino acid. Their livers were then removed and homogenized. Tryptophan pyrrolase antibody was added to the homogenate to precipitate out tryptophan pyrrolase, and the amount of newly synthesized enzyme was determined by measuring the radioactivity of the precipitate. Figure 5.4 shows that hydrocortisone increases the synthesis of tryptophan pyrrolase. This conclusion is based on the observed increase in the amount of radioactive tryptophan pyrrolase in the hydrocortisone treated livers over that in the control livers. There was, however, no increase in tryptophan pyrrolase synthesis in the tryptophan-treated livers. How does tryptophan produce the observed increase in tryptophan pyrrolase activity. This question was resolved in another experiment. Liver proteins were labeled as before with radioactive amino acid. Later, either saline or tryptophan was injected. Still later, the livers were homogenized, and the amount of radioactive tryptophan pyrrolase present was determined after antibody precipitation. Figure 5.5 shows that the amount of radioactively labeled-tryptophan pyrrolase declined steadily after saline injection. Tryptophan, however, prevented this decline. These results suggest that tryptophan keeps tryptophan pyrrolase enzyme active by protecting it from degradative enzymes.

Injected material	*Counts per minute of radioactive amino acid in protein precipitated by antibody against tryptophan pyrrolase*
NaCl	1406
Hydrocortisone	9466
Tryptophan	1954

Fig. 5.4. Stimulation of tryptophan pyrrolase synthesis by hydrocortisone but not by tryptophan.

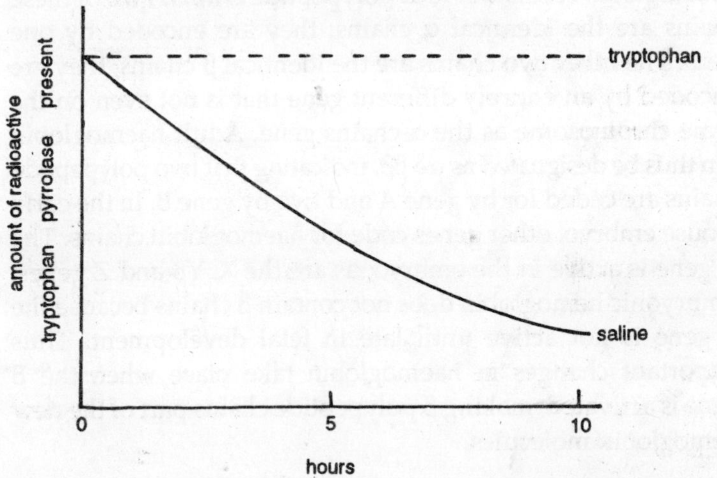

Fig. 5.5. Prevention of tryptophan pyrrolase degradation by tryptophan.

DIFFERENTIATION AND SPECIFIC PROTEINS

Haemoglobin

Haemoglobin is a *tetramer*, a molecule composed of four folded polypeptide chains. Each polypeptide chain carries a non-protein heme group. The heme groups contain iron and bind oxygen. The most important function of hemoglobin is to carry oxygen to different parts of the body. Haemoglobin is the major protein contained in red blood cells and thus is distinct differentiation marker in these highly specialized cell types. There are three different types of haemoglobin: embryonic haemoglobin, fetal hemoglobin, and adult hemoglobin. Stage-specific differences in haemoglobin probably evolved as a result of the different respiratory needs of the embryo fetus, and adult. Mammalian fetuses obtain oxygen by diffusion across the placenta from the maternal blood. Thus the haemoglobin of the fetal red blood cells must have a higher oxygen affinity than that of the mother in order for such an exchange to occur. What is the nature of these stage-specific differences? As mentioned previously, adult mammalian haemoglobin consists of four polypeptide chains. Two of these chains are the identical α chains; they are encoded by one gene. The other two chains are the identical β chains; they are encoded by an entirely different gene that is not even on the same chromosome as the α-chains gene. Adult haemoglobin can thus be designated as αδ ββ, indicating that two polypeptide chains are coded for by gene A and two by gene B. In the early mouse embryo, other genes code for haemoglobin chains. The A gene is active in the embryo, as are the X, Yβ and Z genes. Embryonic hemoglobin does not contain β chains because the B gene is not active until late in fetal development. Thus important changes in haemoglobin take place when the B gene is activated, making β polypeptide chains part of the new hemoglobin molecules.

Abnormal types of adult hemoglobin include hemoglobin S and haemoglobin C. Haemoglobin S is found in individuals who have sickle-cell anemia. In this disease, the red blood cells are sickly shaped. Haemoglobin S differs from normal haemoglobin only in the β subunits; the β subunits are exactly like those in normal adult hemoglobin. The normal β subunit consists of 146 amino acids. The only difference between the B subunits of normal and sickle-cell haemoglobin is that in the

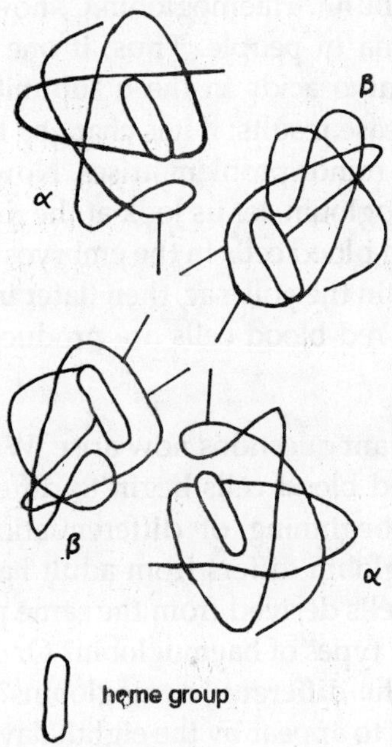

Fig. 5.6. **Diagram of haemoglobin showing four subunits. A haeme group is associated with each subunit.**

latter a single valine residue replaces the number 6 amino acid, glutamic acid. This single amino acid change can be caused by a single base change in the mRNA coding for the b subunit. Glutamic acid is coded for by the base sequence GAA, whereas valine is coded for by GUA. Thus a difference of a single base in a large protein can produce a disease that can be fatal. This change probable arises as the result of a single point mutation in the DNA of the b subunit gene Haemoglobin C, another abnormal adult haemoglobin, also result from a single amino acid change. The same b chain amino acid replaced in haemoglobin S is also replaced in haemoglobin C. Instead of a valine, however, a lysine substitutes for the glutamic acid residue at position 6. The codon AAA codes for lysine, whereas GAA codes for glutamic acid. Again, a single base change, an A for a G, leads to an amino acid change in the b subunit. Haemoglobin C, however, does not cause severe anemia in people. Thus, if one specific amino acid out of 146 amino acids in the b subunit is changed to valine, a major disease results; if it is changed to lysine, on the other hand, only a minor problem arises. Now that we know a little about haemoglobin, let us look at the differentiation of red blood cells. Red blood cells in the embryos of humans and mice originate first in the yolk sac, then (later in development) in the liver. After, red blood cells are produced by the bone marrow.

Several important questions now arise. When do the cells that give rise to red blood cells begin to differentiate? What events mark the beginning of differentiation? Recall that embryonic haemoglobin differs from adult haemoglobin. Do the same cells, or cells derived from the same parent cell, give rise to the different types of haemoglobin? Or do different cell lines give rise to the different haemoglobins? In the mouse, hemoglobin begins to appear by the eighth day of gestation, in red blood cell precursors located in the blood islands of the yolk. sac. On the ninth day, these precursor cells enter the circulation and begin to synthesize embryonic haemoglobin. After they divide four times, their nuclei condense and they become inactive. On the twelfth day of gestation, the liver begins to produce red blood cells. This new cell population synthesizes adult hemoglobin. The yolk sac continues to produce red blood cells through the fourteenth day, so both types of red blood cells (yolk sac and liver) are found in the fetal circulation. Spleen and bone marrow take over the production of red blood cells between day 15 and birth and liver production declines. Thus, there are at least two populations of red blood cells, one that synthesizes embryonic hemoglobin and another than synthesizes adult haemoglobin. No one knows whether a common ancestor cell originally seeded the yolk sac and the liver, spleen, and bone marrow.

In the yolk sac hemoglobin (globin) mRNA is synthesized

before day ten of gestation. After day ten, haemoglobin synthesis is not inhibited by actinomycin D. This suggests that the mRNA made before day ten is stable and used later on for synthesizing the protein chains of haemoglobin. Thus, the main event that triggers the formation of red blood cells is probably the activation of the genes that code for the polypeptide chains of haemoglobin (globin). Messenger RNA that codes for the polypeptide chains of haemoglobin has been isolated. Such mRNA is identified by its ability to direct the synthesis of globin in a cell-free system and by hybridization experiments such as those described in the last chapter. Early fetal mouse liver cells do not contain globin messenger RNA. The glycoprotein hormone, *erythropoietin,* cause these cells to synthesize globin mRNA. Hemoglobin synthesis begins shortly thereafter. Erythropoietin is produced by the kidney in response to decreases in tissue oxygen content. It is a key factor in inducing prospective erythrocytes to differentiate into mature red blood cells. It probably acts by turning on globin genes. Although the exact mechanism is unknown, it may actin in a manner similar to that proposed for estrogen (estradiol). Its final action may be on specific chromosomal proteins that function to regulate gene action. Exactly how certain cells commit themselves to forming red blood cells is unknown. However, the events that immediately lead to differentiation—globin message transcription followed by hemoglobin synthesis—are becoming better understood.

Myosin

The protein myosin is probably present in many cell types. Muscle cells, however, contain a large amount of this protein. Myosin makes up 10-12% of the fresh weight and nearly 50% of the dry weight of adult muscle. Myosin is thus a major marker for muscle cell differentiation. Myosin is a part of the contractile networks present in all eukaryote cells. Since the main function of muscle cells is to contract, these cell types should be enriched in myosin. Myosin is a fibrous protein composed of two identical submits wound around each other.

In muscle, myosin molecules aggregate into filaments. This assembly phenomenon will be examined in the next chapter. Myosin combines with other proteins, such as actin. Units composed of myosin, actin, and certain other proteins are contractile and serve as a cellular contractile apparatus. Proteins associated with actin filaments include α-actin, β-actinin, tropomyosin and troponin. These proteins serve a variety of functions that keep the contractile apparatus intact and allow it to work smoothly.

Let us look first at developing muscle cells and then focus on the differentiation of these cell types. In this chapter we shall stress myosin synthesis. In the next, we shall focus on how muscle proteins assemble into functional contractile units. Experiments using fluorescent labeled myosin antibody have shown that muscle cell contractile proteins are present in many prospective muscle cells before any organized contractile apparatus is visible there. Prospective muscle cells (myoblasts) are derived from the myotome region of the somite and hypomere mesoderm. The prospective muscle cells divide, forming many additional myoblasts. When they stop dividing, they fuse together, forming long muscle tubes (myotubes) containing many nuclei in a common cytoplasm.

In two-day-old chick embryos, somite myotome cells are actively dividing. Some of these cells form spindle shapes containing detectable myosin filaments. By the fourth day most of the myoblasts contain muscle contractile fibrils (myofibrils). All this occurs before the cells fuse into the multinucleated myotubes. In limb muscle, however, myofibrils appear only in the multinucleated myotubes and not in the prefusion myoblasts. We have said that elongated myotubes arise as a result of fusion of single cells rather than continued nuclear divisions without cell division. This was elegantly shown by an experiment in which cells from two different mouse embryos were aggregated. Each embryo strain contained a different form of the enzyme isocitrate dehydrogenase, which is made up of subunits. The mosaic embryos were implanted

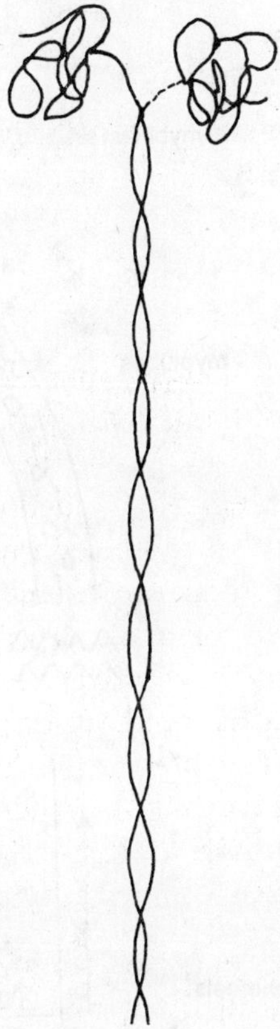

Fig. 5.7. Myosin consists of two subunits wound around units wound around each other.

in the uterus of a foster mother mouse where they developed into normal mice consisting of cells from the two different strains of mice. The skeletal muscle of these mice contained some enzyme like that of each parent, as well as some hybrid enzyme. This hybrid enzyme consisted of some subunits from one parent's enzyme and some from the other parent's enzyme. This type of hybrid enzyme could have been produced only if the myotubes contained nuclei from cells of both parents in a common cytoplasm. Each type of nucleus produced a message

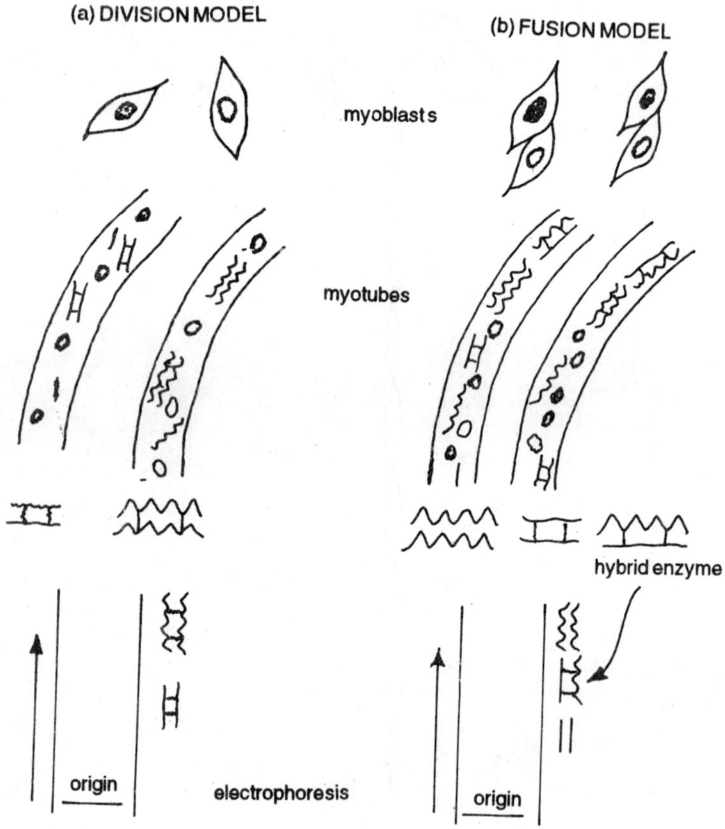

Fig. 5.8. Isocitrate dehydrogenase in myotubes.

that coded for its own enzyme subunits. These subunits, when synthesized, then polymerized into the hybrid enzyme observed. Such enzymes are easily detected using electrophoresis, which separates proteins on the basis of size and charge. Thus skeletal muscle forms when single cells fuse into elongated myotubes. Myoblast fusion has also been studied *in vitro* to learn about the specific factors that influence this phenomenon. Myoblasts in culture do not fuse until six to eight hours after cell division has stopped. Presumably these cells use this time to acquire specific cell-surface properties that facilitate the fusion process.

Limb myoblasts in culture synthesize myosin and assemble myofibrils even if fusion is prevented by chelating agents that bind divalent cations. Thus even in limb muscle, fusion is not absolutely necessary for muscle differentiation to occur.

Yaffe and others have shown that in culture much of the mRNA that codes for myosin is probably synthesized hours before the myoblasts fuse. Actinomycin D inhibits the synthesis of muscle proteins if applied more than six hour before fusion. If applied within six hours of fusion, actinomycin D has little effect. Cell culture experiments have also shown that specific factors may be required for muscle cell fusion and differentiation. Hauschka and Konigsberg found that collagen can stimulate the fusion of myoblasts to form myotubes. Collagen is present in the somite and hypomere where skeletal muscle forms. This substance may play a role in controlling myoblast fusion and differentiation *in vivo*. Exactly what factors cause certain cells to become muscle cells is, as yet, unknown.

Pancreatic Proteins and Differentiation

Before we examine the differentiating pancreas, let us say a few words about one of the pancreatic proteins—the hormone insulin. This hormone is extremely important in metabolizing carbohydrates. Without it, glucose cannot be transformed into glycogen. In its absence, blood sugar builds up, eventually killing the organism. The disease diabetes results from a lack of insulin; however, it can be controlled with insulin injections. Recent work has shown that the human insulin gene can be incorporated into the genome of the bacteria *E. coli*. Bacteria so altered begin to synthesize human insulin. Arber, Smith, and Nathans received Nobel Prizes for their work in discovering enzymes that cut DNA, essential for recombinant DNA studies. Such work is leading to major breakthroughs in medicine. Human insulin, which can now be produced by the insulin gene in *E. coli*, is very useful in treating many difficult cases of diabetes that do not respond well to animal insulin. Virtually unlimited supplies of this hormone are now available.

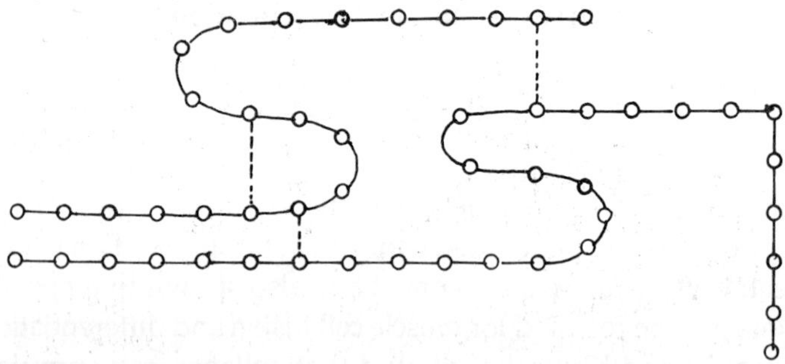

Fig. 5.9. Insulin. The hormone insulin is composed of two polypeptide chains. Disulfide bridges form between the sulfhydryl groups of specific amino acids (cystein residues).

Insulin is a small protein hormone with a molecular weight of only 6000 daltons. It consists of two polypeptide chains. Insulin is one of many important proteins produced by the pancreas. In the mouse embryo, insulin is first detected in the pancreas rudiment nine or ten days after fertilization. The cells that produce insulin pinch themselves off from other pancreas tissue, forming separate masses called Islets of Langerhans. The islet cells differentiate into two distinct cell populations: A-cells, which synthesize the hormone glucagon, and B-cells, which produce insulin. Both hormones are involved in carbohydrate metabolism. B-cells are easily distinguished from A-cells by their dense cytoplasm and beta granules.

The cells that produce digestive enzymes are located at the ends of a branched duct system in pouches called *acini.* The digestive enzymes produced in the acini include lipase, amylase, trypsin, and chymotrypsin. When the mouse pancreas rudiment appears as an evagination from the gut tube, pancreatic enzymes and hormones are detectable, but in very small amounts. A period of rapid cell division follows during which

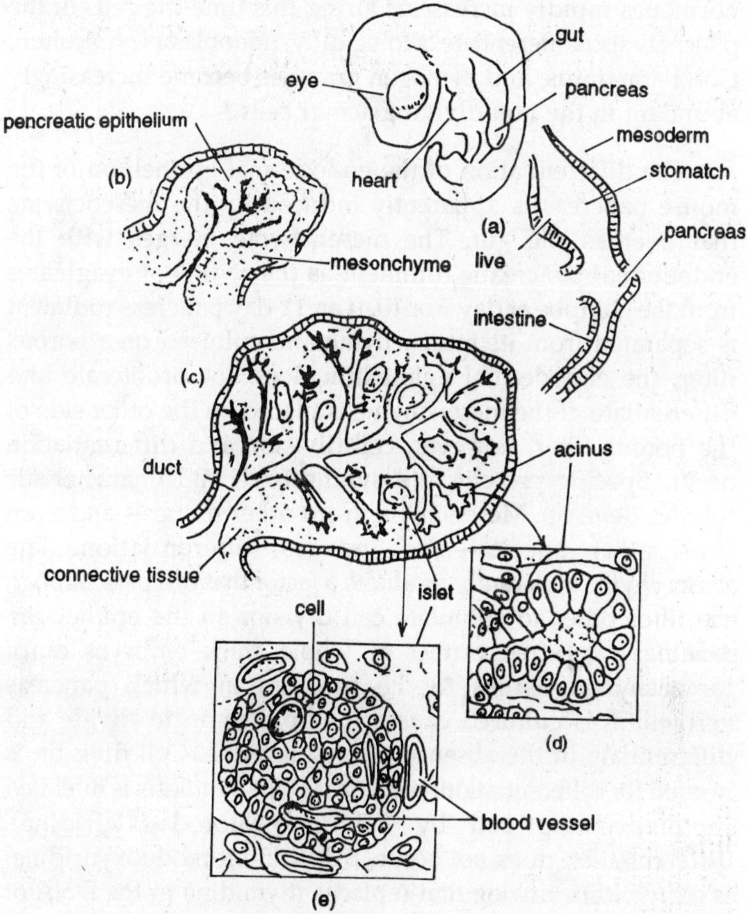

Fig. 5.10. Development of mouse pancreas. (a) Initial elevation of pancreas epithelium, which would have low levels of specific proteins (about 12-day embryo). (b) Acini and islets are now present. The acini composed of exocrine cells with zymogen granules. The islets are endocrine in function. (c). Secretory acini begin to form (about 12-day embryo). (d)—(e) Enlargement of an acinus, containing secretory (zymogen) granules and an islet, respectively.

the enzyme and hormone levels remain low. By about day 15, cell division stop and the synthesis of specific enzymes and hormones rapidly increases. During this time the cells of the pancreas also differentiate cytologically. Endoplasmic reticulum, Golgi apparatus, and zymogen granules become increasingly abundant in the non-dividing acinar cells.

The differentiation of the endodermal epithelium of the mouse pancreas is apparently induced by the mesenchyme that overlies the gut. The mesenchyme merges with the endodermal pancreatic rudiment as the rudiment evaginates from the gut tube at day 9 or 10. If an 11-day pancreas rudiment is separated from its mesenchyme and cultured on a porous filter, the endodermal epithelium does not proliferate and differentiate. If the mesenchyme is placed on the other side of the porous filter, however, cell division and differentiation occur. Specific proteins are synthesized, and characteristic tubules develop. Mesenchyme from other tissues—and even from other species—also induces differentiation. The mesenchyme apparently produces a factor that can pass through the filter pores to stimulate cell division in the epithelium. Adding a cell-free extract of whole chick embryos or of mesenchymal tissues to the medium in which pancreas epithelium is cultured causes the epithelium to divide and differentiate in the absence of mesenchyme. Cell division is needed for differentiation, because if DNA synthesis is inhibited in predividing cell by FUdR (5-fluorodeoxyuridine), differentiation does not occur. BUdR (5-bromodeoxyuridine) is a thymidine analog that replaces thymidine in the DNA of dividing cells but does not prevent DNA synthesis or cell division. The DNA synthesized, however, is abnormal. BUdR also prevents the differentiation of pancreas cells treated during the division period. In the presence of BUdR, mitosis continues and some common cell proteins are synthesized, but levels of pancreas-specific proteins and differentiation do not increase. BUdR may prevent the production of mRNA that codes for cell-specific proteins. The RNA synthesis inhibitor actinomycin

D also inhibits the increase in protein-synthesis if it is applied during the cell division period or before the increase in specific protein synthesis begins. It has no effect one protein synthesis accelerates. This implies that stable messages are produced before the increase in specific protein synthesis. To sum up, period of normal DNA synthesis must occur before a high rate of specific protein synthesis can begin.

The mesenchymal factor (MF) that stimulates cell division and differentiation in pancreas epithelium was isolated by Rutter and colleagues. This factor was covalently bound to inert beads, and the beads were cultured with pancreas epithelium cells. The cells attached to the beads and continued DNA synthesis and cell division. Zymogen granules formed in the cells, indicating that differentiation had occurred. Since it is unlikely that MF attached to the beads entered the cells, these results suggest that MF stimulates mitosis and differentiation by some action at the surface of the pancreas epithelial cells.

In summary, the pancreas develops as an evagination of the gut tube. The pancreatic rudiment joins with the mesenchyme surrounding the gut tube. In the presence of mesenchyme, morphogenesis begins and specific proteins are synthesized at a low rate. Cell division and morphogenesis continue. Finally, specific protein synthesis increases rapidly and cytological differentiation is completed. Stable mRNA coding for the specific proteins is apparently synthesized before the synthesis of specific proteins accelerates. Mesenchyme apparently induces cell division in the pancreas epithelium. Wessels and others have proposed that increased cell mass in the developing pancreas (and in other systems) may be a key factor in the increase in the levels of specific protein synthesis and cytodifferentiation. Large masses of cells could alter their immediate environment, creating conditions that accelerate specific protein synthesis. This idea may suggest additional experiments that will uncover some of the control mechanisms of differentiating systems such as the pancreas.

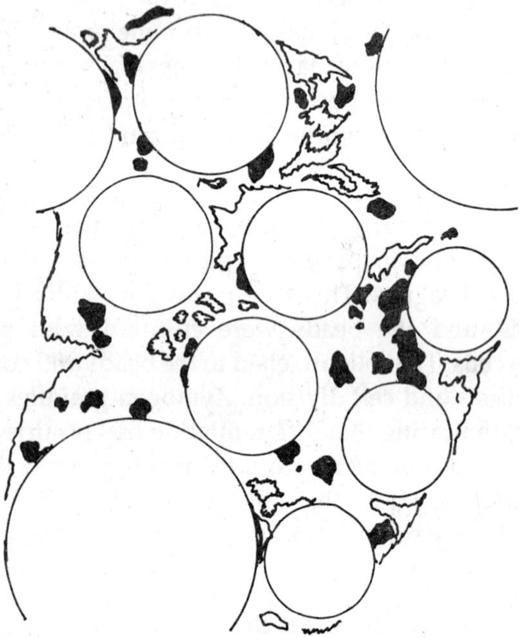

Fig. 5.11. Pancreas epithelial cells attached to sepharose beads coated with mesenchymal factor (MF).

Tissue-and Stage-Specific Isozyme Patterns

We have seen that proteins often consist of subunits coded for by separate genes. Hemoglobin subunits, for instance, are coded for by several genes. The relative activities of these genes vary at different developmental stages, so fetal hemoglobin and adult hemoglobin clearly have different subunit compositions. These different hemoglobins are presumably those best adapted to the different environments of the embryo and the adult. *Isozymes* are enzyme molecules that are functionally similar but that behave differently during electrophoresis because of differences in charge; some forms are more negatively charged than others. These charge differences occur because the enzymes are composed of subunits and, as in hemoglobin, different subunits can combine to form the isozymes. Each different subunit is coded for by a different gene. Thus, if particular genes are more active than others at a given stage or in a given tissue, the resulting proteins will also be different.

Lactate dehydrogenase (LDH) is an enzyme that exists in multiple isozymes. Different isozymes of LDH predominate in different species, in different tissues, and at different developmental stages. LDH catalyzes the interconversion of pyruvate and lactate. Pyruvate is a key intermediate when glucose is oxidized to carbon dioxide and water, releasing energy that is essential for various metabolic activities. LDH makes it possible for cells to function effectively when oxygen is temporarily unavailable. As LDH converts pyruvate to lactate, it generates NAD from NADH. NAD is essential for completion of an early step in a set of metabolic reactions called glycolysis. Like different haemoglobins, different LDH isozymes presumably function best under different metabolic conditions. Thus one isozyme might predominate at a specific stage or in a specific tissue because it is best adapted to function under those cytoplasmic conditions. Before we briefly consider LDH isozymes during development, let us see how LDH was found to be composed of several isozymes in the first place. When crude homogenates of tissues such as mouse skeletal muscle are subjected to electrophoresis, the various proteins present in the homogenate separate on the basis of their net electrical charge. After a time, the gel block is immersed in a staining solution that contains lactic acid (an LDH substrate), the required cofactor NAD, and a tetrazolium salt. If LDH is present in the gel, the lactic acid is oxidized and the tetrazolium salt is reduced to coloured formazan that precipitates at the site of LDH activity. Five of these coloured zones appear after a mouse muscle homogenate is subjected to electrophoresis. Each coloured zone identifies an LDH isozyme. Five isozymes of LDH have been found in most of the mammal and bird tissues examined. Why do five LDH isozymes form?

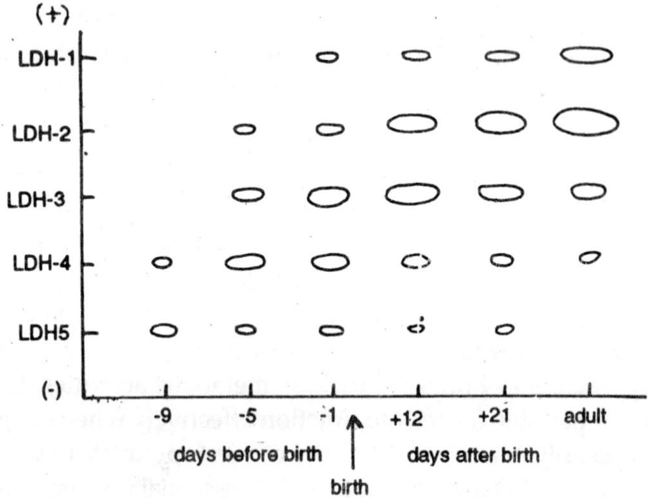

Fig. 5.12. Changing LDH isozyme pattern in heart muscle at different stages of mouse development. LDH-1 migrates toward the anode (+) and LDH-5 migrates to the cathode (-) during electrophoresis.

Markert and Ursprung purified LDH using several methods. The pure LDH also formed five isozyme bands after electrophoresis. By analytical ultracentrifugation in sucrose density gradients, they determined that the molecular weight of each LDH isozyme is about 140,000 daltons. When the pure LDH isozymes were subjected to agents or conditions that separated proteins into subunits, only two, not five, electrophoretic bands appeared that stained for protein. Each band represented a subunit of LDH. Each subunit had a molecular weight of about 35,000 daltons. The two bands were designated LDH subunit A and LDH subunits B.

The separated LDH subunits recombined to form the five LDH isozymes as follows:

LDH-1	BBBB
LDH-2	ABBB
LDH-3	AABB
LDH-4	AAAB
LDH-5	AAAA

Such combinations of subunits probably occur in cells, forming the five LDH isozymes. Different genes code for LDH subunits A and B. Thus, the amount of each isozyme that appears in cells of different tissues or at different developmental stages reflects, in part, the ratio of A subunits to B subunits produced at that time in that tissue. For example, mouse eggs produce mostly LDH-1 (the B tetramer); thus the B gene is the most active subunit during oogenesis. As development proceeds, the A gene becomes activated and the B gene is suppressed. LDH-5 (the A tetramer) begins to predominate. By birth, the B gene has been activated in many tissues of the body. Then, the isozyme pattern shifts again toward LDH-1. Thus shifts in LDDH patterns reflect the shifting relative activities of the A and B gene.

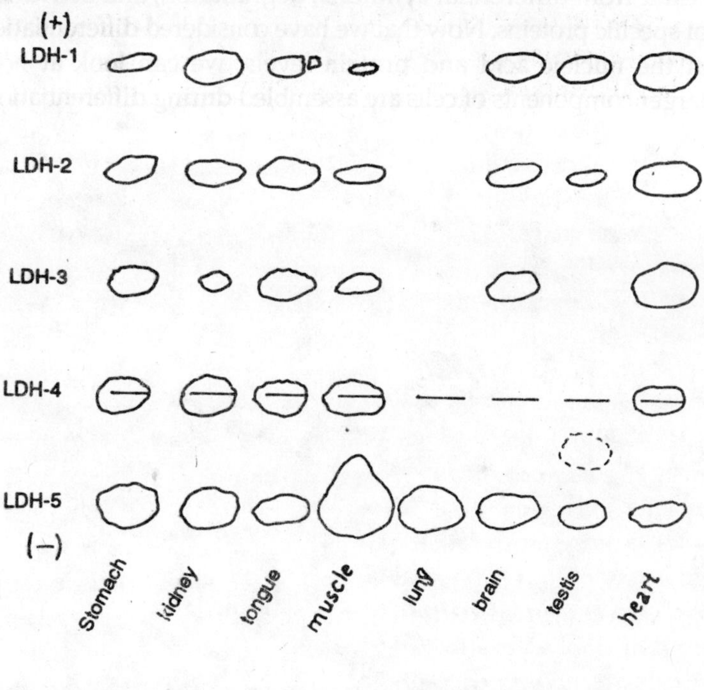

Fig. 5.13. LDH isozyme patterns in eight adult rat tissues. Electrophoretic patterns show tissue specific differences in LDH isozymes. Dotted circle in testis pattern represents an isozyme that may be testis-specific.

Adult rat tissues also differ with respect to LDH pattern. For example, skeletal muscle shows a lot of LDH-5, whereas heart muscle has more LDH-1, LDH-2 and LDH-3. Here, too, shifts occur relative to activities of the A and B genes. Specific cytoplasmic factors such as pH and ionic species probably control the rates of subunit synthesis. Such conditions may also affect subunit accumulation and assembly. Remember that the amount of a given protein present is a function not only of the rate of synthesis of the protein, but also of its rate of degradation (recall the work on tryptophan pyrrolase). Thus the presence of specific LDH isozymes represents a balance between the synthesis and degradation of those isozymes. To sum up, both stage-specific and tissue-specific protein patterns appear during the differentiation process. Such protein patterns result from differential synthesis, degradation, and activation of specific proteins. Now that we have considered differentiation at the nucleic acid and protein levels, we can look at how larger components of cells are assembled during differentiation.

6

Translational Regulation of Development

After a messenger RNA has been transcribed, processed, and exported from the nucleus, it still needs to be translated in order to form the protein encoded in the genome. In this chapter, we shall see that regulation at the level of translation is an extremely important mechanism in the control of gene expression. In such cases, the message is already present but may or may not be translated, depending on certain cellular conditions. Thus, translational control of gene expression can be used when a burst of protein synthesis is needed immediately (as we shall find to be the case in the newly fertilized egg), or it can be used a fine-tuning mechanism to ensure that a very precise amount of protein is made from the available supply of messages (as is the case in haemoglobin synthesis). We shall also see that there are several ways to effect translational control and that different cells have evolved different means to do so.

Translation in Eukaryotes

Translation is the process by which the information contained in the mRNA instructs the synthesis of a particular polypeptide. This process has been divided into three components: initiation, elongation and termination. Initiation consists of the reactions wherein the first aminoacyl-transfer RNA and mRNA are bound to the ribosome. The only transfer RNA (tRNA) capable of initiating translation is a special tRNA (tRNAi), which carries the amino acid methionine. As the first

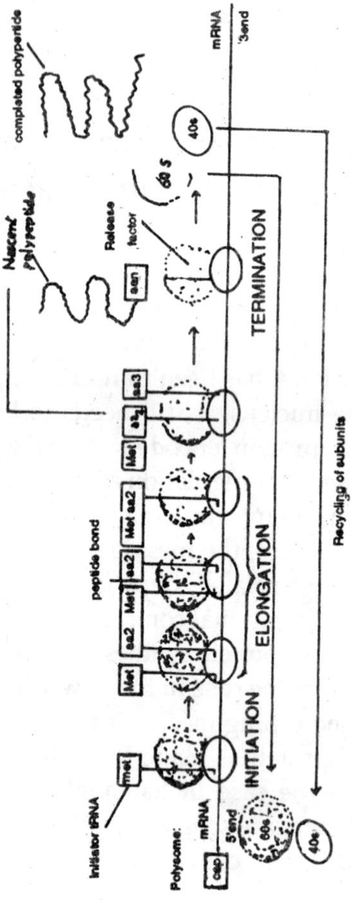

Fig. 6.1. Schematic representation of the events of eukaryotic translation. The initiation step bring together the 40 S (white) and 60S (shaded) ribosomal subunits, mRNA, and the initiator tRNA, which is complexed to the amino acid methionine (Met). During elongation, amino acids are brought to the polysome and peptide bonds are formed between the amino acids. The sequence of amino acids in the growing protein is directed by the sequence of nucleic acid codons in the mRNA. After the last peptide bonds of the protein has been made, the codons UAG, UGA, or UAA signal the termination of translation and the ribosomal subunits and message can be reutilized.

reactions involve the formation of an "initiation complex" consisting of methionyl-initiator tRNA bound to a 40 S ("small") ribosomal subunit. This Met-tRNA, is recognized by *Eukaryotic Initiation Factor* 2 (eIF2), which binds it to the 40S ribosomal subunit. Note that the binding occurs in the absence of mRNA. The mRNA is added next. It first binds its 5' end to the ribosome and then aligns itself so that the methionine-$tRNA_i$ is positioned on an AUG nucleotide triplet (codon). This step involves initiation factors 2, 4A, and 4B as well as the 7-methylguanosine "cap" on the mRNA. Without the cap, the binding of mRNA to the ribosomal subunit is not completed. Only after the mRNA has been complexed to the small ribosomal subunit can the 60S ("large") ribosomal subunit bind. This completes the initiation reaction.

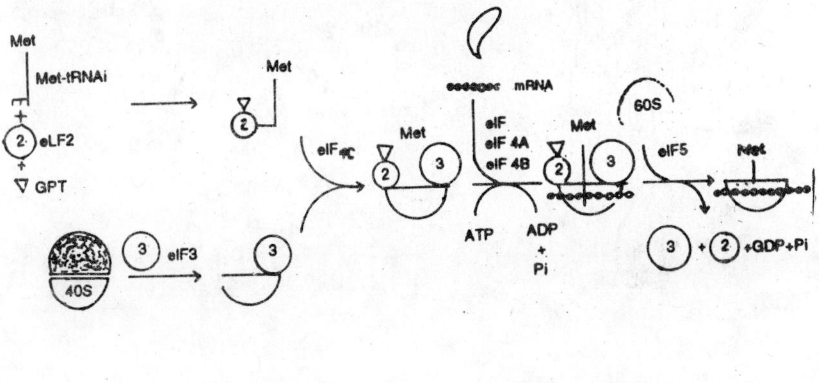

Fig. 6.2. Initiation phase of eukaryotic translation. The first complex is made between an activated initiator-tRNA and the 40S ribosomal subunit, catalyzed by GTP and eukaryotic initiation factors (eIF) 2 and 3. After this complex has formed, the mRNA is positioned and the 60S ribosomal subunit is put into place. These later events are catalyzed by eukaryotic initiation factors 1, 4 and 5. (From Hershey, 1980.)

Elongation involves the sequential binding of aminoacyl-tRNAs to the ribosome and the formation of peptide bonds between the amino acids as they sequentially relinquish their

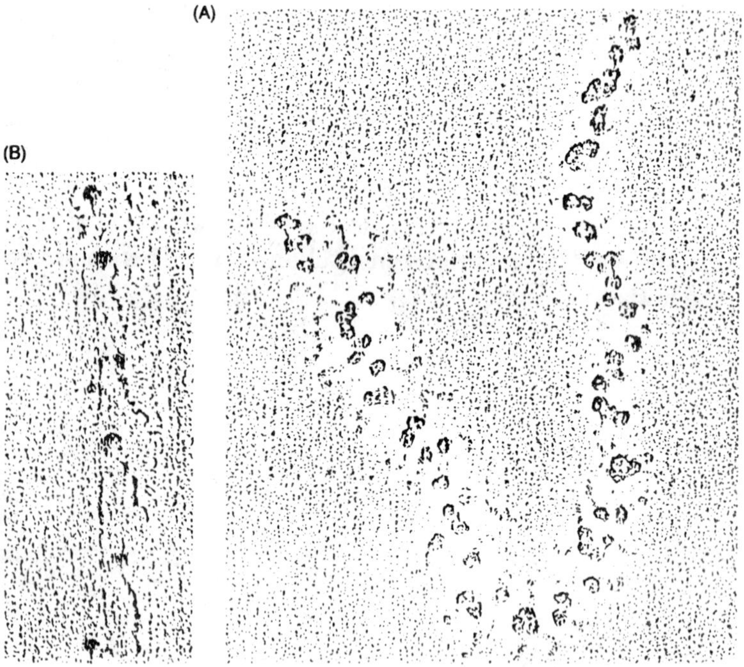

Fig. 6.3. Individual polysome transcribing the giant messenger RNA from the BR2 puff of *Chironomus tentans*. (A) Electron micrograph of a polysome containing 74 ribosomes. The nascent proteins can be seen extending from the ribosomes and growing as the ribosomes move from the 5′ end of the message to the 3' end. Near the 4' end are ribosomes from which the protein has detached. The bar represents 0.5. μm. (B) Higher magnification of such a polysome in which the polysome had been stretched during the preparation of the specimen. The relationship of the mRNA to the ribosomal subunits and the nascent polypeptide can be seen.

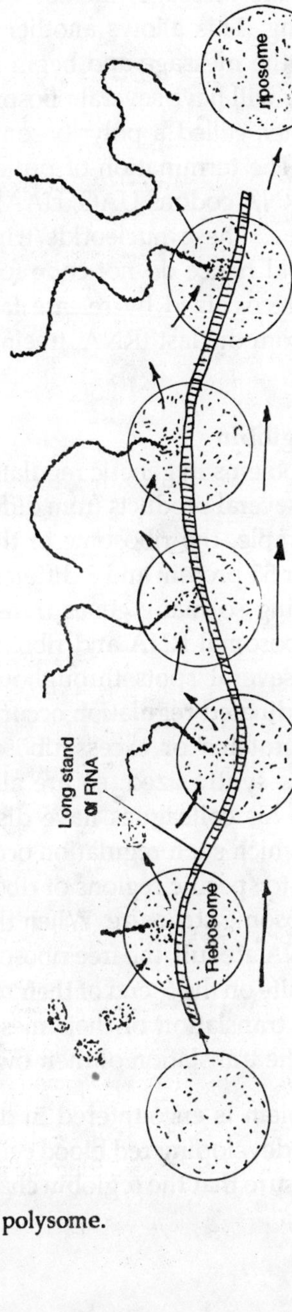

Fig. 6.4. A polysome.

tRNA carriers. As the amino acids are joined together, the ribosome travels down the message, thereby exposing new codons for tRNA binding. This allows another ribosome to initiate on the 5 end of the message and begin its traveling. Thus, any mRNA usually will have several ribosomes attached to it. This structure is then called a polyribosome—or, more commonly, a *Polysome*. The termination of protein synthesis takes place when the mRNA codons UAG, UAA, or UGA are exposed on the ribosome. These nucleotide triplets are not recognized by tRNAs and hence do not code for any amino acids. Rather, they are recognized by release factors, which hydrolyze the peptide from the last tRNA, freeing it from the ribosome.

Ribosomes and Haemoglobin

One of the major problems in genetic regulation is the co-ordinated production of several products from different regions of the genome. For example, the ribosome of the bacterium *Escherichia coli* consists or 55 protein and 3 different ribosomal RNAs. Rather than having so many genes transcribed from the same operon, the ribosomal RNA and ribosomal protein genes are clustered in several spots throughout the *E. coli* chromosome. Still, co-ordinated regulation occurs. One never sees excess ribosomal proteins or excess ribosomal RNAs. When one component is synthesized, so are all the others. Masayasu Nomura and his colleagues have discovered the elegant mechanism by which such regulation occurs. Certain ribosomal proteins bind to specific regions of ribosomal RNA in order to form the ribosome structure. When these binding sites on the ribosomal RNA are full, the free ribosomal proteins bind to a closely related site on the 5' end of their own mRNAs. This blocks any further translation of their messages. Thus, these proteins regulate the translation of their own messages.

A very similar problem is encountered in differentiated eukaryotic cells. When a developing red blood cell synthesizes haemoglobin, it must ensure that the α globin chains, β globin chains, and heme molecules are in a 2:2:4 ratio. Any deviation from this ratio results in severely debilitating diseases.

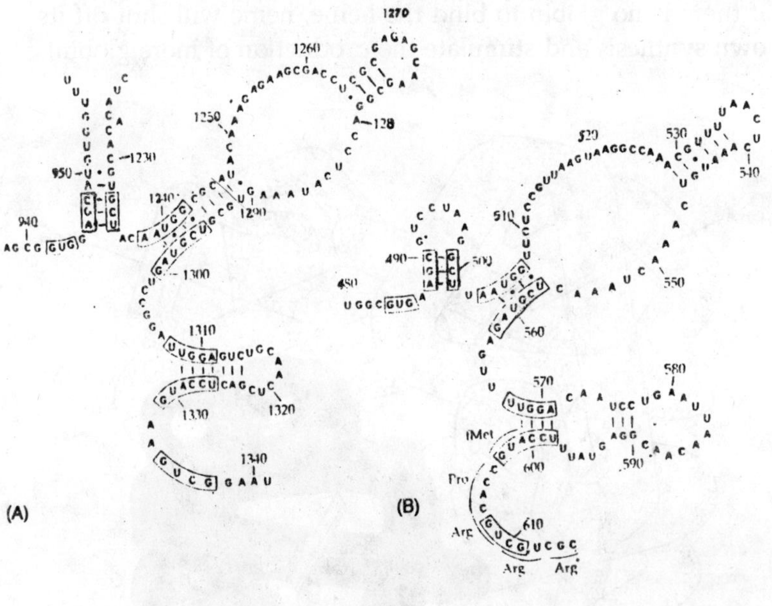

Fig. 6.5. Models of the secondary structures recognized by ribosomal protein S7 in *E. coli*. (A) S7 binding site on the 16 S rRNA of the small subunit. (B) S7 binding site on the leader sequence of its own mRNA. The boxed regions show extensive homologies.

Recent evidence has shown that the heme molecule regulates the proportional synthesis of the haemoglobin components. It accomplishes this feat in two ways. First, excess heme (i.e., heme that has not been bound to protein such as globin) will shut off its own synthesis. It does this by inactivating δ- aminolaevulinate synthase (DALA synthetase), the first enzyme in the pathway for the production of heme. Thus, when there is more heme present than there are molecules to bind it, no more heme will be produced. Second, excess heme is seen to stimulate the production of globin proteins. When heme (as its oxidized form, hemin) is added to a cell-free translation system that includes all the factors needed to translate mRNAs, the synthesis of globin is greatly enhanced. Therefore, if there is no globin to bind the heme, heme will shut off its own synthesis and stimulate the production of more globin!

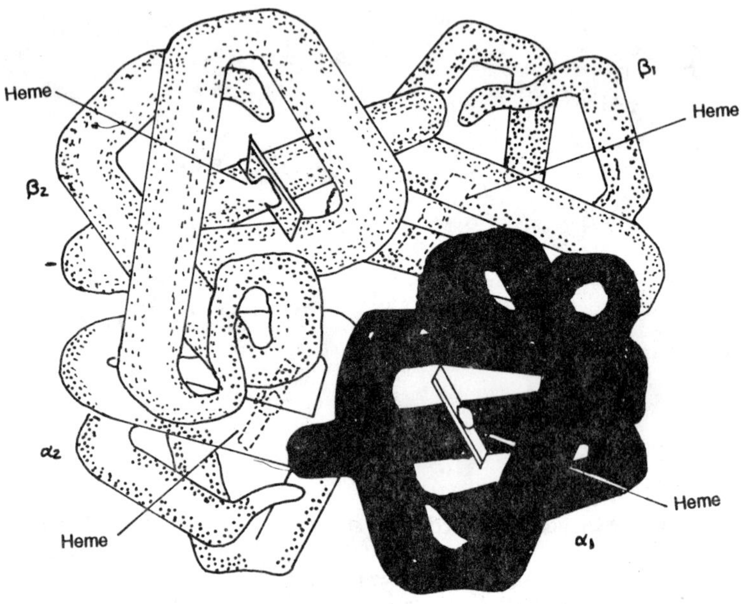

Fig. 6.6. The structure of adult human hemoglobin, which has four polypeptide chains (two α two β) and four heme molecules.

Several laboratories have investigated how such a small molecule as heme could regulate protein synthesis. In 1972, Adamson and colleagues demonstrated that the stimulatory effect of heme on globin synthesis could be mimicked by adding loosely bound ribosomal proteins to the translation system. Because such solutions are rich in translation initiation factors, each factor was tested separately. It was found that eukaryotic initiation factor 2 (eIF2) restored protein synthesis to heme-deficient lysates in the translation system. This initiation factor is responsible for combining with the initiator tRNA and complexing it to the 40 S ribosomal subunit.

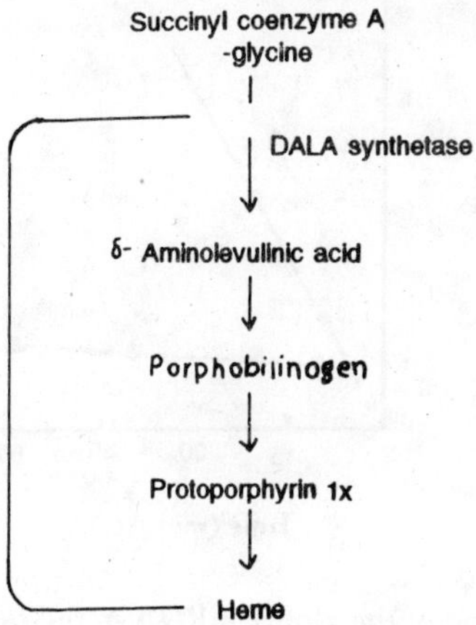

Fig. 6.7. Feedback regulation of heme synthesis.

TABLE 6.1

Components of the in Vitro Translation System Containing rabbit Reticulocyte Lysate

Concentration Component	*Concentration (in 100 μL)*	*Component*	*(in 100 μL)*
Reticulocyte lysate (1:1)	50 μL	KCl	76 mM
Tris-HCl (pH 7.6) buffer	10mM	Proportional amino acid mixture	6-170 μM
ATP	1mM	[$_{14}$C] Leucine	0.8 μ Ci
GTP	0.2 mM	"Cold" leucine	26μM
Creatine phosphate	5mM	Hemin	10-30 μM
Creatine phosphokinase	10 μg	H_2O to bring total reaction	
Magnesium acetate	2mM	to 100 μL volume	

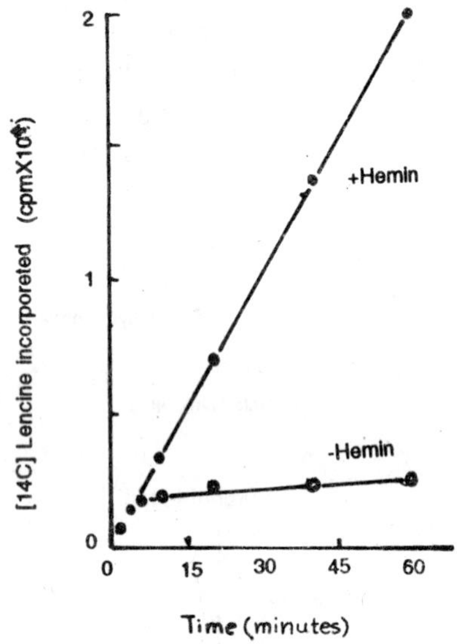

Fig. 6.8. Translation of the globin mRNA in the rabbit reticulocyte in vitro protein synthesis system. Inclusion of hemin causes a dramatic elevation in protein synthesis.

What, then, was the relationship between heme and eIF2? To answer this, London and his co-workers added heme-deficient lysates to heme-supplemented translation systems. They found that a portion of the heme-deficient lysate could actually depress the synthesis of globin in the translation system to which it was-added. This indicated that an inhibitor was present. Furthermore, this inhibiting fraction had enzymatic activity. It was a kinase capable of phosphorylating eIF2.

To summarize, heme appears to regulate globin synthesis in the following manner.

1. In the absence of heme, a specific protein kinase will phosphorylate initiation factor 2.
2. Phosphorylated initiation factor is not active and therefore will not complex Met-tRNA, to the 40 S ribosomal subunit. Thus, no translation occurs.

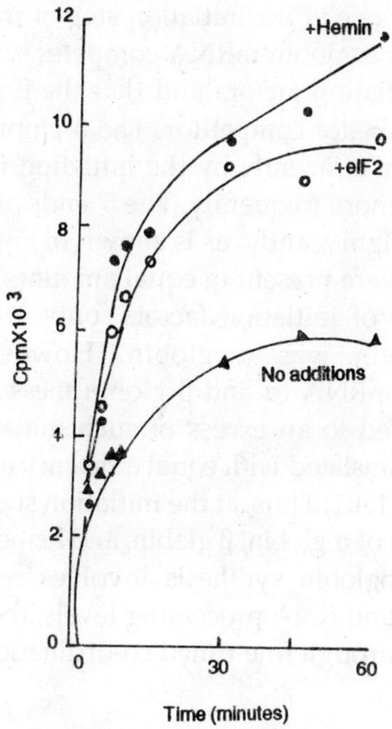

Fig. 6.9. Effect of the addition of extra eukaryotic initiation factor 2 to the rabbit reticulocyte in vitro translation system. The extra eIF2 elevated the level of protein synthesis close to that of the hemin-stimulated system.

3. Excess heme is able to bind to the protein kinase, inactivating it. Inactivated kinase will not phosphorylate eIF2, so translation proceeds. Thus, as long as heme is present, globin synthesis continues.

The story of the translation control of globin synthesis does not end here. As stated already, there are four active α globin genes per diploid cell and only two active β globin genes. If each gene were transcribed and translated at the same rate, one would expect twice as many α globin molecules as β globin molecules. This, of course, is not the case. One finds a 1.4:1 ratio of α:β mRNA, but 1:1 ratio of the proteins. Thus, the mechanism for the equalization has to be translational.

Kabat and Chappell (1977) have suggested that the equalization is done at the initiation step of translation. They showed that the α globin mRNA competes with the β globin message for initiation factors and that the β globin message appears to be a better competitor. The β globin message was recognized more efficiently by the initiation factors and was thus translated more frequently. The 5' ends of the two globin mRNAs differ significantly, as is shown in Figure 6.11. When the two RNAs were present in equal amounts with a severely limiting supply of initiation factors, only 3 per cent of the resulting protein was α globin. However, when the unfractionated mRNA (α and β globin messages from lysed cells) were added to an excess of such initiation factors, all mRNAs were translated with equal efficiency and the resulting α to β ratio was 1.4:1. Thus, at the initiation step of translation, the proper ratios of α globin, β globin, and heme are established. Although hemoglobin synthesis involves regulation at the transcriptional and RNA processing levels, the final molecule is constructed through fine-tuned co-ordination at the level of translation.

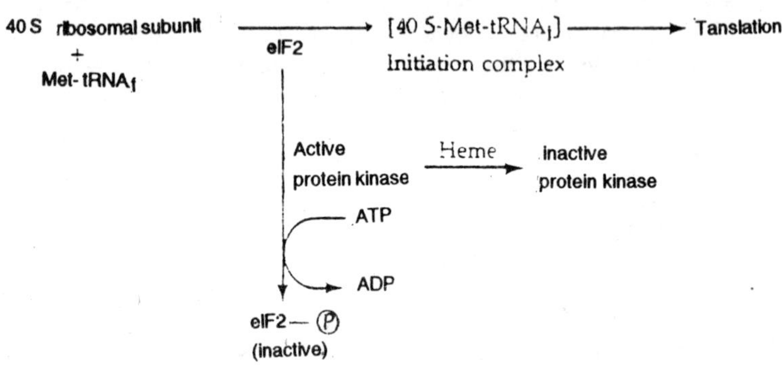

Fig. 6.10. Scheme for the translational control of globin synthesis. Eukaryotic initiation factor 2 is depicted as a result of inactivation by protein kinase, unless heme inactivates the protein kinase.

TRANSLATIONAL CONTROL OF OOCYTE MESSAGES

Evidence for Maternal Regulation of Early Development

In most animal species, the diploid nucleus is not immediately expressed. Evidence that early development is controlled by factors stored in or made by the oocyte came from several experiments at the turn of the century. These experiments clearly demonstrated the dominance of maternal traits during the initial stages of embryogenesis and a switch to paternal or hybrid characteristics only later in development. Such far-reaching maternal effects have already been alluded to in our discussion of the cleavage orientation in snail embryos, in which the oocyte cytoplasm contains a factor that directs the rotations of the cleavage planes in a dextral or sinistral direction.

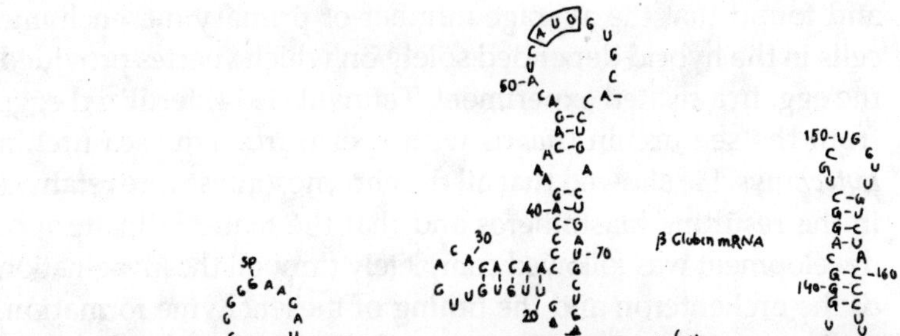

Fig. 6.11. Probable secondary structures for the 5' ends of mouse α globin and β globin chains.

TABLE 6.2

Driesch's data on the Maternal Control of Primary Mesenchyme Cell Number in Hybrid sea Urchin Embryos

Egg		*Sperm*	*Average number of primary mesenchyme cells (range in parentheses)*
Echinus	x	*Echinus*	55(±4)
Spherechinus	x	*Spherechinus*	33(±4)
Spherechinus	x	*Echinus*	35(±5)
Strongylocentrotus	x	*Strongylocentrotus*	49(±3)
Spherechinus	x	*Strongylocentrotus*	33(±3)

In 1898, Hans Driesch crossed two species of sea urchins and found that the average number of primary mesenchyme cells in the hybrid depended solely on which species provided the egg. In a related experiment, Tennant (1914) fertilized eggs from the sea urchin *Cidaris* with sperm from the sea urchin *Lytechinus*. He showed that all the chromosomes were retained in the resulting blastomeres and that the maternal pattern of development was followed completely through the invagination of the archenteron and the timing of mesenchyme formation. The paternal genes were first seen to be involved in the placement of the skeletal cells.

A second type of evidence for maternal regulation of early development comes from enucleation studies. If one could enucleate an egg after fertilization, one should be able to observe how far development could proceed before terminating. This enucleation has been accomplished by physical, chemical, and genetic means. Physical enucleation of the oocyte was accomplished by E. B. Harvey (1940). She placed unfertilized sea urchin eggs into a sucrose solution having the same density as the eggs themselves, and she found that when such eggs were centrifuged, their contents stratified and the egg was eventually split into two spheres. The lighter half contained the nucleus, whereas the heavier half was enucleated. Harvey then parthenogenetically activated the enucleated halves by placing them into hypotonic seawater. These halves ("parthenogenetic merogones") cleaved, formed abnormal blastulae (which lacked a blastocoel), and successfully hatched. No further development was seen. Similar studies involving frog eggs have also shown that cleavage can even occur in the total absence of chromosomes (provided the egg is injected with centrioles to compensate for those normally provided by the sperm). Here, too, an abnormal blastula is created. Physical enucleation studies suggested that certain eggs were capable of development through the midblastula stage, even in the absence of a nucleus.

TABLE 6.3

Tennant's Data on the Control of Early Gastrulation Events in Hybrid Sea Urchin Embryos

Species of sea urchin	*Archenteron invagination (hours)*	*Mesenchyme formation (hours)*	*Site of origin of primary mesenchyme cells*
Cidaris o	20-33	23-26 (follows invagination)	Archenteron tip
Lytechinus o	9	8 (precedes invagination)	Archenterone base and sides
Hybrid (*Cidaris o* x *Lutechinus*) *o*	20	24 (follows invagination)	Archenterone base and sides

Stored Messenger RNA

Evidence that the oocyte controlled early development by storing messenger RNAs was first obtained by Brachet and co-workers (1963) and by Denny and Tyler (1964). The results of these investigations demonstrated that such enucleated and activated sea urchin half eggs contained RNA and that they could synthesize proteins at a rate comparable to that of normally fertilized eggs. Because there could be no transcription from these cells, stored mRNA was implicated. Craig and Piatigorsky (1971) demonstrated that this increase in protein synthesis could not have come from the transcription of mitochondrial

DNA in the oocyte. It seemed that the oocyte had stored messenger RNAs that were not translated until after fertilization.

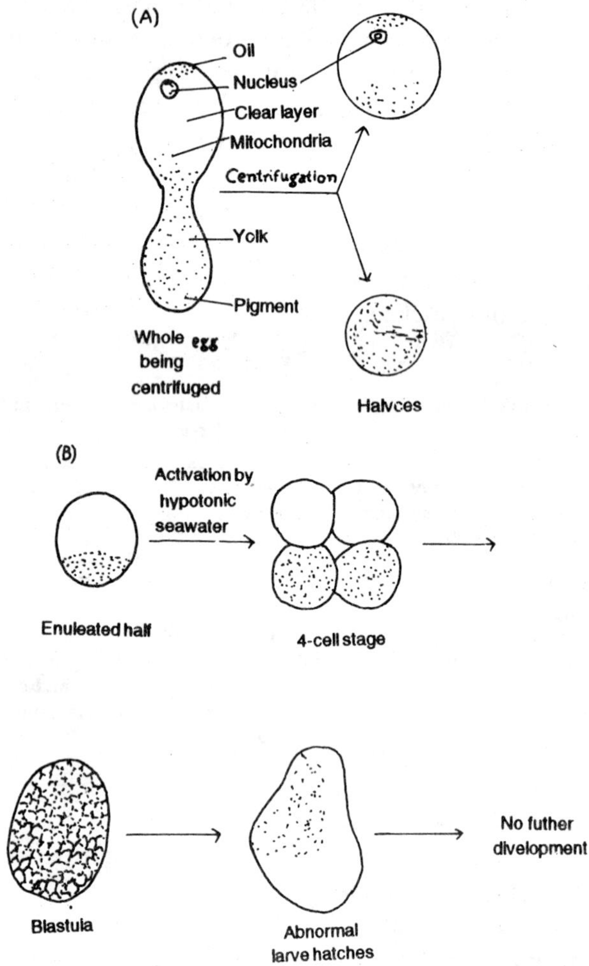

Fig. 6.12. Cleavage in the absence of a nucleus. (A) Physical enucleation of *Arbacia* oocyte by centrifugation in a sucrose solution. (B) Development of enucleated portion of oocyte into "parthenogenetic merogones" having no nuclei. Although the cleavage is often abnormal, without nuclear instruction some enucleated oocytes develop into blastulae and even hatch abnormal larvae.

A gentler means of enucleation was made possible when it was discovered that the drug actinomycin DD inhibited RNA synthesis, and transcription could be eliminated by placing the newly fertilized egg in a solution of this drug. Gross and Cousineau (1964) found that when sea urchin eggs were treated with enough actinomycin D to shut down 94 per cent of their RNA synthesis, the embryos still became blastulae. That this amount of development was dependent on stored messages and not on preformed proteins was shown by treating the fertilized eggs with emetin or cycloheximide. These drugs inhibit translation, and embryos fertilized in the presence of these inhibitors did not develop at all.

Several investigators demonstrated a burst of protein synthesis shortly after fertilization. The actinomycin D experiments of Gross and his co-workers demonstrated that the magnitude of the fertilization burst of protein synthesis in "chemically enucleated" embryos was exactly the same as that in controls. By 6-10 hours, however, a decline in protein synthesis was observed in the treated embryos, and they did not undergo a second burst of protein synthesis at the blastula stage. The message for this second burst of protein synthesis comes from nuclear transcription. Thus, the actinomycin D-treated embryos can develop up to the blastula stage, hatch, and then proceed no further. The stored mRNAs of the oocyte are sufficient to take development only through the hatched blastula stage. Students in amphibians have given similar results, although the stimulus for the increased rate of protein synthesis appears to be ovulation, rather than fertilization, of the egg. In activated enucleated frog eggs, the proteins synthesized shortly after fertilization are not only synthesized in the proper amounts, they are also the same type of protein as is normally made. Here too, it appears that the oocyte cytoplasm is "preprogrammed" to carry out early developmental processes even in the absence of a nucleus. Eventually, transcription is initiated in the nuclei of the embryo; yet even his this transcription of the embryonic nuclei may be activated by the

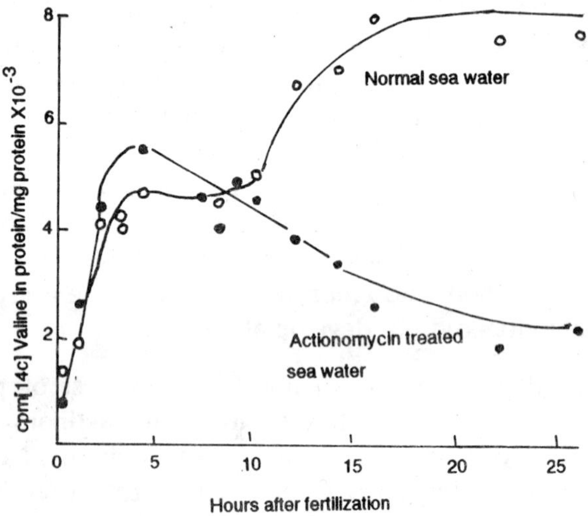

Fig. 6.13. Protein synthesis in embryos of the sea urchin *Arbacia punctulata* fertilized in the presence of actinomycin D. For the first few hours, there is no significant difference in the rate of new protein synthesis. The burst of new protein synthesis beginning in the midblastula stage represents the translation of newly-transcribed messages and therefore is not seen in the embryos growing in achinomycin D.

maternal factors in the oocyte. This concept is supported by the investigations of the *o* mutant of the Mexican axolotl. This is a maternal effect mutation wherein homozygous females produce eggs that are successfully fertilized and are totally normal until the late cleavage and early blastula stages. At midblastula, eggs shed by an *o*/*o* female are seen to have slower mitoses. These eggs will from a dorsal blastopore lip but will always arrest to gastrulation. Malacinski (1971) has shown that in wild-type blastula, new proteins are starting to be synthesized. However, the blastulae from eggs from *o*/*o* mothers do not undergo this burst of protein synthesis and have a pattern f proteins identical to that produced by enucleated

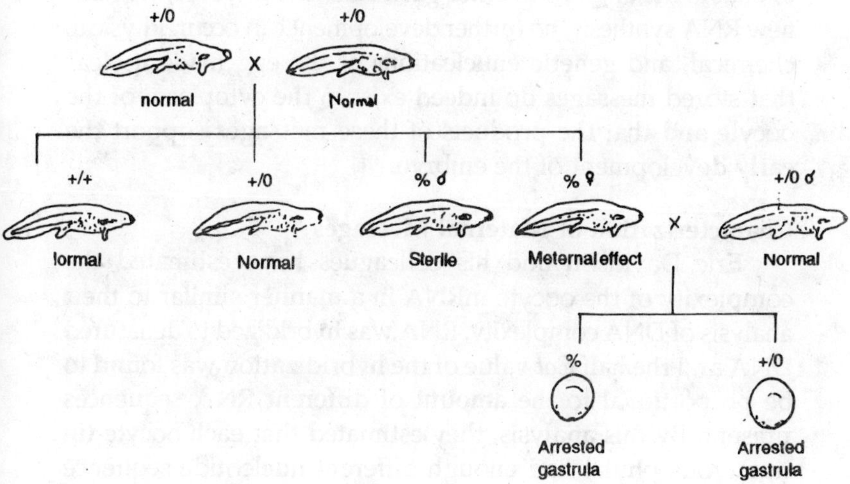

Fig. 6.14. The maternal effect of the *o* mutation in the axolotl. Where heterozygotes are mated, *o/o* individuals can exist because their mothers had a wild-type allele. Male *o/o* individuals are sterile. Female *o/o* individuals, however, produce mutant oocytes that arrest at the gastrula stage even if the wild-type allele is contributed by the male.

zygotes. Carroll (1974) has shown that whereas normal late blastulae show intense RNA synthesis, the mutant embryos show little or no RNA production. Briggs and Cassens (1966) demonstrated that the embryos from *o/o* mothers lacked a factor that activated the nuclear genome during blastula stage. Such a factor could be isolated from normal blastomere cytoplasm or from the nuclear sap of premeiotic oocytes. When this factor was injected into the eggs of *o/o* mothers, it repaired the gastrula arrest and allowed the embryo to develop normally. Thus, a specific factor is needed to activate the amphibian nucleus. In the absence of such a factor, the only development that occurs is that which can be supported by the stored oocyte mRNA. In amphibians, there is enough stored oocyte material to enable the embryo to enter gastrulation. However, without new RNA synthesis, no further development can occur. Physical, chemical, and genetic enucleations of the egg made it clear that stored messages do indeed exist in the cytoplasm of the oocyte and that the products of these messages support the early development of the embryo.

Characterization of Maternal Messages

Eric Davidson and his colleagues have estimated the complexity of the oocyte mRNA in a manner similar to their analysis of DNA complexity. RNA was hybridized to denatured DNA and the half *Cot* value of the hybridization was found to be proportional to the amount of different RNA sequences present. By this analysis, they estimated that each oocyte (in numerous phyla) had enough different nucleotide sequence to account for roughly 1600 copies each of 20,000-50,000 RNA types. This is the greatest message complexity of any known cell type, and it reflects the enormous developmental potential of the oocyte. Even though this figure represents nearly 4 per cent of the total RNA of the oocyte, only a few of these messages have been characterized. The use of cell-free translation systems and cDNA probes has enabled the identification of histone mRNA and tubulin mRNA in the oocyte cytoplasm. The products of these messages are of obvious importance for the formation of chromatin and the mitotic spindle during cleavage. Biochemical assays have permitted the identification of the messages for nucleotide reductase (an enzyme essential for the formation of deoxyribose nucleotides from stored ribonucleotides) and for the hatching enzyme, which permits the blastulae to digest their fertilization membranes. The utilization of recombinant DNA techniques discussed earlier is also being applied to maternal messages and has enabled investigators to isolate stored mRNA for actin and other proteins.

Interestingly enough, stored mRNAs are not uniformly distributed throughout the oocyte. We have already seen that certain RNA sequences show cytoplasmic localization in snail and tunicate eggs, but such localizations have also been observed in sea urchin eggs. Rodgers and Gross (1978) separated blastomeres at the 16-cell stage, cultured them in actinomycin D, and found that the mesomeres had more different types of mRNA than did the micromeres. Thus, cytoplasmic localization of preformed oocyte messages can be seen in both *"mosaic"* and *"regulative"* eggs.

Mechanisms for Translational Control of Oocyte Messages

There are, at present, at least four hypotheses for the regulation of oocyte mRNA translation. Three of them involve the availability of messenger RNAs; the fourth involves the efficiency of mRNA translation. Although these hypotheses may be seen as competing with one another, it is probable that each is used by different species and that some species use more than one of these mechanisms to regulate oocyte mRNA translation.

Masked Messages

The first hypothesis contends that the oocyte messages are physically masked by proteins so that the mRNA cannot attach to ribosomes. Messenger RNA is never found devoid of proteins. However, the type of protein associated with the RNA can vary 1966. Spirin proposed that the mRNA of the oocyte was stored in Informosomes, ribonucleoprotein complexes wherein the mRNA was masked. The masked messages would be unable to bind to the ribosomes and thus would not be translated. At fertilization, the proteins masking the message would be released (possibly because of the ionic changes occurring during fertilization) and the message would be free to start translation. Support for this hypothesis followed shortly. In 1968, Infante and Nemer found in the sea urchin oocyte ribonucleoprotein (RNP) particles that sedimented more slowly than ribosomes, and Gross and co-workers (1973) found that these particles contained various messages. Other studies showed that oocyte ribosomes and initiation factors were perfectly capable of translating exogenously supplied message, so that the control of translation did not seem to be due to the number or capabilities of the ribosomes.

A series of experiments in Rudolf Raff's laboratory demonstrated that oocyte mRNA is indeed stored in a non-translatable form that is susceptible to ionic changes such as those occurring during fertilization. Nonribosomal RNP particles can be isolated from mature sea urchin oocytes and are found to contain RNA with poly(A) tails. Thus, these RNP particles seem to contain messenger RNA. These RNP particles were isolated in two different ionic solutions. Some were isolated "oocyte" butter containing 0.35 mM K^+ and 5 mM Mg, an ionic condition approximating that of unfertilized oocytes. Other particles were isolated into 0.35 mM Na^+, an ion concentration that should remove the proteins bound non-covalently to the mRNA. The mRNA-containing particles prepared in the "oocyte" buffer were very poorly translated in a cell-free system. However, the RNA extracted into the sodium-containing buffer were above to support protein synthesis almost as well as isolated egg mRNA. It thus appears that the influx of sodium during fertilization may destabilize the RNP particle, allowing its mRNA to be translated. Recently certain oocyte-specific proteins have been implicated in the masking of stored messages. When globin mRNA is injected into *Xenopus* oocytes, it is readily translated. However, if this message is first mixed with proteins isolated from *Xenopus* oocyte RNP particle, very little translation is observed. Other RNA-binding proteins isolated from other tissues did not affect the translatability of the injected globin messages, so the oocyte-specific proteins of the RNP particles are implicated in the translational regulation of stored maternal messages.

It is known that whereas the unfertilized sea urchin oocyte has only 0.75 per cent of its ribosomes incorporated into polysomes, the cleavage-stage blastomeres have nearly 20 per

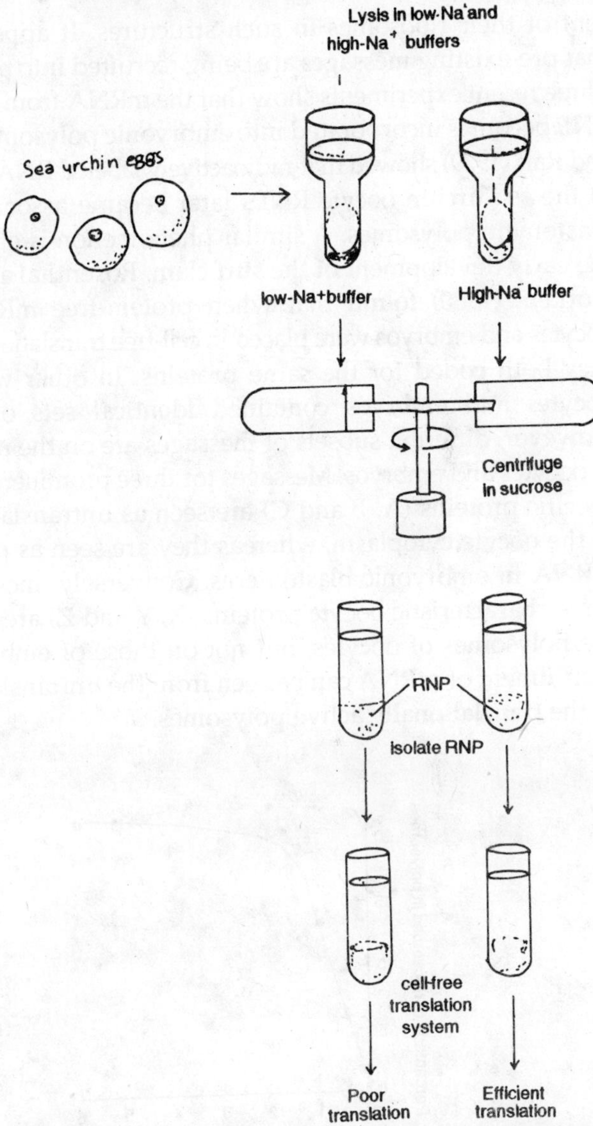

Fig. 6.15. Evidence for the presence of RNA in RNP and its inability to be translated while in the particles. Protein-bound RNA cannot be translated in low sodium buffer, but can be translated into proteins in buffer containing high concentrations of sodium ions.

cent of their ribosomes in such structures. It appears, then, that pre-existing messages are being recruited into polysomes. Three recent experiments show that the mRNA from the oocyte RNP becomes incorporated into embryonic polysomes. Young and Raff (1979) showed that radioactively labeled RNA originally in the sea urchin oocyte RNPs later became associated with blastomere polysomes. A similar phenomenon is observed in the early development of the surf clam. Rosenthal and his co-workers (1980) found that when protein-free mRNA from oocytes and embryos were placed in cell-tree translation systems, they both coded for the same proteins. In other words, the oocytes and embryos contained identical sets of mRNA. However, different subsets of messages are on the polysomes in oocytes and embryos. Messages for three prominent embryo-specific proteins (A, B and C) are seen as untranslated RNPs in the oocyte cytoplasm, whereas they are seen as polysomal mRNA in embryonic blastomeres. Conversely, messages for three characteristic oocyte proteins (X, Y and Z) are found on the polysomes of oocytes, but not on those of embryos. The recruitment of mRNA can be seen from the untranslated RNP to the translationally active polysomes.

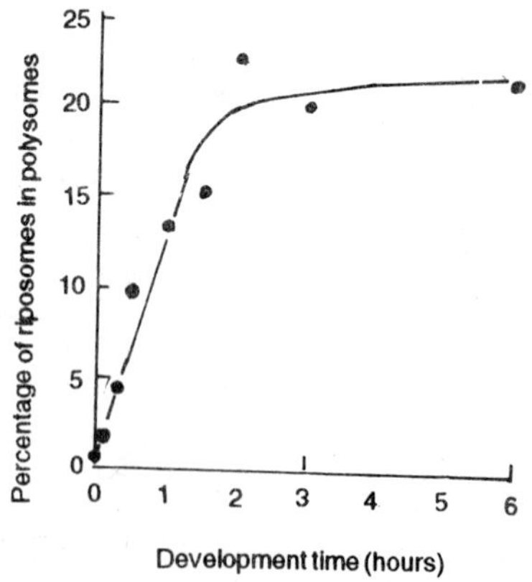

Fig 6.16. Increase in the percentage of ribosomes incorporated into polysomes during early sea urchin development.

Message recruitment is also seen *Drosophila* embryos. Mermod and co-workers (1980) isolated poly(A)-containing RNA from oocyte RNP,oocyte polysomes, and embryonic polysomes. This mRNA was then translated in a cell-free system containing radioactive amino acids. They found that the mRNA from *Drosophila* oocyte RNP can made certain proteins that cannot yet be synthesized by the mRNA isolated from oocyte polysomes. However, these proteins can be made from the mRNA isolated from *embryonic* polysomes. Thus, mRNA formerly stored in the oocyte RNP particles has been recruited into the embryonic polysomes. In this manner, gene expression is controlled by the initiation of translation.

Uncapped Messages

Certain moths use a different mechanism for translational control within the oocyte. In order to be translated, most all eukaryotic messages need to have a "cap" on their 5' ends (Shatkin, 1976). These eukaryotic messages are characterized by 5' 7-methylguanosine. It is believed that this structure is necessary for the recognition of the mRNA with the 40 S ribosomal subunit. The stored messages of the tobacco hornworm moth oocyte have a non-methylated cap. The guanosine is present, but the methyl group has not been added to it. Such messages are not translated into proteins in a cell-free system. However, at fertilization, there is a burst of methylation in these oocytes and the cap is completed. The mRNAs with the completed caps are then able to bind to the ribosomes and initiate translation. Therefore, in two different ways, the oocyte keeps its mRNA and ribosomes separate. Whether by masking the message or not completing its 5′ end modifications, the initiation of translation is inhibited. Thus, mRNAs can be stored without being translated, and fertilization triggers the events that enable their translation the oocyte.

Sequestered Messages

Some studies have disputed the evidence that oocyte RNPs are untranslatable. Rather, it is possible that the protein synthetic apparatus is compartmentalized such that mRNA (within the

RNP) does not have a chance to get close to the ribosomes. The histone mRNAs of sea urchin oocytes seem to be regulated by this type of restriction. The histone messages of the oocyte are not found in cytoplasmic RNPs. Rather, they are localized in the pronucleus. It is only when the pronucleus breaks down, at the end of fertilization, that the histone mRNA gets into the cytoplasm. This may not be the case for other messages. Less than 0.1 per cent of the total oocyte messenger RNA is found in pronuclei, and those RNPs containing the messages for actin and tubulin are predominantly cytoplasmic.

Changes in Translational Efficiency

In one model of translational regulation in the oocyte the physical or chemical separation of mRNAs need not be postulated. Rather, the initial low pH of the oocyte is able to impede protein synthesis. As stated already, there is an enormous release of hydrogen ions during sea urchin fertilization, resulting in an elevation of cytoplasmic pH from 6.9 to 7.4. Winkler and Steinhardt (1981) prepared an in vitro cell-free translation system from unfertilized sea urchin oocytes. The pH of the resulting suspension was then retained or altered by dialysis, and protein synthesis was measured by the incorporation of radioactive valine into proteins. No exogenous mRNA was added. Figure 6.17 shows the results of one such experiment. Increasing the pH from oocyte levels (pH 6.9) to zygote levels (pH 7.4) occasioned a burst of protein synthesis mimicking that seen during fertilization. It is not known whether this increase in protein synthesis is due solely to the increased efficiency of protein synthesis or whether increasing the pH may cause other conditions (such as the unmasking of messages) that then allow protein synthesis to occur. This increased efficiency alone would not account for the specificity of translation as seen in the surf clam (where messages formerly on the polysomes are replaced by others). In any event, the oocyte is able to exercise translation-level control to regulate the expression of preexisting mRNAs.

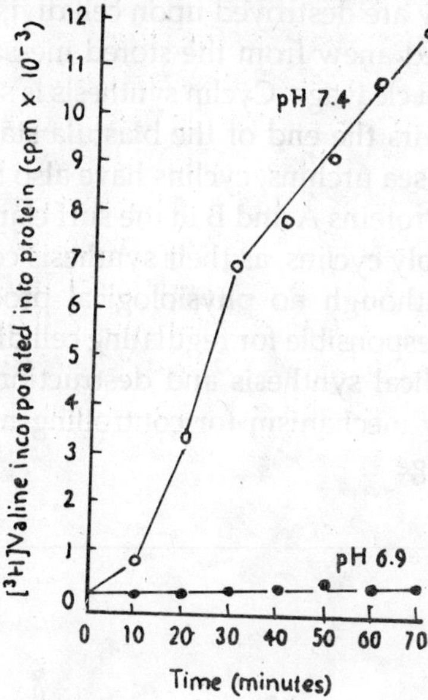

Fig. 6.17. Evidence for the inefficiency of protein synthesis at prefertilization pH levels. The in vitro translation system made from unfertilized eggs is kept at pH 6.9 or dialyzed to pH 7.4. Endogenous messages are translated much more efficiently at the postfertilization pH.

Maternal mRNA and Embryonic Cleavage

Cleavage in sea urchins requires continued protein synthesis from stored maternal mRNAs. If sea urchins are fertilized in the presence of an inhibitor of translation, no cleavage results, even though the two pronuclei fuse. No mitotic spindle forms, the chromosomes do not condense, and the nuclear membrane does not break down. Thus, it is probable that some of the proteins specified by material messages are involved in cell division during cleavage. This hypothesis has gained support from the discovery of a class of proteins called *Cyclins*. These proteins are coded for by maternal mRNAs, and they are not detectable in the unfertilized egg. However, their synthesis increases dramatically after fertilization. What is striking about cyclins is that they are destroyed upon cell division and have to be resynthesized anew from the stored messages after the completion of each cleavage. Cyclin synthesis is seen to decline as the embryo nears the end of the blastula stage. Although originally seen in sea urchins, cyclins have also been detected in other species. Proteins A and B in the surf clam (mentioned earlier) are probably cyclins, as their synthesis correlates with the cell cycle. Although no physiological proof exists that these cyclins are responsible for regulating cell division during cleavage, the cyclical synthesis and destruction of a protein would be a likely mechanism for controlling mitosis during embryonic cleavage.

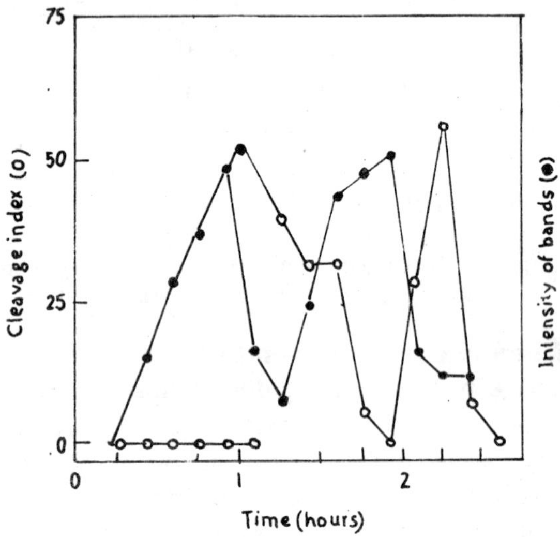

Fig. 6.18. Correlation of cyclin levels with cleavage divisions in the sea urchin embryo. Eggs were fertilized in the presence of radioactive amino acids and were analyzed every 10 minutes for the percentage of dividing cells and for the presence of a particular protein that exhibited cyclicity. This protein was at its highest levels during the middle of each cell cycle and was then degraded and resynthesized. Other proteins increased linearly during this period.

Translation Control of Casein Synthesis

Differentiated gene products are often synthesized in response to hormonal induction. In some of these cases, the hormones do not increase the transcription of certain messages but act at the level of translation. One such case involves the

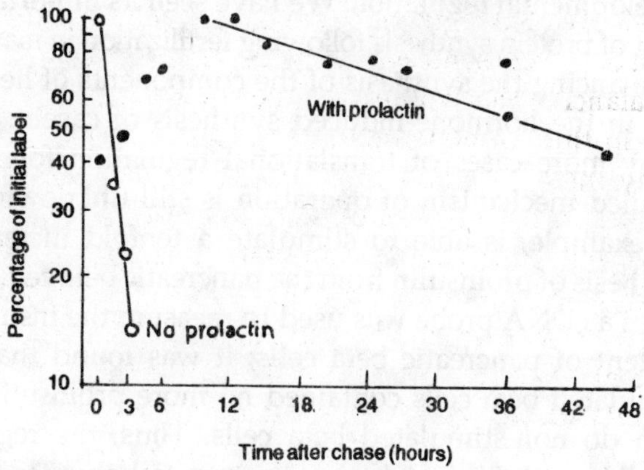

Fig. 6.19. The degradation of casein mRNA in the presence and absence of prolactin. Cultured mammary, cell were given radioactive RNA precursors (pulse) and, after a given time, were washed and fed non-radioactive precursors (chase). The casein RNA synthesized during the pulse time was then isolated and counted. In the absence of prolactin, the newly-synthesized casein mRNA decayed rapidly, with a half-life of 1.1 hours. When the same experiment was done in medium containing prolactin, the half-life was extended to 288.5 hours.

synthesis of casein by lactating mammals. Casein is the major phosphoprotein of milk and is therefore a differentiated product of the mammary gland. The mammary gland is prepared by the sequential actions of several hormones. Prolactin, however, is the hormone responsible for lactation, that is, actual milk production. Prolactin augments the transcription of casein messages only about twofold; its major effect appears to be the stabilization of the casein mRNA. Prolactin somehow increase the longevity of the casein message such that it exists 25 times longer than most other messages in the cell. This means that each casein mRNA can be used for many more rounds of translation. In this way, more casein molecules can be synthesized from each casein message.

The Widespread use of Translational Regulation

Translational control is probably a major mechanism of developmental regulation. We have seen its importance in the burst of protein synthesis following fertilization in many species, in balancing the synthesis of the components of hemoglobin, and in the hormone-induced synthesis of casein. There are many more cases of translational regulation for which the detailed mechanism of operation is still unknown. Glucose, for example, is able to stimulate a tenfold increase in the synthesis of proinsulin from the pancreatic beta cell. However, when a cDNA probe was used to measure the insulin mRNA content of pancreatic beta cells, it was found that glucose-stimulated beta cells contained no more proinsulin message than do non-stimulated beta cells. Thus, the regulation of insulin synthesis by glucose appears to be under translation control. The reactivation of dehydrated brine shrimp embryos also seems to be regulated at the level of translation. These embryos (often sold as *"brine shrimp eggs")* begin development and can then lie dormant for long periods of time. During this dormancy, no protein synthesis occurs. It has been found that an inhibitor of protein synthesis exists in these embryos and that upon reactivation, a 20-fold increase in the level of functional eIF2 occurs. Moreover, Sierra *et al.* (1977) showed that this inhibitor may be a protein kinase. Thus, the same mechanism that regulates haemoglobin synthesis may also be used by brine shrimp embryos.

Even sperm appear to regulate the expression of certain genes by translational control. Just like eggs, sperm can store messages for later use. This can be seen in the case of lactate dehydrogenase-X in avian and mammalian sperm. Although this *protein* is only synthesized in the last stages of sperm development, the *gene* for LDH-X is transcriptionally active throughout all stages of sperm development. The synthesis of

LDH-X protein probably occurs by initiating the translation of stored mRNAs. Translational control, then, is an important and widely used mechanism for regulating gene expression in development.

POSTTRANSLATIONAL REGULATION OF GENE EXPRESSION

Once a protein is made, it becomes part of a larger level of organization. It may become part of the structural framework of the cell or it may become involved in one of the myriad enzymatic pathways for the synthesis or breakdown of cellular metabolites. In either case, the individual protein is now part of a complex *"ecosystem"* which integrates it into a relationship with numerous other proteins. Thus, we do not abandon the regulation of gene expression after synthesizing a protein, for several changes can still take place. First, some newly synthesized proteins are inactive without further modifications. These modifications can involve the cleaving away of certain inhibitory sections of the protein or may involve the binding of a small compound to enhance its activity. Second, some proteins may be selectively inactivated. In some cases, inactivation involves the degradation of the protein itself; in other cases, inactivation may be brought about by the binding of an inhibitory ligand. Third, some proteins must be *"addressed"* to their specific cellular destinations. The cell is not merely a sack of enzymes; and proteins will often be sequestered in certain regions such as membranes, lysosomes, or mitochondria. Fourth, some proteins need to assemble with other proteins to form a functional unit. The haemoglobin protein, the microtubule, and the ribosome are all examples of numerous proteins joining together to form a functional unit. Therefore, the expression of genetic information can still be influenced at the posttranslational level. Let us now review each of these of these in turn.

Activation of Proteins by Posttranslational Modifications

Many newly synthesized polypeptides are inactive unless certain amino acid residues are cleaved away. This is especially

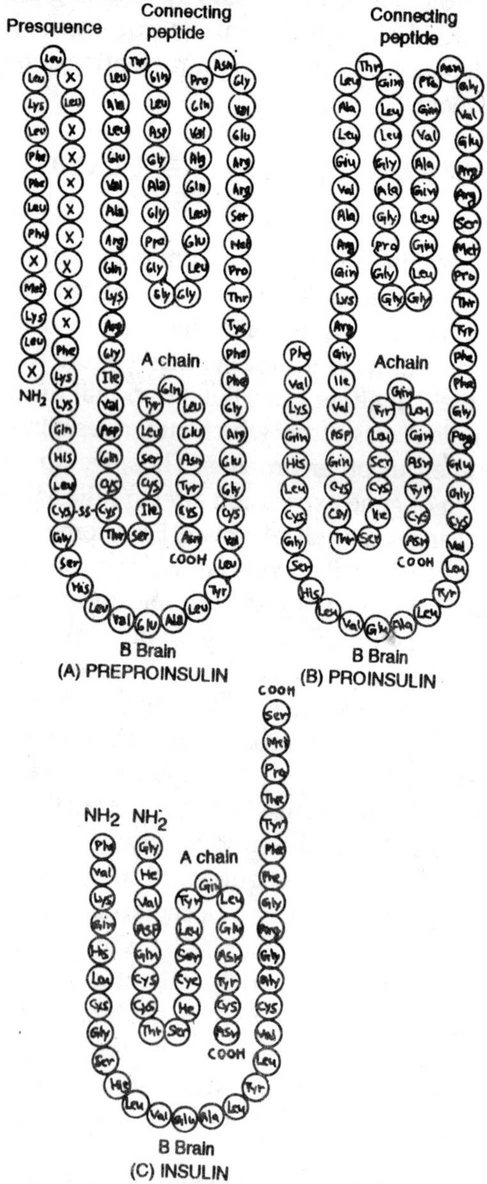

Fig. 6.20. Amino acid sequences of fat insulin II (shaded circles) and its precursors. The insulin molecule is inactive until the presequence and the connecting middle sequence have been cleaved away. (X indicates unknown amino acids).

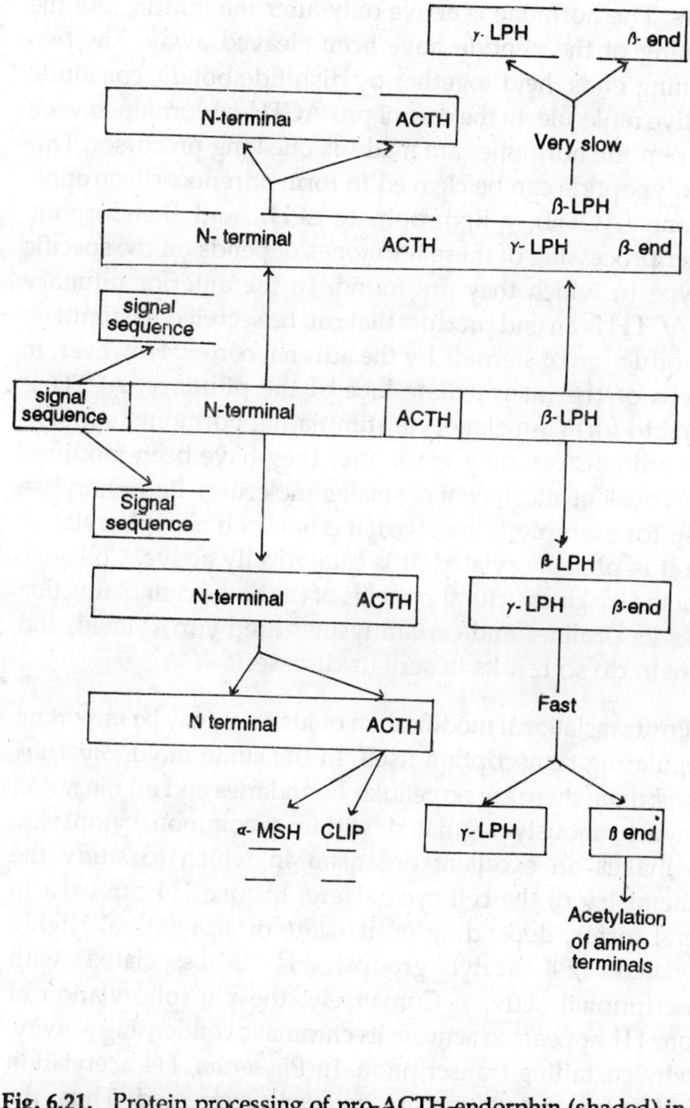

Fig. 6.21. Protein processing of pro-ACTH-endorphin (shaded) in the intermediate and anterior lobes of the rat pituitary gland.

evident in the case of peptide hormones. The precursor of insulin, for instance, is a long polypeptide with three disulfide bridges. The hormone is active only after the middle and the beginning of the peptide have been cleaved away. The two remaining ends, held together by disulfide bonds, constitute the active molecule. In the case of pro-ACTH-endorphin, several small peptide hormones are made as one long precursor. This one polypeptide can be cleaved to form adrenocorticotrophic hormone (ACTH), g-lipotropin (g-LPH), and b-endorphin. Further processing of these hormones depends on the specific cell type in which they are found. In the anterior pituitary cells, ACTH is an end product that can be secreted to stimulate the production of steroids by the adrenal cortex. However, in the cells of the intermediate lobe of the pituitary, ACTH is cleaved to form a-melanocyte stimulating hormone (a-MSH). Some proteins can only work after they have been modified by the covalent attachment of smaller molecules. Phosphorylase kinase, for example, is inactive if it is not itself phosphorylated. Once it is phosphorylated, it is catalytically active. Collagen, one of the major structural proteins of the body, cannot function unless its prolines and certain lysines are hydroxylated, and failure to do so results in serious diseases.

Posttranslational modification of histones may be important in regulating transcription itself. In the slime mold *Physarum polycephalum*, there are no cellular boundaries and all the nuclei are synchronously regulated within a common cytoplasm. This makes an excellent organism in which to study the biochemistry of the cell cycle. Here, histone H4 can exist in several forms, depending on its state of acetylation. Highly acetylated (2-4 acetyl groups) H4 is associated with transcriptional activity. Conversely, the phosphorylation of histone H1 appears to activate its chromatic condensing activity, thereby curtailing transcription. In *Physarum*, H4 acetylation and H1 phosphorylation are inversely expressed. Thus, the gross transcriptional activity of chromatin may be regulated by the phosphorylation or acetylation of histone proteins.

Inactivation of Proteins by Posttranslational Modifications

Alterations to existing proteins may also by used to *inactivate* them. Earlier in this chapter we have seen how phosphorylation inactivated initiation factor 2 in the regulation of haemoglobin synthesis. Glycogen synthetase and pyruvate kinase are similarly inactivated by phosphorylation. Glutamine synthesis is inactivated when an AMP moiety is transferred to it. Since these reactions are all reversible, the protein's catalytic activity can be regulated to a very precise level by regulating the activities of the enzymes that add or cleave the modifiers.

Another way of regulating gene expression posttranslationally is to degrade specific polypeptides after translation. One such example is lactate dehydrogenase (LDH), which catalyzes the reversible interconversion of pyruvate and lactate. LDH is coded by two different genes: one that encodes LDH-A subunits and one that encodes LDH-B subunits. These genes are on different chromosomes. The functional LDH enzyme is a tetramer, meaning that it is composed of four subunits. These submits can be of either the A or the B variety. If both genes are expressed, one expects to find five different forms of LDH: A4, A3B1, A2B2, A1B3 and B4. (Alternative forms of an enzyme are called *Isozymes.)* The LDH isozymes have different catalytic efficiencies under different cellular conditions. Moreover, if the genes are being expressed equally and recombination is random, one would expect the ratio of types to be 1(A4): 4(A3B1): 6(A2B2): 4(A1B3): 1(B4). These isozymes have different electrical charges: the B subunit is more negatively charged than the A subunit. Therefore, the five isozymes can be separated by electrophoresis and stained for clatalytic activity. When equal quantities of the subunits are combined artificially, one indeed finds the 1:4:6:4:1 ratio. However, when one looks at individual organs, large deviations from this ratio are found. Muscle LDH is predominantly made of A subunits, whereas testes, heart, and kidney isozymes are composed mostly of B subunits.

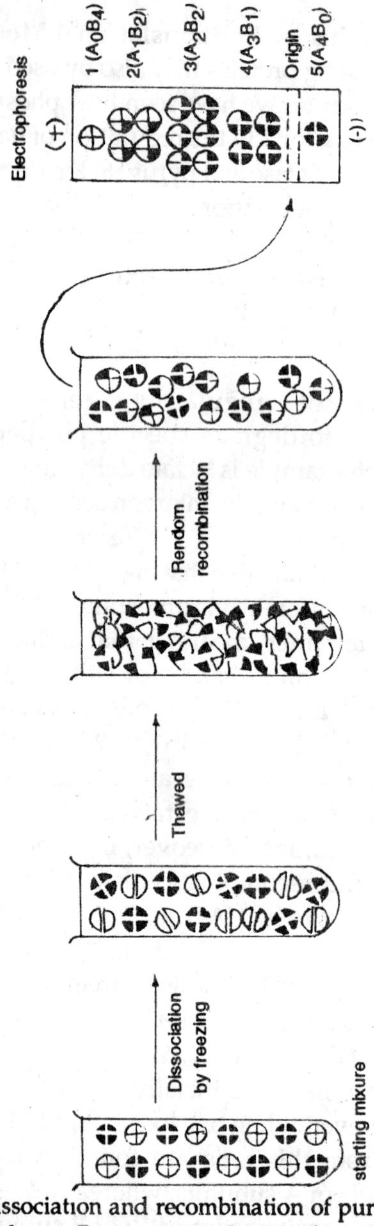

Fig. 6.22. Dissociation and recombination of pure A tetramers (black) and pure B tetramers (white) to form the five elctrophoretic isozymes of LDH.

Using anti-LDH-antibodies to precipitate any newly synthesized A subunits, Fritz and co-workers (1969) discovered that the synthesis of LDH-A does not vary greatly from tissue to tissue. However, the rate of its degradation varies enormously. The degradation of LDH-A is 22 times greater in heart muscle than in skeletal muscle. So the control of LDH isozymes is not at the level of LDH transcription or translation, but at the level of enzyme degradation.

Subcellular Localization of Proteins by Posttranslational Modifications

After they are synthesized, proteins have to be properly positioned. Thus, the protein may ultimately be located in the soluble cytoplasm, the mitochondria, the lysosomes, the endoplasmic reticulum, or the nucleus; it may even be secreted from the cell.

Those proteins destined for lysosomes, endoplasmic reticula, or secretion are passed into the lumen of the rough endoplasmic reticulum which they are still being translated. These proteins (including collagen and the peptide hormones discussed earlier) contain a signal of approximately 30 amino acids, which are recognized by a *Signal Receptor Complex* that floats in the cytoplasm. When this complex (which contains six different peptide chains and a small RNA molecule) binds to the growing peptide of a cytoplasmic ribosome, translation stops. The complex then joins to a docking protein on the endoplasmic reticulum and translation resumes. The growing polypeptides pass through the reticular membrane and once inside the lumen, this signal sequence is removed and the remaining protein can be modified.

One such modification is the covalent addition of mannose-6-phosphate to the protein. This sugar acts as an address label that packages the glycoprotein for lysosomes. The mannose-6-phosphate is recognized by receptors at specific areas of the endoplasmic reticulum, which thereby concentrate those enzymes having this address label. This region of the reticulum

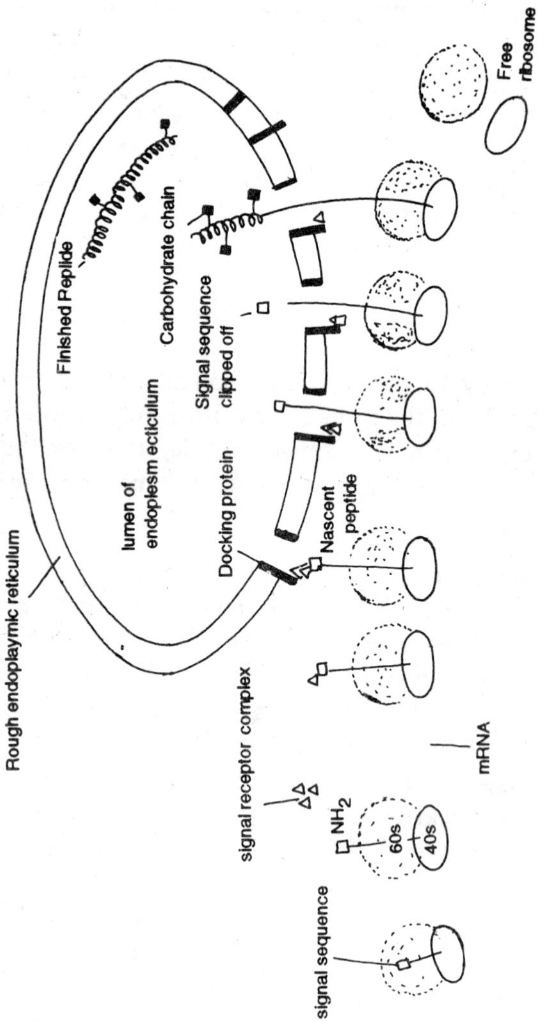

Fig. 6.23. Diagrammatic model for the translation, glycosylation, and transmembrane insertion of secreted proteins. The first 50 or so amino acids of the polypeptide usually code for a "signal sequence" which identifies the protein as one to be inserted into the membranes of the rough endoplasmic reticulum. The sequence is recognized by a signal receptor complex and then by sites on the RER. Once within the lumen of the RER, part of this sequence is clipped off. As the protein continues to elongate, carbohydrates are added to it.

then buds off to form the lysosome. When binding does not occur, these proteins are secreted from the cell. Evidence for this comes from a lethal genetic syndrome, I-cell disease, where patients are unable to put the mannose-6-phosphate address label on the proteins. The lysosomes of these individuals lack the appropriate enzymes, which have been mistakenly secreted into the blood. Some proteins have specific binding proteins, which localize them in certain regions of the cell. In mouse liver, b-glucuronidase is found both in the lysosomes and on the endoplasmic reticulum. Although the mannose-6-phosphate label can get this enzyme into the lysosomes, b-glucuronidase is bound to the liver endoplasmic reticulum by the *protein Egasyn,* which specifically recognizes this protein. In mutant mice lacking egasyn, there is no b-glucuronidase found on the liver endoplasmic reticulum. The localization of b-glucuronidase is different in various cell types. Tissues such as liver and kidney are rich in egasyn and have b-glucuronidase on their endoplasmic reticula. Brain and spleen cells, however, appear to be entirely deficient in egasyn, and these have all their ß-glucuronidase in their lysosomes. So the intracellular localization of proteins is posttranslationally regulated and may differ from tissue to tissue.

Supramolecular Assembly

The last form of posttranslational regulation that we will consider involves the assembly of these newly synthesized proteins into a functional unit. We have already discussed several proteins that can spontaneously assemble themselves into functional units. Hemoglobin and lactate dehydrogenase, for example, both contain four polypeptide chains, which are non-covalently jointed to form the functional protein. On a slightly higher level, the contractile proteins tubulin and actin are both found to exist in polymerized or non-polymerized forms. We have seen that the polymerization of actin from globular monomers into microfilaments is essential for the extension of the acrosomal process during fertilization. Similarly, tubulin is assembled into the microtubules of the mitotic apparatus, only to be converted into monomers as mitosis finishes. These tubulin units can then be reassembled into new microtubules, which are important in establishing cell shape. There appears to be a class of proteins (microtubule-associated proteins) that are necessary for regulating the polymerization and function of microtubules.

Several proteins form assemblies with nucleic acids. Histones, for instance, are useless unless assembled with DNA to form the nucleosome particle. The eukaryotic 80 S ribosome consists of approximately 70 proteins and 4 different ribosomal RNAs. The proteins and nucleic acids have to assemble together to be functional, and it is remarkable that this assembly occurs quickly and efficiently within all cells. It should not be a total surprise, then, to discover that whole infectious viruses can be obtained by incubating the component proteins and nucleic acids together. Thus, spontaneous assembly is another important form of posttranslational control.

Collagen : An Epitome of Posttranslational Regulation

By looking at the synthesis of collagen, one of the most important structural proteins of the body, we can demonstrate the importance and variety of posttranslational control mechanisms. First, the collagen mRNA sequence is translated into a procollagen peptide. The amino terminal of the procollagen contains a signal sequence that enables it to enter the lumen of the rough endoplasmic reticulum. This signal peptide gets cleaved from the remainder of the collagen molecule. Within the endoplasmic reticulum, the collagen molecules also encounter three enzymes that hydroxylate it. Two of these enzymes hydroxylate proline residues to 3-hydroxyproline and 4-hydroxyproline; the third converts lysines to hydroxylysyl residues. Only certain proline and lysine residues are in the correct positions to be recognized by these enzymes. As the collagen chains are being hydroxylated, sugar residues can be added to the hydroxylysines. The first enzyme, galactosyltransferase, adds a galactose to the hydroxylysine; the second enzyme, glucosyltransferase, adds a glucose unit

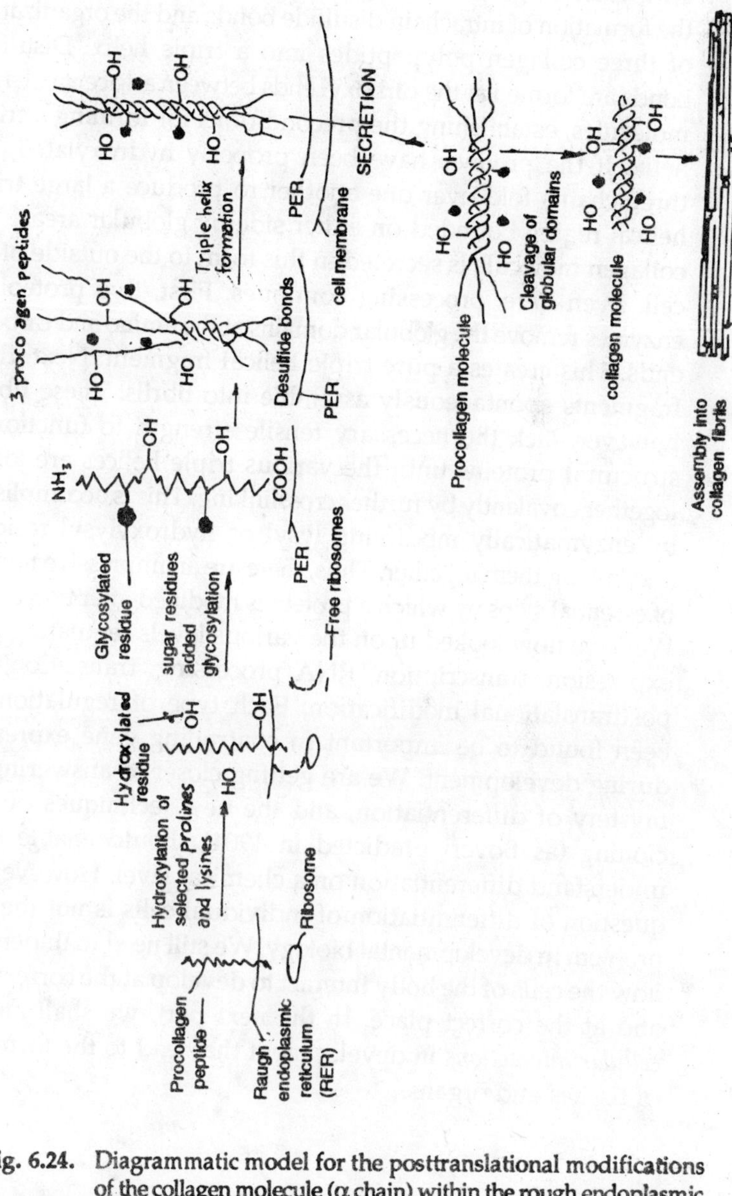

Fig. 6.24. Diagrammatic model for the posttranslational modifications of the collagen molecule (α chain) within the rough endoplasmic reticulum and after secretion.

to the hydroxylsylgalactose. The region near the carboxyl end is modified by glucosamine and mannose. The next step involves the formation of intrachain disulfide bonds and the organization of three collagen polypeptides into a triple helix. Disulfide bonds are formed at the carboxyl ends between adjacent collagen molecules, establishing the preconditions for forming a triple helix. If the prolines have been properly hydroxylated, the three chains fold over one another to produce a large triple helical region bounded on either side by globular areas. The collagen molecule is secreted in this form to the outside of the cell. Even here, processing continues. First, two proteolytic enzymes remove the globular domains at the amino and carboxyl ends. This creates a pure triple helical fragment. Next, these fragments spontaneously assemble into fibrils. These fibrils, however, lack the necessary tensile strength to function as structural proteins until the various triple helices are joined together covalently by further crosslinking. This is accomplished by enzymatically modifying lysyl or hydroxylysyl residues and linking them together. Thus, there are an impressive number of essential steps by which a protein is modified after translation. We have now looked upon the various levels regulating gene expression: transcription, RNA processing, translation, and posttranslational modification. Each type of regulation has been found to be important in controlling gene expression during development. We are getting closer to answering the mystery of differentiation, and the new techniques of gene cloning (as Boveri predicted in 1904) should enable us to understand differentiation on a chemical level. However, the question of differentiation of individual cells is not the only problem in developmental biology. We still need to understand how the cells of the body interact to develop at the correct time and at the correct place. In the next part, we shall analyze *cellular interactions* in development that lead to the formation of tissues and organs.

7

Differentiated Somatic Cells and Genomes

The process of cellular differentiation in developing multicellular organisms culminates in the formation of specialized somatic cell types. Once cell types acquire a specific cellular phenotype for performing a specialized function, no further changes usually occur. We call this the stability of the differentiated state. It is now widely accepted that the process and stability of cell specialization are under genetic control. But the frequency and extent to which this genetic control involves irreversible genetic changes among eukaryotic animals remain unanswered. Irreversible genetic changes involving DNA losses or whole chromosomes, so far as we know, occur as a regular event in only a few animal species. Examples of such losses are present in some protozoans, nematodes, crustaceans, insects, and mites of the lower animal phyla. Among vertebrates, DNA losses occur in the holocephalan fish, the mammalian Marsupialia, as well as in the genetic rearrangement of the immunoglobulin genes in chicken (Weill and Reynaud, 1987) and mammalian B lymphocytes and of certain receptor genes in mammalin T lymphocytes. Most of the evidence available today indicates that a large portion of the genome is retained in specialized somatic cells. For example, the activation of dormant genes in specialized cells is a relatively common phenomenon, consistently demonstrated in several experimental systems (DiBerardino *et al.* 1984), including cell cultures, transdifferentiation, heterokaryons, cell hybrids (H. Harris, 1985; *this volume*), cancer and nuclear transplantation into oocytes and eggs. In these cases, the activation of dormant genes is accompanied by changes in the cell phenotype of specialized cells. Therefore, the evidence indicates that many silent genes are maintained in the genome in the absence of expression and that under favourable conditions they are stimulated to function. Nevertheless, a fundamental problem concerning the developmental process in multicellular organisms still remains unsolved, i.e., whether most differentiated somatic cell types retain the genomic totipotency of the zygote nucleus.

The genomic potential of differentiated cells can best be evaluated by the transplantation of nuclei oocytes and eggs, because this procedure has the potential to test the entire genome. The idea that the gene content of a cell might be tested by injecting a living nucleus into an enucleated egg was suggested by Hans Spemann (1938). The rationale was that the type of development that results is a reflection of the gene content of the introduced nucleus. This idea materialized when Briggs and King (1952, 1960) succeeded in obtaining metamorphosed frogs following transplantation of blastula and early gastrula nuclei into enucleated eggs of *Rana pipiens*. Later, McKinnell (1962), Gurdon (1962), and subsequently others obtained fertile adult frogs from embryonic nuclei, demonstrating, for the first time, the genomic totipotency of embryonic nuclei. These studies were extended to the insect *Drosophila* and the teleost fish; their young embryonic nuclei were also shown to be totipotent. Recent studies in mice indicate that totipotency is restricted to early cleavage stage nuclei. Tsunoda *et al.* (1987) obtained several fertile mice from nuclei of the 8-cell stage, but no development ensued from nuclei of the inner cell mass of blastocysts; however, Illmensee and Hoppe (1981) reported fertile mice from these latter-stage nuclei. Among the larger mammals, three lambs and two calves have resulted from 8-cell and 9- to 15-cell-stage nuclei, respectively. So far, the success of nuclear transfers in mammals is limited to stages that are much earlier than that in amphibians however, mammalian nuclear transfer studies are still in their infancy, and we await future developments in this area.

This chapter deals mainly with nuclear transfers from terminally differentiated somatic cells, i.e., cells in which the genome is imprinted to specify a specific and final cellular phenotype. So far, successful nuclear transfer studies from terminally differentiated somatic cells have been performed in frogs. The results of these studies summarized below demonstrate that the genomes of several differentiated somatic cell types can (1) undergo widespread activation, (2) direct the formation of a multiplicity of cell types, and (3) display genetic multipotentiality, but (4) evidence for genomic totipotency of differentiated somatic cells in still lacking.

Nuclear Transfers from Differentiated Somatic Tissues

The critical experiment posed by Spemann (1938) concerned the genetic repertoire of differentiated somatic cells. The term differentiated cell has undergone multiple uses. For example, the unfertilized egg is considered by some to be a highly differentiated cell (gamete), yet at fertilization it becomes totipotent. The critical test of nuclear equivalence can best be assayed on genomes that have been imprinted to specify a specific and final cellular phenotype, i.e., genomes of cells that normally have no other potential in the organism. Here we consider such cell types as definitive tests of the genomic potential of differentiated cells.

Nuclear transfer experiments that assess the genetic potential of differentiated cells must include evidence for two conditions: (1) the donor cells are differentiated, and (2) the nuclear transplants are derived from the donor nucleus, and not from the egg nucleus. We shall consider two series of experiments that differ in the certainty that the donor cells were derived from differentiated cells. The first series comprises the initial attempts to work with differentiated cells at a time when the techniques usually precluded distinction between stem cells and differentiated cells in the tissues employed. In

Table 7.1A.
Nuclear Transfers from Differentiated Somatic Tissues[a]

Donor		Identification of donor cells	Total nuclei tested	Total transfers[b] at stage								Verification	Reference
				Postneurula		Larva		Metamorphosis		Adult			
Stage	*Tissue*			*N*	*%*	*N*	*%*	*N*	*%*	*N*	*%*		
Larva	Intestine[X]	Size	726	36	5.0	31	4.3	-	-	4[2F]	0.6	1-nu	Gurdon and Uehlinger (1966)
	Intestine[X]	Size	522	8	2.0	8	2.0	1	0.2	-	-	98% haploids[c]	Marshal and Dixon (1977)
	Epidermis[X]	Nonciliated	440	2	0.4	2	0.4	-	-	1	0.2	1-nu	Kobel et al. (1973)
	Minced tad-poles[X]	Cell culture	3686	23	0.6	9	0.2	3	0.1	-	-	1-nu[d]	Gurdon and Laskey (970)
Adult	Liver[X]	Cell line	365	2	0.6	2	0.6	-	-	-	-	-	Kobel et al. (1973)
	Skin, lung, kidney[X]	Cell culture	2322	26	1.1	7	0.3	-	-	-	-	1-nu[d]	Laskey and Gurdon (1970)
	Intestine[X]	Size	1112	10	0.9	6	0.5	-	-	-	-	95% haploids[c]	McAvoy et al. (1975)

a Superscript letters: X, *Xenopus laevis;* F, fertile.
b Includes the results of serial transfers.
c Denotes efficiency of UV irradiation in controls.
d Chromosome number determined in nuclear transplants. 1-nu, one nucleolus.

Table 7.1B
Nuclear Transfers from Differentiated Somatic Cells[a]

Donor				Total transfers[b] at stage							
				Tadpole							
				Postneurula		Prefeeding		Feeding			
Stage	Tissue/cell	Differentiated state of donor cells	Total nuclei tested	N	%	N	%	N	%	Verification	Reference
Larva	Melanophores cell culture[X]	Pigmented	257	2	0.8	-	-	-	-	1-nu	Kobel et al. (1973)
Juvenile frog	Erythrocytes[R]	Shape, hemoglobin	51	4	8.0	4	8.0	3	5.9	Triploid[d]	DiBerardino et al. (1986)
Adult	Erythrocytes[R]	Shape, hemoglobin	130	11	8.5	6	4.6	-	-	Exovate[d]	DiBerardino and Hoffner (1983)
	Erythroblasts[X]	Shape, hemoglobin	442	8	2.0	8	2.0	-	-	1-nu[d]	Brun (1978)
	Skin, cell culture[X]	Keratin antibody	129	6	4.6	4	3.1	-	-	1-nu[d]	Gurdon et al. (1975)
	Spleen[X]	Immunogen antibody	100	6	6.0	6	6.0	-	-	1-nu[d]	Wabl et al. (1975)

a Superscript letters: X, *Xenopus laevis:* R. *Rana pipiens.*
b Includes the results of serial transfers.
c Stages of hindlimb bud.
d Chromosome number determined in the nuclear transplants.

the second series the specialized properties of the donor cells were documented. In both series only those nuclear transplants that developed to postneurula stages, larval (tadpole) stages, and adults will be considered. Postneurula embryos have developed primitive organ systems, which in the latter stages display muscle and heart function. During the early larval stages, the major cell types, tissue, and organ systems have differentiated and are functional. Thus, transplanted nuclei capable of programming eggs to develop to these stages are considered genetically multipotent. Nuclei that direct the formation of fertile frogs are interpreted to be totipotent.

Verification that the nuclear transplants developed from the transplanted test nucleus and not form the egg nucleus that remains due to faulty enucleation have been performed in the following ways. In *Xenopus*, the egg nucleus is inactivated by ultraviolet (UV) irradiation but, since this procedure is successful in only 61-92% of cases, a genetic marker is required. The most used marker in *Xenopus* has been the 1-nu (nucleolar) mutation that can be seen in cells from the gastrula stage and beyond. However, as pointed out by DuPasquier and Wabl (1977), it is necessary to verify that one nucleolus is accompanied by diploid set of chromosomes and does not result from haploidy or aneuploidy. In *Rana pipiens*, triploidy has served as a convenient cytogenetic marker because the triploid set of chromosomes can be identified throughout the life cycle of the frog. Another verification procedure has also been used in *Rana pipiens* because microsurgical removal of the egg nucleus can be accomplished in 98-100% of cases. The exovate formed at the time of enucleation adheres to the vitelline membrane outside of the egg. This membrane is later removed, sectioned, and stained with the Feulgen procedure that specifically stains DNA and exhibits the presence of the egg nucleus in the exovate. This information together with the determination of chromosome number of the nuclear transplant and the time of the first cleavage insures that development ensued from the test nucleus.

Differentiated Somatic Tissues

The advanced development of nuclear transplants obtained from differentiated somatic tissues of *Xenopus* larvae and adults is summarized in Table 7.1. Five adult frogs were obtained from five original larval nuclei, four from nuclei of the larval intestine, and one from a nucleus of larval epidermis. Three of these adult frogs were fertile — two from intestinal nuclei, and one from a non-ciliated cell of the epidermis. The donor cells from the intestine were selected on the basis of large size. The epidermal cell from the experiments of Kobel *et al.* (1973) was identified as non-motile (non-ciliated). Nuclei from epidermal cells that were motile (ciliated), and therefore specialized, did not give advanced development. The authors concluded that the single case showing totipotency was derived from a non-specialized cell. In addition to the studies tabulated in Table 7.1, endodermal nuclei of the primitive gut taken from postneurula embryos and initial larval stages before tissue differentiation directed *Xenopus* eggs to develop into 20 fertile frogs (~3%), most of which bore the 1-nu genetic marker of the donor nucleus.

At least four metamorphosed frogs developed from eggs injected with larval nuclei—one from an intestinal cell selected on the basis of large size and three from cell cultures. The latter three were obtained from cell cultures, initiated by mincing whole tadpoles that were in the stage of beginning circulation in the gills, but it is known if all the donor cells were differentiated. The above tissues (intestine, epidermis, and cell cultures of tadpoles) also contained nuclei capable of directing eggs to develop into 69 postneurula embryos, and 50 of these proceeded into larval stages. Nuclear transfers of adult somatic nuclei into *Xenopus* eggs resulted in 38 postneurula embryos, and 15 of these developed into larvae. The somatic nuclei were derived from a cell line established from liver from cell cultures of skin, lung, and kidney, and from *in vivo* intestinal cells selected on the basis of size. Several conclusions can be made from the above studies. First, cells from both larval and adult tissues contain some nuclei that can direct the formation of the diverse cell types found in postneurula embryos and tadpoles. Therefore, these nuclei display genetic multipotentiality. Second, genetic totipotency has been shown for a few nuclei from advanced embryonic and young larval stages, since these nuclei directed the formation of fertile frogs. Third, the interpretation of these results remains equivocal, because with the exception of the non-ciliated epidermal cells, no valid criteria were used to distinguish between stem cells and differentiated cells in the tissues employed.

Differentiated Somatic Cells

In this section we consider the advanced development obtained from nuclei of donor cells that were unequivocally differentiated. Among the cell types listed in Table 7.2, melanophores, erythrocytes, and erythroblasts were directly identified at the time of nuclear transfer. For tests of melanophore nuclei, short term monolayer cultures from prefeeding *Xenopus* larvae were used, and the melanophore donor cells were identified on the basis of pigment inclusions. Their nuclei promoted enucleated eggs to develop into postneurula embryos s(0.8%) that contained the 1-nu genetic marker of the donor cells. Nucleated erythroid cells of amphibians are particularly ideal for nuclear transfer experiments: (1) they are unequivocally differentiated; (2) they can be conveniently isolated from the peripheral blood by low speed centrifugation; (3) the cells exist singly in the plasma, and therefore, no chemicals are required to achieve cell dissociation as is the case when using donor cells from solid tissue; and, most importantly, (4) individual erythroid cells can be directly identified with the stereomicroscope at the time of nuclear transfer because of their oval shape and presence of hemoglobin. Erythrocyte nuclei from juvenile *Rana* frogs promoted the highest yield of tadpoles and expressed the greatest genomic potential among the differentiated somatic cell types tested: 8% of the original nuclear population promoted the formation of prefeeding tadpoles and 5.9% directed the development of *feeding* tadpoles that advanced to hindlimb bud stages, and in the best cases survived up to a month. All the tadpoles bore the triploid marker of the donor nucleus. Nuclear transfer studies performed on adult erythrocyte nuclei from *Rana* resulted in 4.6% prefeeding tadpoles. The origin of the latter tadpoles was verified by recovery of the egg nucleus in the exovate and determination of chromosome number. Another study of adult erythroid nuclei was performed in *Xenopus* (Brun, 1978). In the latter study, erythroblast nuclei (2.0%) promoted development as far as the first tadpole stage at which hatching is initiated, and all these nuclear transplants bore the 1-nu genetic marker.

Two other studies examined the genomic potential of adult skin and spleen cells. The donor cells were derived from a nearly homogeneous population. In the case of the skin cells, correlative studies showed that over 99% of the skin cells from short term culture were producing immunoreactive keratin. Nuclear transfers from similarly cultured skin cells resulted in 3.1% prefeeding tadpoles that carried the 1-nu genetic marker. In the case of the spleem cells, their nuclei directed the development of prefeeding tadpoles (6.0%) that bore the 1-nu genetic marker of the donor cells of which 96.1-98.7% were judged to be producing immunoglobins (Ig). In summary, nuclei from five documented differentiated somatic cell types promoted the formation of 37 postneurula embryos. With the exception of cultured melanophores, nuclei from the other 4 cell types promote the formation of 28 tadpoles. Thus, there is evidence for genomic multipotentiality of differentiated somatic cells from the larva, juvenile frog and adult. *However, to date, no nucleus of a documented differentiated somatic cell nor of any adult somatic cell (stem or specialized) has yet been shown to be totipotent.*

Nuclear Transfers from Germ Cells

The genomic totipotency of germ cell nuclei certainly does not require formal proof via the nuclear transplantation assay, because the germ cells of an individual retain the repertoire of zygotic genes, except for an occasional mutation. Nevertheless,

Table 7.2
Nuclear Transfers from Germ Cells[a]

Donor			Total nuclei tested	Total transfers at stage								Reference
				Postneurula		Feeding larva		Metamorphosis		Adult		
Stage	*Gonad*	*Cell*		*N*	*%*	*N*	*%*	*N*	*%*	*N*	*%*	
Larva	Genital ridge	Primordial germ cells[R]	410	35	8.5	31	7.6[b]	-	-	-	-	Smith (1965)
Larva to metamorphosis	Genital ridge, undifferentiated and diferentiated gonads	Primordial germ cells, gonocytes[P]	2986	46	1.5	12	0.4	9	0.3	6	0.2[2F]	Lesimple *et al.* (1987)
Adult	Testis	Enriched for spermatogonia[R]	116	4	3.5	1	0.9	-	-	-	-	DiBerardino and Hoffner (1971)

a Superscript letters: R, Rana Pipiens; P. Pleurodeles waltl; 2F, two adults were fertile.
b Ten larvae raised, final stage not indicated, one in the climax of metamorphosis; see text.

nuclear transfer tests of germ cells have been informative because they showed that their genome after transplantation into eggs expresses different degrees of developmental potential depending on the stage of differentiation of the donor cells. Smith (1965) tested nuclei of primordial germ cells taken from *Rana pipiens* tadpoles at the initial feeding stage. The primordial germ cells are large and yolky and can be easily distinguished from the small somatic cells of the genital ridge. The investigator's efficiency in microsurgical removal of the host nucleus was 99.5%, as judged by the production of androgenetic haploids. Among the 31 normal tadpoles that developed from 7.6% of the injected eggs, 10 were reared for periods of 1-3 months, were then autopsied and found to be normal. Although the precise larval stages were not reported, a photograph of one metamorphosing tadpole was illustrated.

Lesimple *et al.* (1987) tested germ cell nuclei of the salamander *Pleurodeles walti* taken from progressive stages of larval development up to metamorphosis. The larger germ cells were distinguished from the smaller somatic cells. Among the total injected eggs, 12 (0.4%) attained the feeding larval stage, 9 (0.3%) metamorphosed and 6 (0.2%) developed into adults. Two of the adults were mated and produced normal progeny. The adults arose from experiments in which the donor nuclei were not genetically marked. The authors argue that it is unlikely that the adults arose from the egg nuclei, because in *Pleurodeles* gynogenesis results in 50% ZZ males and 50% WW females, but they obtained 3 ZZ males and 3 ZW females and no WW females. Nevertheless, proof of this interpretation is lacking. One study examined the developmental potential of adult germ cell nuclei, since their nuclei are derived from differentiated germ cells. The cell suspension of donor cells from adult testes was enriched for spermatogonia, and the donor cells were selected on the basis of size. Enucleated *Rana pipiens* eggs injected with these nuclei yielded 4 (3.5%) postneurulae and 1 (0.9%) feeding tadpole. Recovery of the egg nucleus in the exovate indicated that development was directed by the transferred nucleus. Failure to reveal the genomic totipotency of differentiated germ cells in nuclear transplantation experiments indicates that their genome and/or chromatin undergo modifications required for specifying the phenotype of a differentiated cell. Under the experimental conditions, the modifications are stable and prevent the full expression of their genomic potential, but certainly they are not irreversible. Lesimple *et al.* (1987) reported a decrease in the percentage of hatching larvae when the germinal nuclei were taken after the mid-larval stages. They suggested that differentiation of germinal stem cells into spermatogonia or oogonia might account for the decrease in successfully nuclear transplantations, while the retention of a few stem cells (gonocytes) might account for the successful nuclear transplants.

Imprinting of the Genome

Recent nuclear transfer studies in the mouse have shown that normal development requires the presence of both the female and male pronuclei. Embryos with two female or two male pronuclei arrest early in embryogenesis. Those embryos with solely maternal or paternal chromosomes exhibited different phenotypes. Embryos with a diploid set of maternal chromosomes developed into well formed but small 25-somite embryos; however, the extraembryonic tissues were underdeveloped. By contrast, embryos with a diploid set of paternal chromosomes differentiated poorer embryonic structures but better extraembryonic tissues than the embryos with maternal chromosomes. These results indicate that the maternal and paternal genomic contributions to embryogenesis are functionally different and that both are essential form normal development. The investigators have concluded that a differential imprinting of the genome occurs in the male and female gametes during gametogenesis and suggest that different subsets of genes are inactivated in the parental genomes but in a complementary pattern. This imprinting of the genome is reversible, however, because homozygous uniparental mouse embryos in chimeric combination with normal embryos can develop into fertile adults that produce normal gametes derived from the uniparental genome. These latter investigators interpret the cases of totipotency to stem from reimprinting the genome during gametogenesis.

In amphibians, gynogenetic diploids produced by suppression of the second polar body can develop into fertile adults. These cases include *Rana pipiens* and *Xenopus laevis.* Thus, if differential imprinting of amphibian parental genomes occurs during gametogenesis, such imprinting must be easily reversed in amphibian embryos.

Amphibian embryos also differ from mouse embryos in their onset of zygotic RNA transcription. In *Xenopus,* the first detectable RNA transcription occurs at the sixth cleavage stage, whereas in murine embryos RNA synthesis is initiated during the first cleavage stage. The later stage of onset of RNA transcription in frog embryos would permit the genome of a transplanted frog nucleus additional cell cycles to adjust to a new programme of gene expression compared with the one cell cycle in the mouse. This differences in the onset of RNA synthesis might account for successful nuclear transplantations from advanced stages of frog cells, whereas successful nuclear transfers in mammals are so far limited to cleavage stage nuclei. Whether the temporal programme of gene expression, parental imprinting of genomes or other factors account for the difference in nuclear reprogramming between mammals and frogs remains to be elucidated.

Although the successful cases of amphibian nuclear transplants are those that are frequently emphasized, many frog nuclei from determined regions of embryos before overt tissue differentiation displayed developmental restrictions after transfer to enucleated eggs. Somatic nuclei tested from different embryonic stages, germ layers and primitive organs from various anuran and urodelen amphibians showed a progressive decrease in the percentage of normal nuclear transplants when the donor nuclei were taken from progressively more advanced stages of embryogenesis. Other main points derived from these studies were (1) a few of nuclei still displayed genetic totipotency, and this was especially evident in the endodermal nuclei of hatched larvae of *Xenopus*; (2) but most nuclei did not promote normal development of the test eggs; (3) the developmental restrictions analyzed in endodermal and neural nuclear transplants were stable, since they could not be reversed through serial transplantation; (4) furthermore, the restrictions expressed by endodermal nuclei were intrinsic since they were not corrected by parabiosing nuclear transplant embryos with normal ones, nor could they be corrected when a haploid set of egg chromosomes was combined with the diploid set from the endodermal nucleus.

Among the *Rana* nuclear transplants derived from endodermal and neural nuclei, a small number displayed a pattern of developmental abnormalities consistent with the origin of their nuclei. During postneurula stages some nuclear transplants derived from late gastrula endodermal nuclei exhibited deficiencies and degenerative changes in ectodermal and mesodermal organs, but not in endodermal tissues. This phenotype was not reversed by serial cloning, alleviated by parabiosis with normal embryos, or associated with obvious chromosomal abnormalities. Similar studies of nuclei from the presumptive medullary plate regions of late gastrula and the definitive medullary plate of early neurula and mid-neurula embryos showed a small group with a phenotype different from the endodermal nuclear transplants. The neural nuclear transplant group (13%) displayed during larval stages cellular deficiencies mainly in mesodermal and endodermal derivatives but showed good differentiation in the organs and tissues of ectodermal origin and did not present detectable chromosomal abnormalities. These distinct phenotypes suggested that by late gastrulation some endodermal and neurula nuclei have acquired stable properties for specific pathways of cell differentiation. Although these phenotypes were described more than 20 years ago, the mechanism probably is analogous to the imprinting of the genome recently described in murine nuclear transplants. In the mouse differential imprinting of parental genomes, presumably during gametogenesis, leads to two distinct phenotypes, depending on whether the conceptus developed from two male or two female pronuclei. In some of the amphibian nuclear transplants, we probably observed the effect of genomic imprinting initiated during embryogenesis. The limitation of these studies is that the data are not extensive. Perhaps, an analysis of the pattern of DNA methylation in the endoderm and neural nuclear transplants would clarify this interesting problem. Presumably, the normal frog nuclear transplants were derived from stem cells present among the other differentiating donor cells and would differ from the abnormal nuclear transplants in their pattern of DNA methylation.

Reversal of Genomic Imprinting

The genome of differentiated cells has been imprinted by molecular mechanisms to specify a particular cellular phenotype. Evidence available indicates that this imprinting results from the interaction of chromosomal proteins with the genome and modification of its DNA by methylation. Here we consider the extent to which the genomic imprinting in differentiated somatic frog cells can be erased under differing cytoplasmic environments.

Nuclear Transfer into First Meiotic Metaphase Oocytes

Nuclei of *Rana* erythrocytes have expressed the greatest genomic potential among the genomes of differentiated somatic cells so far tested. The main difference in the experimental procedure was that the erythrocyte nuclei were initially injected into oocytes, whereas the other nuclear types were injected directly into eggs. The standard host for nuclear transplantation has been the egg, the host originally proposed by Spemann (1938). However, we hypothesized that the expression of the genetic potential of differentiated somatic cells might be enhanced if the donor nuclei were first exposed to the molecular components in the oocyte that normally prepare the oocyte chromosomes to participate in fertilization. Initially, we tested whether somatic nuclei of embryonic cells would, in fact, function properly in the egg after residing in oocyte cytoplasm. Oocytes at the stage of first meiotic metaphase were used for hosts. Such oocytes display in the animal pole the first black dot, indicative of the first meiotic metaphase. A single broken embryonic cell, together with it nucleus, was injected into the animal hemisphere near the equator region. At this time, the oocytes are not activatable, but becomes so about 24 hrs. later when they have matured *in vitro* into eggs (18^{0}C). The matured oocytes (eggs) were activated by penetration of a glass microneedle. Within 10 min., the second black dot, indicative of the second meiotic metaphase of the egg, appeared and was removed microsurgically with a glass microneedle. We found that embryonic somatic nuclei exposed to cytoplasm directing meiotic events could still support development through embryogenesis. Cytological studies performed on a parallel series of nuclear transplants showed that transplanted embryonic nuclei formed metaphase chromosomes on newly induced spindles in concert with the behaviour of the oocyte nucleus during first meiotic metaphase. When the matured oocytes were activated, the injected nuclei transformed into pro-nuclei. Thus, somatic nuclei can reversibly respondent to the molecular cues directing meiotic and mitotic events in a sequential manner and still direct the host to develop through embryogenesis. These findings indicated that nuclei of differentiated cells could now be tested in oocytes to determine whether the imprinting of their genome could be erased by exposure to the molecular components of maturing oocytes.

Genomic Potential of Erythrocytes

The erythrocyte satisfies the most stringent criterion of a differentiated cell, it is non-cycling and terminally differentiated cell, one that is almost transcriptionally quiescent. Its oval shape and the presence of hemoglobin (Hb) permit its direct identification under the stereomicroscope at nuclear transfer.

Thus, any development ensuing from the transplantation of an erythrocyte nucleus would unequivocally result from a differentiated genome.

The developmental potential of the adult erythrocyte genome was tested in two ways. In one series, the nuclei were injected into enucleated eggs; in another series, the nuclei were injected into oocytes at the stage of first meiotic metaphase. Some of the blastulae derived from the hosts of each series were used to provide donor nuclei for a second transplant generation. The blastula nuclei were injected singly into enucleated eggs and nuclear clones were produced. None of the transplants from the egg series developed beyond the early gastrula stage. However, among the 12 clones derived from the original oocyte series, eight clones (67%) contained members that developed into postneurula embryos, and six clones (50%) had members that differentiated into prefeeding tadpoles. When erythrocyte nuclei from juvenile frogs were tested under the same conditions, they directed the development of feeding tadpoles that formed hind limb buds in 75% of the four nuclear clones (DiBerardino *et al.*, 1986). These tadpoles survived for up to 1 month and are the longest surviving and most advanced nuclear transplants derived from documented differentiated somatic cells. The interval from zygote to metamorphosis is approximately 3 months. Thus, the month old tadpoles developed and survived for about one third of this interval.

Several conclusions resulted from the nuclear transfer studies of the non-cycling and terminally differentiated erythrocytes of *Rana*. First, the erythrocyte genome retained the genes for directing the differentiation of the cell types, tissues, and organ systems of tadpoles. In fact, the specification of the various cell types in these tadpoles requires the activation of dormant genes that never functioned in the cell lineage of an erythrocyte. Second, exposure of the erythrocyte genome to the cytoplasm of maturing oocytes induced the most widespread activation of the erythrocyte genome ever obtained in an

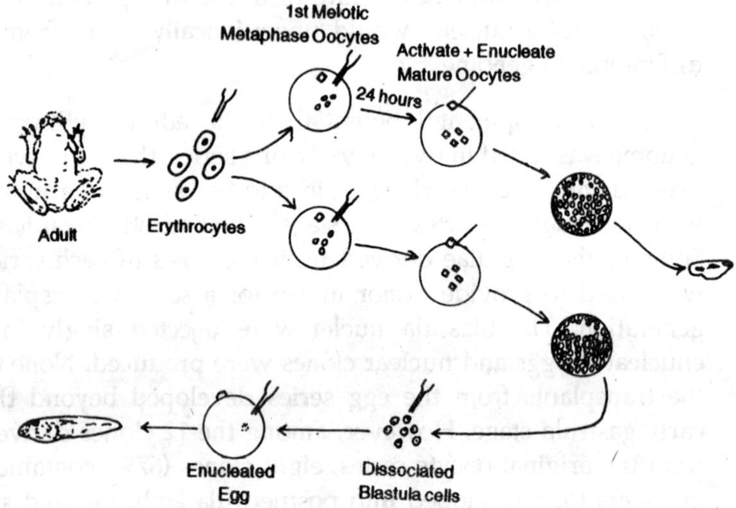

Fig. 7.1. Erythrocytes obtained by intracardiac puncture of adult *Rana pipiens* frogs were broken in distilled water by osmotic shock and microinjected into oocytes near the equator at first meiotic metaphase. Approximately 24 hrs later (at 18°C), when the oocyte matured, the matured oocyte (egg) was activated by pricking with a glass needle, and the egg nucleus was removed microsurgically. (Top) Original transplant generation; prehatching tadpoles resulted. In some cases, nuclear transplant blastulae were dissected and their animal hemisphere nuclei were transplanted singly into activated-enucleated eggs. (Bottom) First retransfer generation; swimming tadpoles resulted. Artwork by Betti Goren.

experimental system. In the regard, the imprinting of the genome that formerly specified the phenotype of an erythrocyte, was erased to the extent of specifying the cell phenotypes of tadpoles. Third, serial transplantation of erythrocyte nuclei through eight transplant generations showed that once the erythrocyte genome is activated by its progression through the cytoplasm of the oocyte and activated egg, it maintains its potential not only for replication, but also for directing the complex functions of organogenesis in excess of 100 cell cycles. In addition, there was no evidence that the mitotic progeny of the erythrocyte nuclei lost there ability to replicate their genomes and continue cell cycling. Fourth, some tadpoles fed and therefore functioned as independent organisms. Thus, the developmental block previously exhibited in prefeeding tadpoles from differentiated somatic nuclei was surmounted. The progression of tadpoles to the feeding stage is a critical developmental event, because at this stage maternal (egg) components are largely used. Thereafter, continued development, further differentiation, and survival of the tadpoles are dependent on an external source of nourishment.

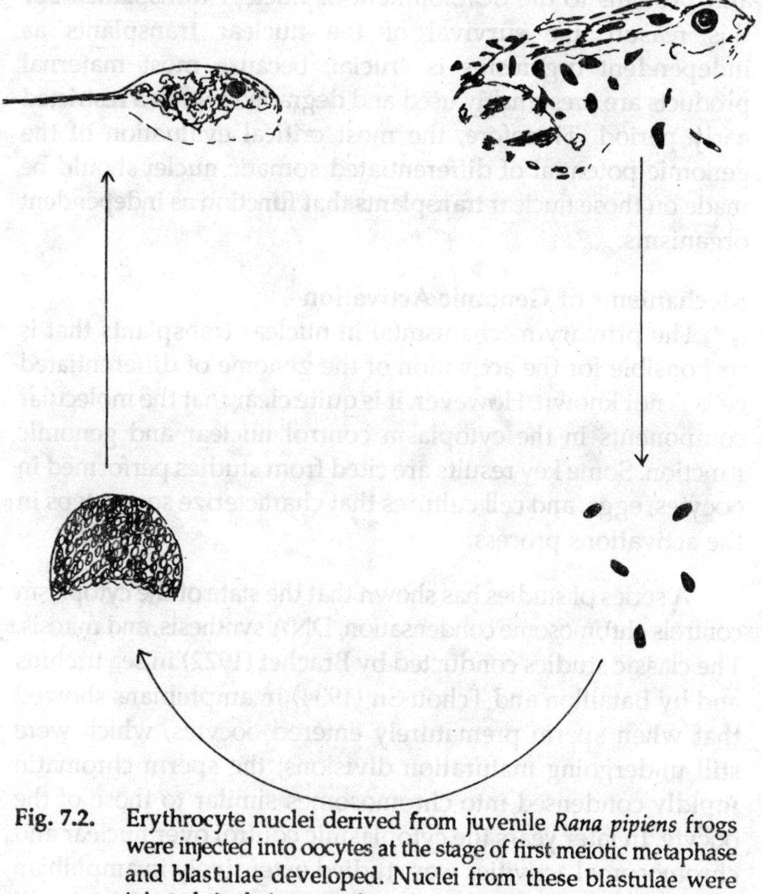

Fig. 7.2. Erythrocyte nuclei derived from juvenile *Rana pipiens* frogs were injected into oocytes at the stage of first meiotic metaphase and blastulae developed. Nuclei from these blastulae were injected singly into enucleated eggs that developed into feeding tadpoles and advanced to hind limb bud stages.

An important factor that must be considered in evaluating the formation of nuclear transplants is the contribution of maternal RNAs and proteins to their development. It is certain that the transplanted nucleus contributes a significant role to the development of nuclear transplants, because eggs lacking a functional nucleus but containing a functional centriole develop at best only into partially and abnormally cleaved blastulae. Nevertheless, there is not enough information to evaluate exactly the relative contribution of maternal RNAs and proteins to the development of nuclear transplants. For this reason, the survival of the nuclear transplants as independent organisms is crucial, because most maternal products are presumably used and degraded within a restricted early period. Therefore, the most critical evaluation of the genomic potential of differentiated somatic nuclei should be made on those nuclear transplants that function as independent organisms.

Mechanisms of Genomic Activation

The primary mechanism(s) in nuclear transplants that is responsible for the activation of the genome of differentiated cells is not known. However, it is quite clear that the molecular components in the cytoplasm control nuclear and genomic function. Some key results are cited from studies performed in oocytes, eggs, and cell cultures that characterize some steps in the activations process.

A series of studies has shown that the state of the cytoplasm controls chromosome condensation, DNA synthesis, and mitosis. The classic studies conducted by Brachet (1922) in sea urchins and by Bataillon and Tchou-Su (1934) in amphibians showed that when sperm prematurely entered oocytes, which were still undergoing maturation divisions, the sperm chromatin rapidly condensed into chromosomes similar to those of the oocyte. In later years the cytoplasmic control over nuclear and chromosomal activities was studied extensively in amphibian eggs and oocytes by nuclear transplantation. Nuclei injected into activated eggs behave like pronuclei; they enlarge, undergo chromatin decondensation and synthesize DNA. During nuclear activation, some nonhistone proteins leave the donor nuclei, while both histone and nonhistone proteins migrate from the recipient cytoplasm into the injected nuclei. At the diplotene stage of the oocyte, when its nucleus (germinal vesicle) is greatly enlarged and is engaged in RNA but not DNA synthesis, transplanted nuclei also enlarge and synthesize new RNA products directed by the regulatory mechanisms of the oocyte. Here, too, an exchange of nuclear proteins between nucleus and cytoplasm is observed. When the nuclear envelope of the germinal vesicle breaks down, the oocyte chromosomes condense and become aligned on the spindle of the first meiotic metaphase. Similarly, injected nuclei also transform into metaphase chromosomes aligned on spindles. The cytoplasmic control of nuclear activity also operates in the mouse oocyte as well as in oocytes of other species. The same phenomena were reported in cultured mammalian cells when cells from different phases of the cell cycle were fused together. Cells in S phase induced cells in G1 phase to undergo DNA synthesis, whereas mitotic cells fused with G1, G2, or S cells caused premature chromosome condensation. The main conclusion drawn from the above studies is that a new program of genomic expression is induced in test nuclei by a specific set of molecular factors residing in the oocyte, egg and different phases of the cell cycle, respectively. The availability of cytoplasmic extracts from activated eggs that can induce sperm to transform into pronuclei and mitotic chromosomes *in vitro* will ultimately permit the biochemical analysis of the factors in eggs that induce DNA synthesis and chromosomal condensation. With respect to extracts from oocytes, Korn *et al.* (1982) identified a positive regulator that activates dormant 5S RNA genes of erythrocytes. The activating factor is sensitive to protease and heat treatment suggesting that protein(s) are involved.

Heterokaryons have also provided clues into the process of gene activation. An outstanding example is the activation of certain dormant genes in the hen erythrocyte. The classic experiment by H. Harris (1974) involved the fusion of hen erythrocytes with HeLa cells. In the heterokaryons, the nuclei of the parent cells remain separate, and each nucleus can be identified and monitored. The erythrocyte nuclei enlarged, their highly condensed chromatin dispersed, they resumed synthesis of RNA and DNA, and later, a series of chick proteins were synthesized. During the activation process, there is an uptake of mammalian nuclear proteins by the chick erythrocyte nucleus and a loss of histone 5 (H5) from the chromatin of the erythrocyte. Blau *et al.* (1985) fused mouse muscle cells with a variety of nonmuscle cell types of human origin. Such heterokaryons containing muscle cells do not undergo cell division, and therefore, the heterokaryons are stable and retain the complement of chromosomes. In these heterokaryons, various specific mRNAs and proteins, normally made by muscle cells were also made by the nonmuscle genomes. In addition to extending previous studies on heterokaryons, the authors suggest that their system will be amenable to isolating the trans-acting regulators that induce muscle gene activation.

In the case of *Rana* erythrocyte nuclei, we have shown that exposure of erythrocyte nuclei to the components in the cytoplasm from the stage of first meiotic metaphase and subsequent maturation stages leads to widespread activation of the genome. This activation involves in some manner the ability of noncycling and nonreplicating erythrocyte nuclei to undergo significant DNA replication in activated eggs. Autoradiographic studies by Leonard *et al.* (1982) demonstrated that only 24% of adult erythrocyte nuclei transplanted to eggs synthesized DNA and, in these cases, only a small portion of the genome engaged in DNA synthesis. If, however, adult erythrocyte nuclei were first injected into oocytes that later matured and were activated, more then 75% synthesized DNA, and more than one half of these nuclei synthesized DNA in amounts similar to those of the egg pronucleus. Thus, exposure of erythrocyte nuclei to oocyte cytoplasm leads not only to induction of DNA synthesis in more nuclei, but also to significant enhancement of the extent of replication of the genome. We have proposed that a remodeling of the proteins in the erythrocyte chromatin occurs in the oocyte cytoplasm that renders the chromatin capable of responding more normally to the signals in activated egg cytoplasm that induce the synthesis of DNA. Such remodeling of chromosomal proteins could lead to changes in chromatin structure and in patterns of DNA methylation that eventually permit the genome to respond later to the appropriate transcription factors and synthesize the appropriate RNAs during embryogenesis. Other autoradiographic studies examined the nucleocytoplasmic exchanges of nonhistone proteins when endodermal nuclei were transplanted to eggs. These studies showed that during the first cell cycle there is an accumulation of cytoplasmic egg nonhistone proteins by transplanted nuclei and also a major loss of nonhistone proteins from the transplanted nuclei into the cytoplasm. Perhaps a similar but more effective nucleocytoplasmic exchange occurs in erythrocyte nuclei placed in the oocyte, not only because of a longer exposure to the cytoplasm, but also because the nuclei are exposed to additional factors not present in the egg. With respect to chromosomal proteins there is evidence that changes occur in the proteins of chromatin in germ cells during spermatogenesis, oogenesis, and after fertilization. Thus, there are precedents that substantiate our hypothesis that remodeling of chromosomal proteins may have occurred in erythrocyte nuclei exposed to oocyte cytoplasm. This hypothesis can be tested by examining appropriate chromosomal proteins, as well as the state of DNA methylation and the chromatin structure of specific genes in test nuclei before and after exposure to oocyte cytoplasm. Such studies, together with the identification of specific extracts of oocyte cytoplasm that control molecular changes in the chromatin of test nuclei, might ultimately indicate the primary mechanisms involved in the activation of an entire genome of differentiated somatic cells. This information might also indicate how the germ cell chromatin of oocytes is prepared for participation in fertilization and embryogenesis.

In addition to the remodeling of chromosomal proteins, an essential replication factor may be required. For example, BLow and Laskey (1988) have shown that sperm chromatin can rereplicate in a cell-free system extracted from *Xenopus* eggs, if the nuclear membrane is permeabilized or breaks down. They suggest that an essential replication factor gains access to DNA when the nuclear envelope breaks down during mitosis and that one role of the nuclear envelope is to limit DNA to one replication per cycle. In our transfers of nuclei into oocytes at first meiotic metaphase, the nuclear membrane breaks down and the interphase chromatin is transformed into metaphase chromosomes aligned on a newly induced spindle. These events occur by 1 hr or earlier after transfer to the oocytes. It is likely that noncycling erythrocyte nuclei lack such an essential replication factor or have negligible amounts, but acquire it from the oocyte cytoplasm after nuclear membrane breakdown. Nuclei injected initially into eggs retain their membranes until after the period of DNA synthesis and would therefore be deficient in the replication factor. The above speculation is consistent with our findings that adult erythrocyte nuclei incubated in oocyte cytoplasm undergo significant DNA synthesis in activated eggs and support tadpole development, whereas those injected in eggs undergo little DNA synthesis and support development only to the early gastrula stage.

8

Morphogenesis

Cell contact and adhesion take place at the cell surface. Cells in embryos move over one another and stick to one another at their surfaces. Thus in order to examine embryonic cell rearrangements and associations, we must look at the cell surface—the place where these associations begin. It has already described the cell surface as a lipid bilayer in which proteins and glycoproteins are imbedded. The cell surface has fluid properties that allow the proteins and glycoproteins to move laterally in the lipid. This movement is sometimes restricted by rod-shaped proteins, such as microtubules and microfilaments, that anchor some of the surface proteins from the cytoplasmic side of the membrane. Singer and Nicolson's fluid-mosaic model of the cell surface, currently the most popular and widely accepted model, describes the cell surface as having fluid properties and being mosaic in nature—that is, it consists of various proteins interspersed in the lipid.

Fluid-Mosaic Model of the Cell Surface

An elegant experiment by Frye and Edidin provided the basis for determining that the cell surface has fluid properties. Using immunological techniques, these workers developed complex stains that specifically bind to certain cell-surface proteins. Thus stains specific to the surface proteins of given cell types could be produced and coupled to fluorescent dyes. Two different cell types (a human cell and a mouse cell) were fused together, with Sendai virus used to promote cell fusion. The surface proteins specific for each cell type were then traced using the complex stains and a fluorescence microscope.

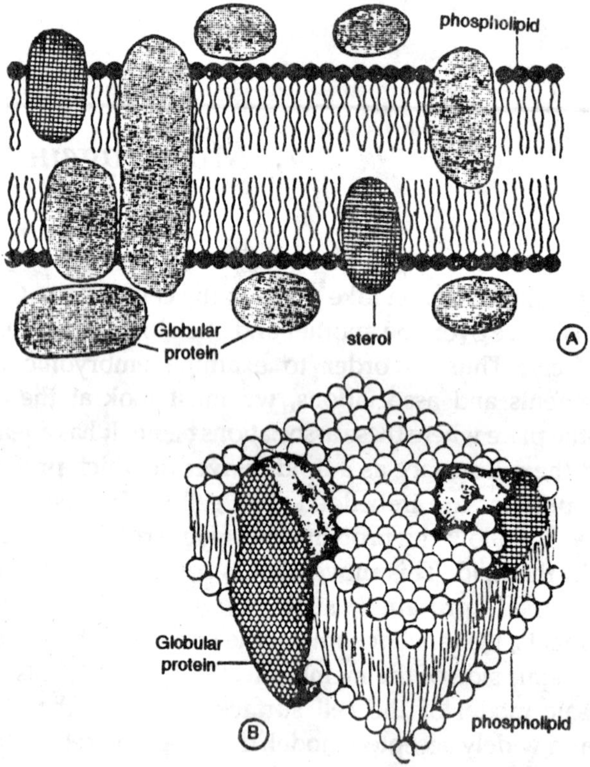

Fig. 8.1. (A) Schematic representation of a section through a eukaryote fluid-mosaic membrane.

(B) The fluid-mosaic model of Singer and Nicolson (1972): bacterial membrane

By watching the movement of each fluorescent stain bound don the surface of the hybrid cell, Frye and Edidin showed that the surface proteins of the human cell and mouse cell intermixed. The rate at which these proteins intermixed suggested that the proteins diffused laterally (moved sideways) in the plane of the cell surface). Numerous subsequent experiments supported these results. Thus investigators concluded that the cell surface is fluid in nature. We must stress, however, that under certain conditions (for example, low temperatures that solidity the lipid phase of the membrane) the movement of membrane proteins becomes restricted. In addition, there is increasing evidence that certain cell-surface proteins are anchored by cytoskeletal elements (microtubules and microfilaments) that restrict their movement.

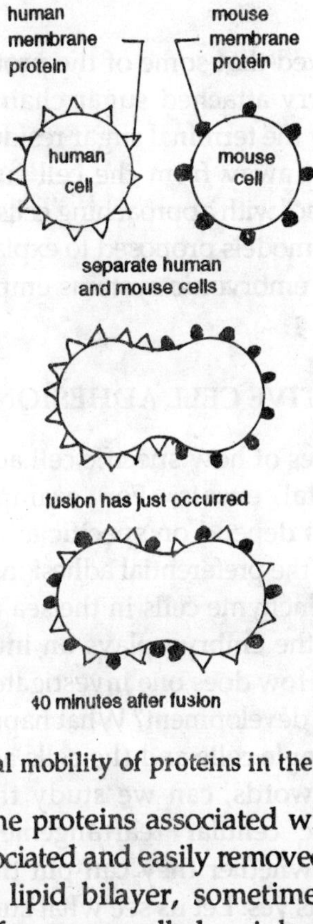

Fig. 8.2. Lateral mobility of proteins in the membrane lipid.

Some of the proteins associated with the cell membrane are loosely associated and easily removed; others are imbedded deeply in the lipid bilayer, sometimes spanning its entire thickness. The former are called peripheral membrane proteins, and the latter are called integral membrane proteins. Integral membrane proteins can often be viewed using the freeze-fracture technique. In this technique the membrane is split down the middle and one of the lipid bilayers is peeled away. The surface of the cleaved membrane is coated with a heavy metal that forms a replica of the fractured surface. The replica is examined with the electron microscope. Integral membrane proteins are observed as particles on the replica surface. To sum up, the cell surface has fluid properties and is a mosaic consisting of lipid interspersed with proteins, many of which can move laterally in the lipid.

Surface Sugars

It has been observed that some of the proteins and lipids of the cell surface carry attached sugar chains. These sugar chains—and especially the terminal sugar residues of the sugar chains—reach farthest away from the cell surface and thus offer a first line of contact with approaching cells and molecules. Some of the molecular models proposed to explain the adhesive recognition of cells in embryonic systems emphasize the role of the surface sugars.

SELECTIVE CELL ADHESION

We have seen examples of how specific cell adhesion plays a role in developmental events. For example, sperm-egg recognition appears to depend on specific adhesion between the gametes. Similarly, the preferential adhesion of the filopodia of the secondary mesenchyme cells in the sea urchin gastrula to the animal end of the embryo plays an important role in forming the gut tube. How does one investigate the role of cell adhesion in embryonic development? What happens if embryos are dissociated into single cells and the cells are permitted to reassociate? In other words, can we study the mechanisms controlling embryonic cellular rearrangements by taking embryos apart to see whether they can put themselves back together? The answer is yes. Let us see what such experiments can tell us about the mechanisms of morphogenesis. We shall begin with sponges and then move onto embryonic systems.

In Sponge Cells

The Regenerator in sponges was studied by H.V. Wilson. He found that sponges could be disaggregated into single cells by pressing them through silk cloth. When the single sponge cells were allowed to settle in a dish of sea water, they moved back together to form a new sponge. When cells from the red sponge (*Microciona prolifera*) were mixed with cells from the purple sponge (*Haliclona occulata*), the red cells sought out other red cells and formed a red sponge and the purple cells likewise formed a purple sponge. The red cells seldom stuck to purple cells. Thus, sponge cells can specifically recognize cells of their own species. This specificity has served as a model for studying adhesive recognition in developing systems.

Moscona and Humphreys continued to investigate specific reaggregation in the sponge system. Using a rotary shaker method developed by Moscona, these workers found that when mixtures of red and purple sponge cells were rotated together, they still formed aggregates consisting mainly of only one type of cell. Red stuck to red, purple to purple. These results suggested that the major reason sponge cells form species-specific aggregates is that each type of cell preferentially sticks to its own kind. The shaker method allowed cell aggregates to form as a result of adhesive stability only without the directed movement (migration) that could occur in a stationary (nonrotated) system. Moscona and Humphreys tried to determine the molecular nature of this species-specific adhesion. They found that sponges could also be disaggregated into single cells by immersing them in seawater free of calcium and magnesium. These cells, however, did not reaggregate if returned to sea water and rotated at low temperature (5^{0}C).

When calcium was added to the disaggregation supernatant (the solution left over after the disaggregated cells were removed) and the supernatant was added back to the cells, they did reaggregation of only red sponge cells. Similarly, the supernatant from the purple sponges promoted reaggregation of only purple sponge cells. Moscona and Humphreys concluded that a molecule responsible for species-specific sponge cell adhesion was released when the sponge disaggregated in sea water free of calcium and magnesium. If these molecules were added back to the cells from which they were removed, the cells would stick back together. Apparently, cells can also resynthesize this molecule, because they reaggregated to some extent at higher temperatures (24°C) but not at low temperatures (5°C). At higher temperatures, cellular synthetic reaction could occur.

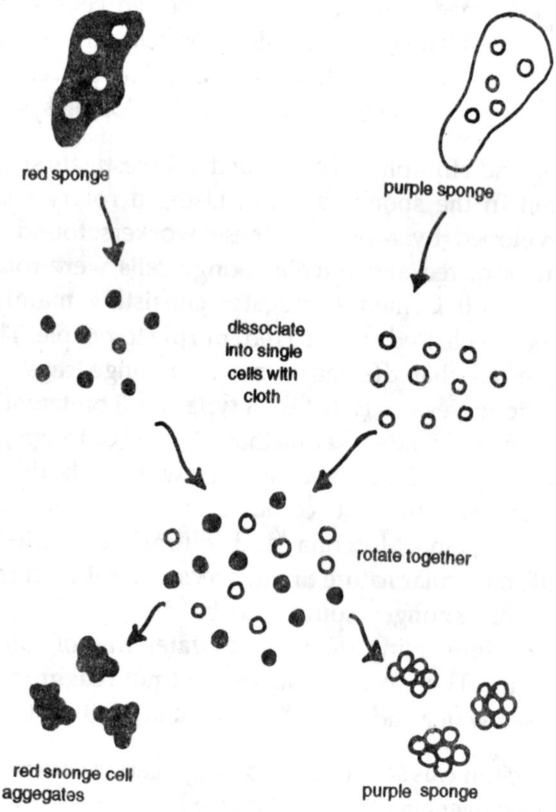

Fig. 8.3. Species-specific sponge cell reaggregation.

The molecules promoting adhesion in sponge cells were purified both by Moscona's group and by Humphrey's group. They appeared to be high-molecular-weight glycoproteins or proteoglycans (proteins associated with carbohydrate). Turner and Burger found that the sugar D-glucuronic acid was on the

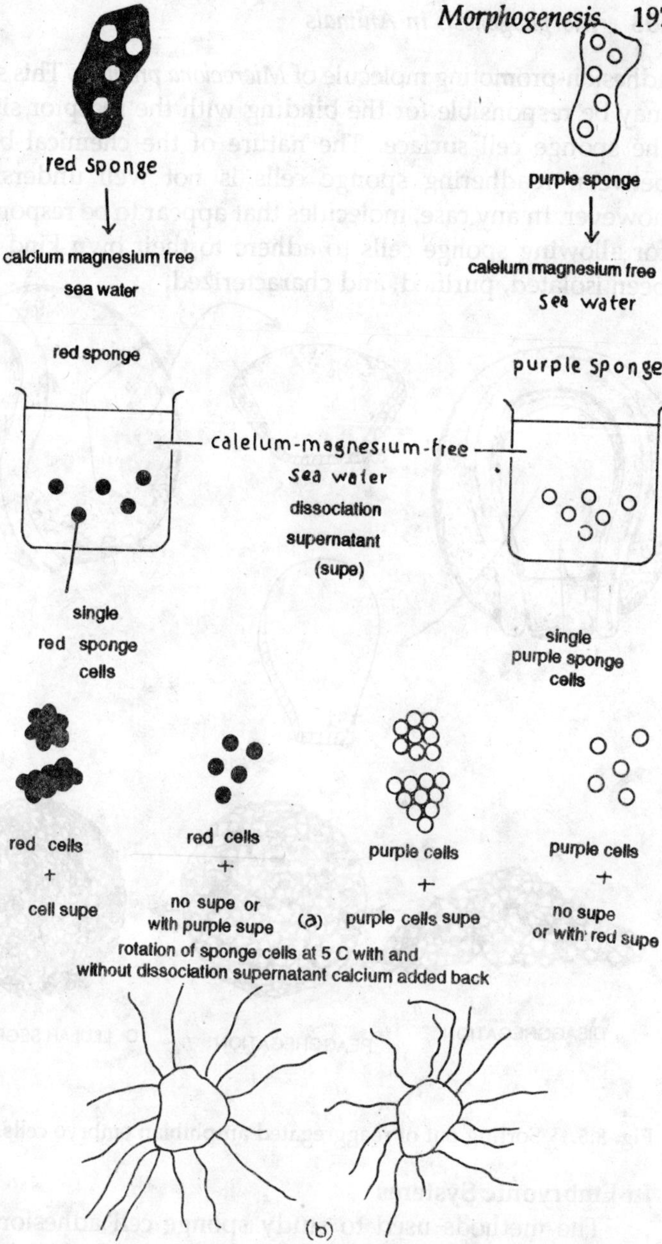

Fig. 8.4. (a) Species-specific sponge aggregation factors. From experiments of Moscona and Humphreys. (b) "Sunburst" configuration of *Microciona* sponge aggregation factor.

adhesion-promoting molecule of *Microciona prolifera*. This sugar may be responsible for the binding with the receptor site on the sponge cell surface. The nature of the chemical bonds between readhering sponge cells is not well understood, however. In any case, molecules that appear to be responsible for allowing sponge cells to adhere to their own kind have been isolated, purified, and characterized.

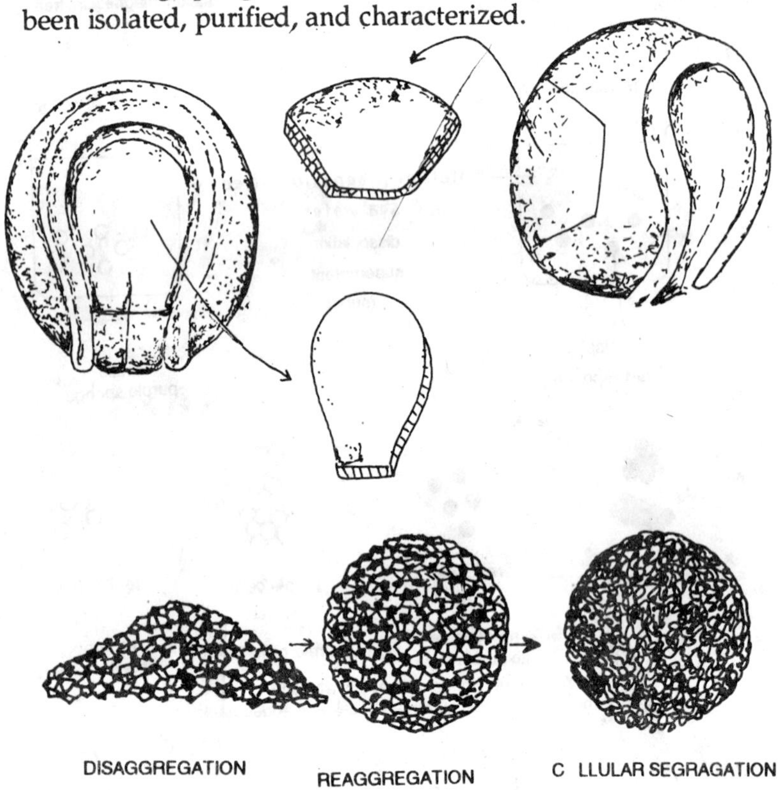

Fig. 8.5. Sorting out of reaggregated amphibian embryo cells.

In Embryonic Systems

The methods used to study sponge cell adhesion have also been used to study embryonic cell adhesion. Townes and Holtfreter, Moscona, and Steinberg performed some pioneering adhesion experiments with embryonic cells. Embryos of all types can be disaggregated into single cells using alkaline solutions, solutions free of calcium and magnesium, or solutions containing proteolytic (protein-degrading) enzymes such as trypsin. Embryonic cells disaggregated by such means can be rotated in the medium using a gyratory shaker. The cells from a whole embryo or from a variety of embryonic tissues recombine into aggregates of many cell types. After a day or two, however, cells tend to "sort out" into groups of only one specific cell type. The mixed-cell aggregate is gradually transformed into an aggregate containing several areas, each composed of only one cell type. For example, ectoderm cells stick to ectoderm cells, retina cells stick to retina cells, and preheart cells stick to preheart cells. The nature of this "sorting out" phenomenon is unclear, but it appears to involve the relative adhesiveness of the different cell types. The most tightly adhesive cells squeeze together toward the center of the aggregate, pushing the less adhesive cells to the periphery. McGuire and coworkers found that if embryonic cells were disaggregated from the embryos very gently, they behaved like the sponge cells. That is, embryonic liver cells mainly adhered immediately to other liver cells and not to cells from other embryonic tissue cell types. Thus tissue-specific recognition occurs between like embryonic cells. This adhesive recognition may govern embryonic cell rearrangements. Let us now turn to some experimental evidence suggesting that specific adhesive recognition plays an important role in morphogenesis. Roth's group used the chick embryo retina-tectum system in their study. Nerves grow from the retina of the eye to specific regions of the optic tectum, the brain center that interprets visual impulses. Nerves from the dorsal portion of the retina always grow to the ventral portion of the optic tectum, not to the dorsal portion. Roth and coworkers sought to determine whether adhesive recognition plays a major role in these specific nerve connections. They first cut retinas and tecta into dorsal and ventral halves. They then labeled retina cells from each half of the retina with radioactive precursors (molecules that become incorporated into cells) or with retinal pigment. Single

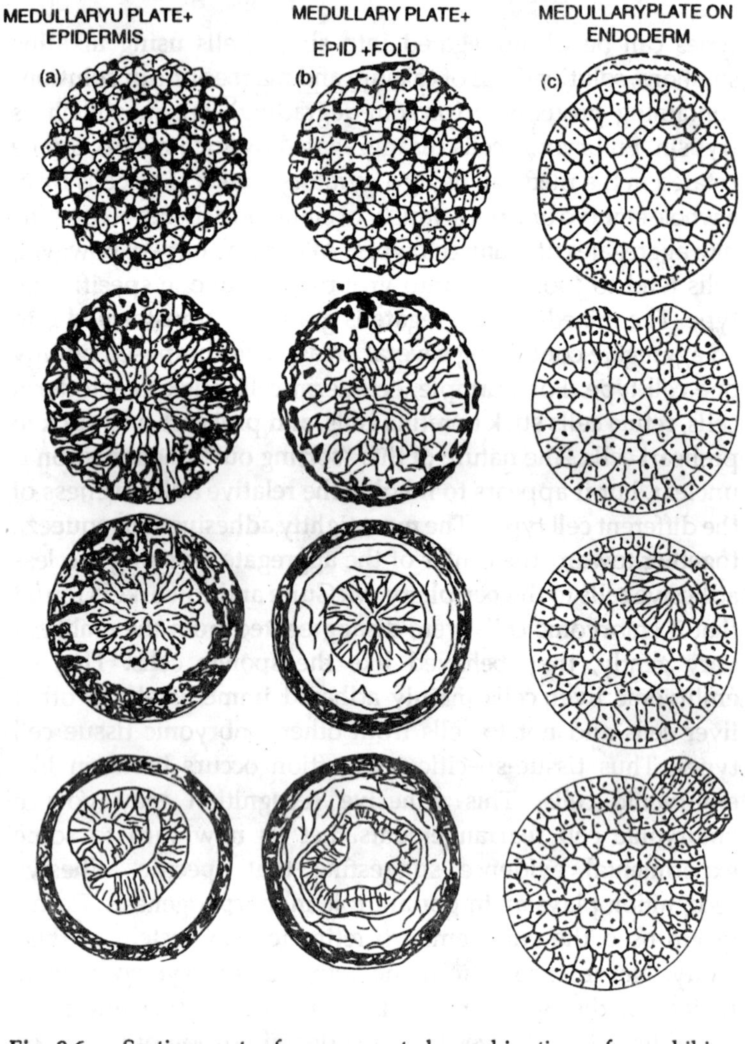

Fig. 8.6. Sorting out of reaggregated combination of amphibian embryonic cells. Sections through successive stages of composite reaggregates. (a) Randomly arranged cells of epidermis (black) and medullary plate (white) move in opposite directions and reestablish homogeneous tissues. (b) When cells from the neural fold are added, they move to occupy the space between the neural tissue and the epidermal covering. (c) A piece of medullary plate or of larval forebrain moves first into and then out of a matrix of endoderm.

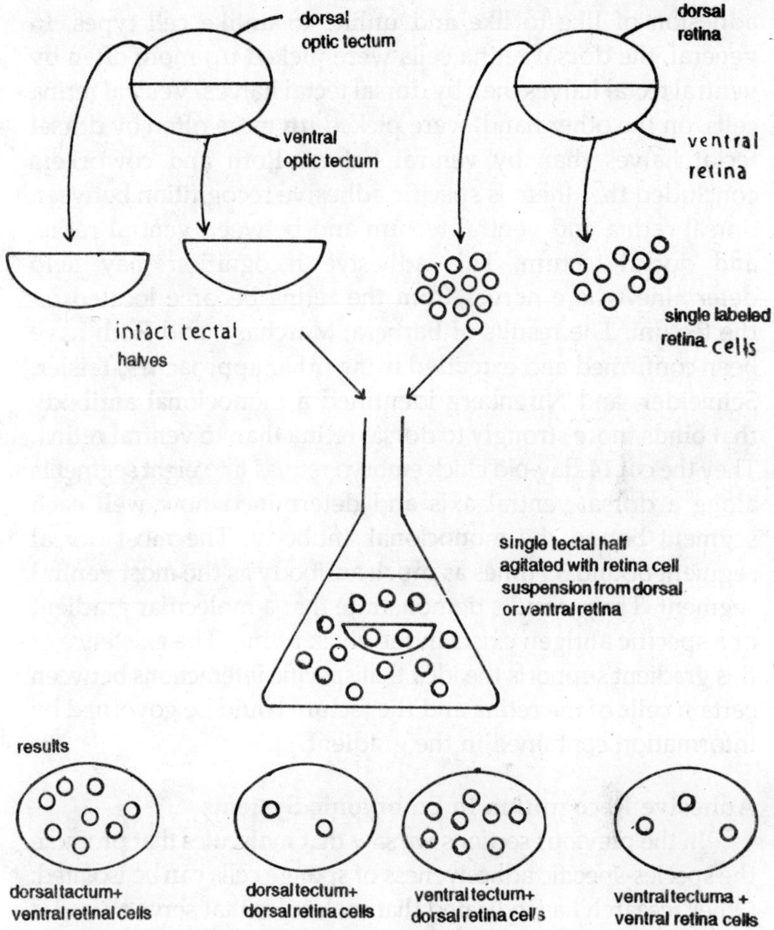

Fig. 8.7. Retinotectal adhesive recognition. Barbera, Marchase, and Roth found that dorsal retina cells adhere best to ventral tectum and ventral retina cells to dorsal tectum. The time needed for adhesion varied in these experiments.

labeled dorsal or ventral retina cell suspensions were agitated with either dorsal tectal halves or ventral tectal halves. This technique of shaking chunks of tissue with more than one cell type was developed by Roth and Weston to measure the adhesion of like to like and unlike to unlike cell types. In general, the dorsal retina cells were picked up more often by ventral tectal halves than by dorsal tectal halves. Ventral retina cells, on the other hand, were picked up more often by dorsal tectal halves than by ventral halves. Roth and coworkers concluded that there is specific adhesive recognition between dorsal retina and ventral tectum and between ventral retina and dorsal tectum. This adhesive recognition may help determine where nerves from the retina become located on the tectum. The results of Barbera, Marchase, and Roth have been confirmed and extended using other approaches. Trisler, Schneider, and Nirenberg identified a monoclonal antibody that binds more strongly to dorsal retina than to ventral retina. They the cut 14-day-old chick embryo retinas into eight segments along a dorsal-ventral axis and determined how well each segment bound the monoclonal antibody. The most dorsal segment bound 35 times as much antibody as the most ventral segment. These results demonstrate that a molecular gradient of a specific antigen exists in the chick retina. The existence of this gradient supports the idea that specific interactions between certain cells of the retina and the tectum could be governed by information contained in the gradient.

Adhesive Recognition in Embryonic Systems

In the previous sections we saw that molecules that promote the species-specific adhesiveness of sponge cells can be isolated. Other research has indicated that molecules that serve a similar function can also be isolated from embryonic cells. Moscona and Lilien have isolated a molecule from a culture medium in which chick embryo neural retina cells were grown. this molecule promoted the adhesion of embryonic neural retina cells, but not that of other cells. When chick embryo retina cells were rotated with the adhesion-promoting factor, the cell aggregates formed over time were much larger than when the adhesion factor was absent. Similar adhesion-promoting factors have been isolated from mouse teratoma cells be Oppenheimer and Humphreys, from cellular slime molds by Rosen and

Barondes, and from sea urchin sperm by Vacquier and colleagues.

Edelman and colleagues have identified a number of *cell-adhesion molecules* (CAMs) in specific tissues of various vertebrates. One of these molecules controls neuron-neuron and neuron-muscle adhesion. During embryonic development, this neuron cell-adhesion molecule (N-CAM) loses much of it sialic acid content. Such a molecular change could account for the changing adhesive properties of developing cells and tissues. In the cerebellums of mice with the staggerer mutation, a mutation in which nerve-cell interactions are defective, the N-CAMs do not change. This observation is consistent with the hypothesis that changing CAMs help to control cellular interactions during embryonic development. Embryonic nerve cells actually produce N-CAMs. These cells grow along the outer surface of the brain, guided by a trail of N-CAMs contained on projections of glial cells. N-CAMs also appear to guide formation of synaptic connections between nerve and muscle cells. Antibodies to N-CAMs alter the route taken by the optic nerve and disrupt synapse formation. Glial cells make N-CAMs just when the nerve reaches them, apparently to guide the nerve in the right direction. Other molecules called *cadherins* also appear to play an important role in morphogenesis. These are calcium-dependent cell-cell adhesion molecules. at least three cadherins are known. E-cadherin (epithelial cadherin), N-cadherin (neural cadherin), and P-cadherin (placental cadherin). These are cell-surface proteins with molecular weights ranging from 118,000 to 127,000 daltons. Cadherins help maintain associations between cells of the same type and foster separation of cells of different types during tissue and organ formation. When the ectoderm separates from the mesoderm, the cells that form mesoderm stop producing E-cadherin and make N-cadherin. Also, when nervous tissue forms from ectoderm, the ectoderm switches from producing E-cadherin to producing N-cadherin. This switch suggests that cadherins may play a role in breaking old cellular connections so that a new layer or organ can be formed. a number of other cell adhesion molecules have also been discovered, including very large particles such as *adherons* and *toposomes*. These cell-surface particles have been found in a wide variety of organisms ranging from sea urchins and fruit flies to chick embryos, and they may play a role in embryonic cellular interactions and morphogenesis.

Oppenheimer and Meyer isolated an aggregation-enhancing protein from blastulas of the sea urchin *Strongylocentrotus purpuratus* by disaggregating them in sea water free of calcium and magnesium. The isolated protein dramatically promoted the reaggregation of *Strongylocentrotus purpuratus* blastula cells, but not the corresponding gastrula cells. It had no effect whatsoever on blastula cells from *Lytechinus pictus*, a different species of sea urchin. Evidence suggests that this protein binds cells together at their surfaces, possibly by attaching to carbohydrates. This molecule, like the sponge aggregation factor, is clearly species specific. It is also stage specific in its action. The fact that this material, isolated from intact embryo cells, promotes the adhesion of specific embryo cells at specific stages in their natural medium, sea water, supports the contention that the molecule helps mediate cell adhesion in the embryo. Continued work with this material and with other factors isolated in different systems should illuminate the nature of specific cellular adhesiveness during embryonic development. Isolating, purifying, characterizing, and determining the functional groups of these adhesion factors should help elucidate the nature of specific cell adhesion at the molecular level. The mechanisms controlling selective cell association are probably not the same in all systems. A couple of models have been proposed to explain cell adhesion in dynamic systems such as developing embryos and tumors. One model, the cell-surface glycosyl transferase—carbohydrate acceptor model, was proposed by Roseman and developed further by Roth and others. This model suggests that embryonic cell adhesion and "de-adhesion" result from cell-surface enzymes called *glycosyl transferases*, which attach to or detach from the carbohydrate

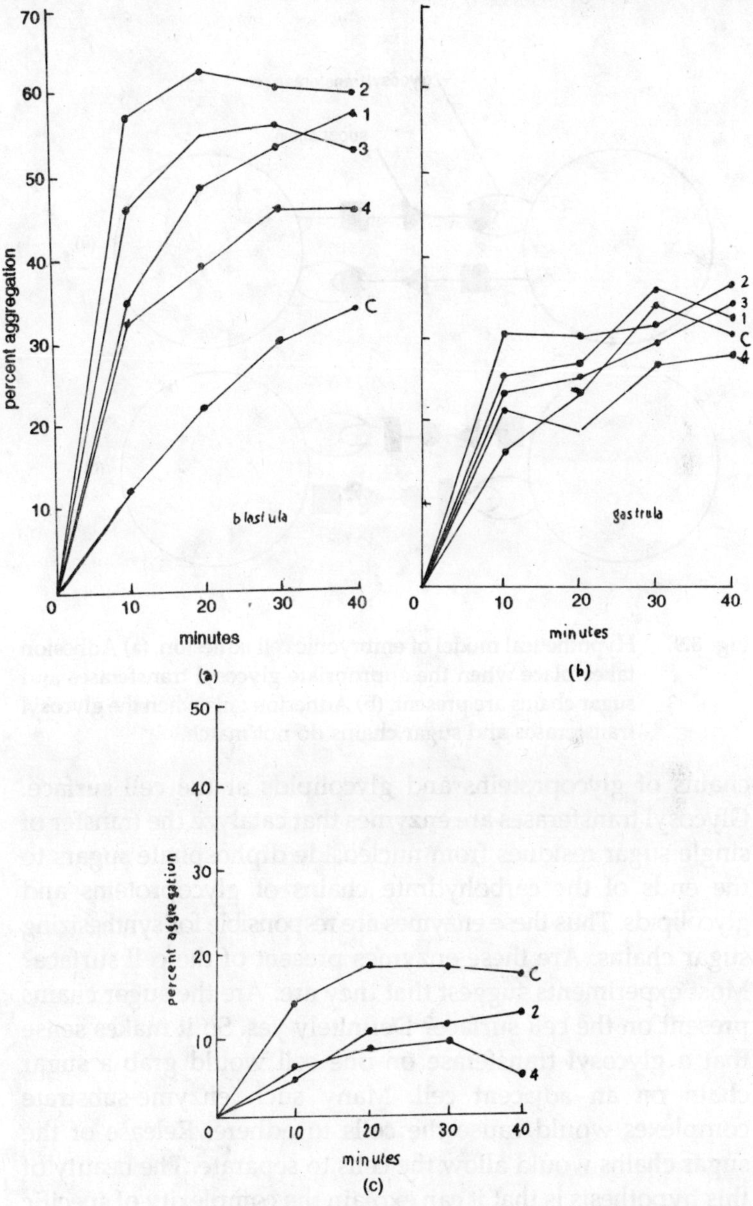

Fig. 8.8. (a-b) Developmental stage-specificity of *Strongylocentrotus purpuratus* blastula aggregation protein. (c) Species specificity of *Strongylocentrotus purpuratus* blastula aggregation protein.

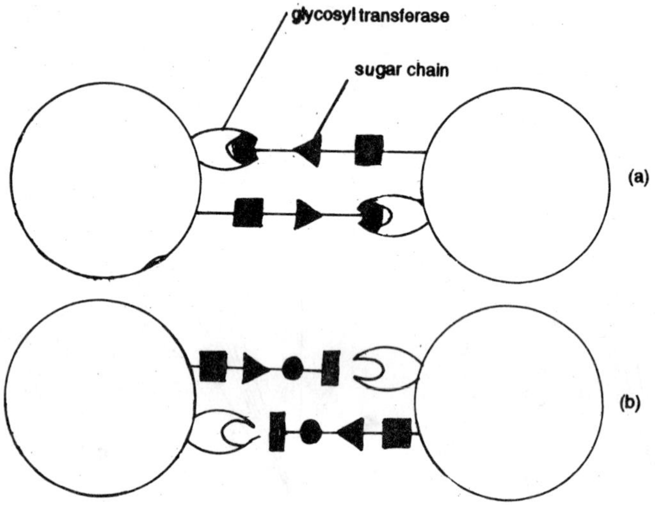

Fig. 8.9. Hypothetical model of embryonic cell adhesion. (a) Adhesion takes place when the appropriate glycosyl transferases and sugar chains are present. (b) Adhesion fails when the glycosyl transferases and sugar chains do not match.

chains of glycoproteins and glycolipids at the cell surface. Glycosyl transferases are enzymes that catalyze the transfer of single sugar residues from nucleoside diphosphate sugars to the ends of the carbohydrate chains of glycoproteins and glycolipids. Thus these enzymes are responsible for synthesizing sugar chains. Are these enzymes present of the cell surface? Most experiments suggest that they are. Are the sugar chains present on the cell surface? Definitely yes. So it makes sense that a glycosyl transferase on one cell would grab a sugar chain on an adjacent cell. Many such enzyme-substrate complexes would cause the cells to adhere. Release of the sugar chains would allow the cells to separate. The beauty of this hypothesis is that it can explain the complexity of specific cell associations and disassociations in embryos. Each transferase is specific for a given receptor chain and for the sugar it transfers. Thus certain cells could stick together if the right transferases and the right sugar chains were present. It should be emphasized that this model explains only initial cell contacts. It does not deal with the more stable associations that result from the secretion of cementing substances in differentiated tissues.

Glycosyl Transferase

Experimental evidence is accumulating in support of the hypothesis advanced by roseman and colleagues that cell-surface glycosyl transferases mediate cell-cell adhesion by binding to cell-surface molecules containing carbohydrate. A number of studies have suggested that glycosyl transferase-acceptor interaction mediates cellular interactions in several developmental systems.

In one such study, Shur and Hall showed that mouse sperm bind to mouse eggs be means of a specific glycosyl transferase located on the surface of the sperm cell. The specific enzyme involved is a galactosyl-transferase that binds to *N*-acetylglucosamine groups on the zona pellucida of mouse eggs. Findings that support this mechanism include the following:

1. Two reagents that interfere with surface galactosyltransferase activity inhibit sperm-egg binding. The milk protein alpha-lactalbumin changes the ability of the transferase to bind to *N*-acetylglucosamine and also inhibits sperm-egg binding; UDP-dialdehyde inhibits sperm-egg binding and sperm-surface galactosyltransferase activity to identical degrees.
2. Sperm-surface galactosyltransferase can add sugar to add cell surfaces.
3. Enzymatic removal of egg-surface *N*-acetylglucosamine reduces sperm binding by 86%.

The molecular basis of sperm-egg binding is indeed becoming clearer!

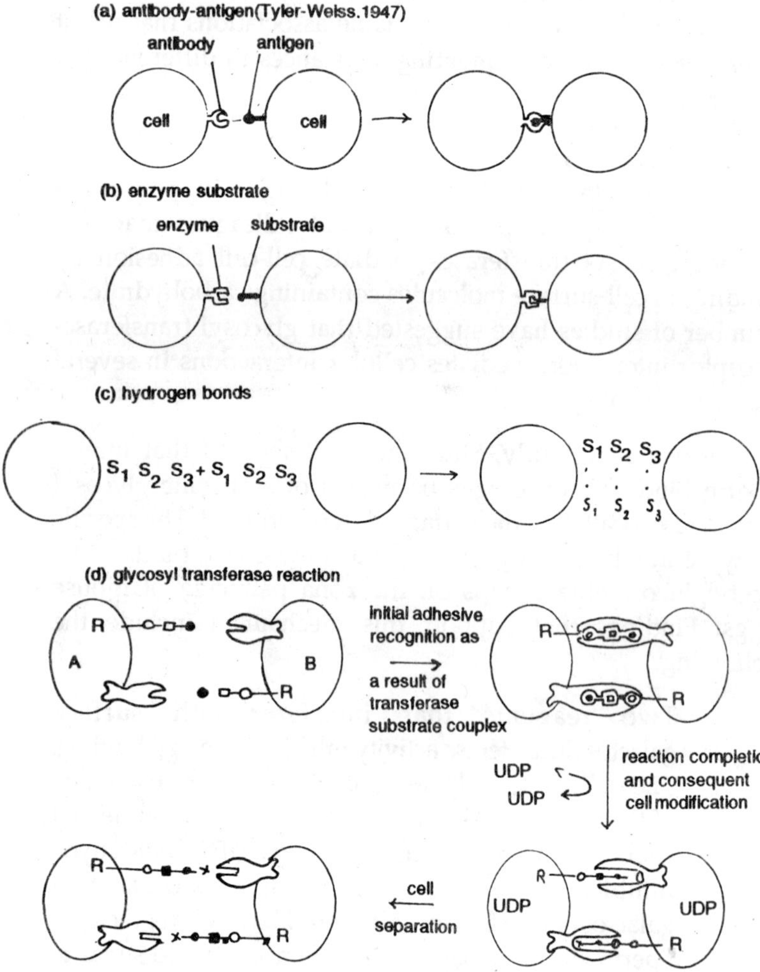

Fig. 8.10. Hypotheses of cell-cell adhesion: antibody-antigen, enzyme-substrate, hydrogen bonds, modification of cell-cell adhesion by glycosyl transferase reaction.

Although glycosyl transferases had been found on the cell surface, until lately their role in mediating adhesion had been suggested only indirectly (as in the previously mentioned experiment). Recently a pure glycosyl transferase (a fucosyl transferase) became available in quantities sufficient to test its role in cellular adhesion. A variety of cell types, such as embryonic human skin cells (fibroblasts), were found to attach and spread on surfaces on which the pure transferase was attached. The adhesion of cells to the transferase was specific in that the cells would not attach to other proteins and specific alterations of the transferase prevented adhesion. Adhesion of the cells to the enzyme-covered surface was inhibited by compounds that specifically bind to the transferase. Thus evidence suggests that protein-sugar interactions, such as glycosyl transferase-sugar or lectin-sugar binding play a role in mediating cellular adhesion. Substantial evidence points to a role for cell-surface carbohydrates in cell associations; experimenters whose work supports this view include Roseman, Roth, Oppenheimer, Steinberg, Moscona, Burger, Rosen, and Barondes. When embryonic cells are treated with purified glycosidases (enzymes that catalyze the removal of sugar residues from carbohydrates), their adhesion and aggregation rates change. Molecules that promote aggregation rates change. Molecules that promote aggregation are often inactivated when treated with glycosidases. In addition, specific carbohydrate-binding proteins have been isolated from slime mold and embryo cells. These proteins promote cell association when the cells are rotated with these molecules. The proteins apparently bind specific sugar residues on the cell surface. Thus protein-carbohydrate interaction may be of key importance in controlling cell-cell associations.

In summary, the molecular nature of cell associations in embryos is not well understood. Cell adhesion, however, plays an important role in the cellular rearrangements that occur in developing embryos. Carbohydrate-protein interaction may control embryonic cell associations in some systems. Because the carbohydrate chains of cell-surface glycoproteins and glycolipids reach farthest away from the cell, these chains must play a role in at least the initial contact with approaching cells and substrates.

OTHER MECHANISMS

Specific adhesive recognition among cells is clearly not the only mechanism that controls the development of form in embryos. We have treated cell adhesion at length because this area has been intensively investigated in recent years, and because specific testable models of the molecular nature of adhesive recognition among cells have been developed. Now let us briefly examine some of the other specific mechanisms important in controlling embryonic morphogenesis.

Cytoskeletal Elements in Morphogenesis

We have mentioned that cytoskeletal elements such as rod-shaped proteins, microtubules, and microfilaments—the "microskeleton" of the cell—are involved in a variety of embryonic cell activities. For instance, cytoplasmic microtubules are associated with elongating cells. Cell elongation often occurs during embryonic cellular rearrangements and during morphogenesis of embryonic organs. When cell elongation takes place, microtubules become visible, lined up in the long axis of the cell. Microtubules are observed in elongating cells around the amphibian blastopore, in the neural plate cells, in cells moving into the primitive groove in bird embryos, and in outgrowing nerve cells. Cell elongation is prevented by colchicine, which specifically disrupts microtubules. Colchicine prevents nerve cell outgrowth and inhibits cell elongation in the chick primitive streak and in the neural plate. Colchicine is not known to affect anything other than microtubules. Microtubules thus play a role in embryonic cell elongation. They may directly force the cells to elongate, or they may serve as tracks for cytoplasmic flow in the direction of elongation. Cytoplasmic microfilaments cause cell narrowing. Micro-

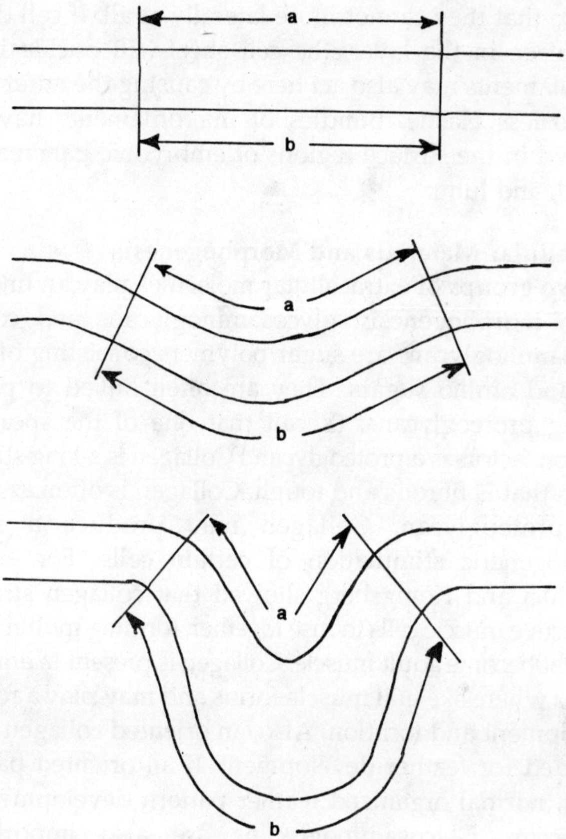

Fig. 8.11. Epithelial folding: a hypothetical mechanism. Surface a (outer surface) is tightly adherent. Cell division takes place within the layer, and the sheet buckles.

filaments serve as a simple cellular contractile system. The morphogenesis of tubular glands, for example, may be aided by bundles of microfilaments that contract and exert force on the sides of cells, narrowing the cells at one end. If a sheet of cells narrows at one side, the sheet will bend. Other hypotheses, however, have been advanced to explain how certain cell layers (epithelia) bend or fold. For example, suppose that the outer surface cells of the layer are tightly attached to each other so that they cannot more laterally at all. If cell division takes place in the layer, the cell sheet-will buckle inward. Microfilaments may also act here by causing the outer cells to become less elastic. Bundles of microfilaments have been observed in the surface regions of embryonic pancreas, lens, thyroid, and lung.

Extracellular Materials and Morphogenesis

Two groups of extracellular molecules play an important role in morphogenesis: glycosaminoglycans and collagen. Glycosaminoglycans are sugar polymers consisting of uronic acids and amino sugars. They are often linked to proteins, forming proteoglycans. (Recall that one of the sponge cell adhesion factors is a proteoglycan.) Collagen is a long structural protein that is fibrous and tough. Collagen is often associated with proteoglycan. Collagen may produce a specific morphogenetic stimulation of certain cells. For example, Hauschka and Konigsberg showed that collagen stimulates prospective muscle cells to fuse together, forming multinucleated cells that become adult muscle. Collagen is present in embryonic regions where skeletal muscle forms and may play a role in its development and location. Also, an oriented collagen pattern is needed for feather development. If an oriented pattern is absent, normal organized feather pattern development does not occur. Glycosaminoglycans are also important in morphogenesis. These molecules apparently help maintain the shape of branching epithelial structures. If embryonic salivary gland, lung, or kidney epithelia is treated with enzymes that remove surface glycosaminoglycans, normal morphogenesis is inhibited and the structures tend to round up and form a ball.

The branching or lobulation of epithelial layers in structures such as the salivary gland, mammary gland, lung, kidney, and thymus is a morphogenetic event that has been carefully studied. Although the mechanisms involved are not completely understood, a well-documented picture is emerging. Branching glands consist of an epithelium associated with mesenchyme. The entire organ rudiment can be removed from the organism and studied in culture. In this way the roles of the different parts have been studied, helping us understand the nature of branching morphogenesis.

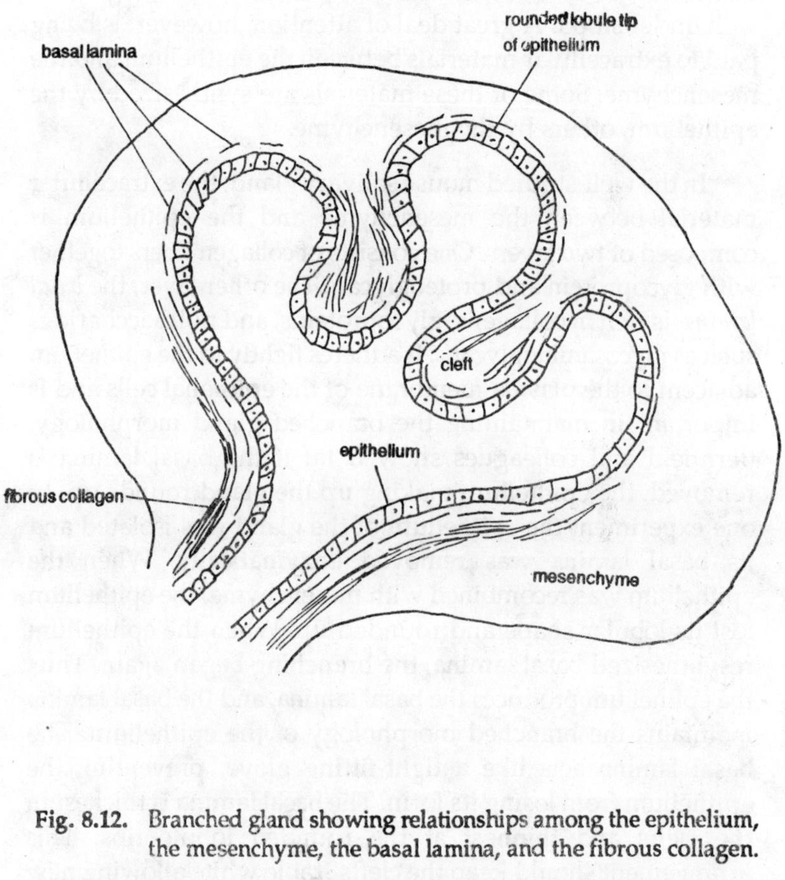

Fig. 8.12. Branched gland showing relationships among the epithelium, the mesenchyme, the basal lamina, and the fibrous collagen.

Branching begins when the epithelial layer folds inward, forming clefts in the epithelium. One of the mechanisms that control this infolding is the contraction of cytoplasmic microfilaments at one end of the cells in the folding layer.

Also, the outer surface cells of the epithelial layer may be tightly attached to each other, so that when cell division occurs within the layer the cell sheet buckles inward. Cleft formation in epithelial layers is only one step in the formaation of a lobulated gland. Another component involved in epithelial lobulation is the mesenchyme that envelops the epithelium. The epithelium will not branch in the absence of mesenchyme. The influence of mesenchyme on epithelial branching is not well under stood. A great deal of attention, however, is being paid to extracellular materials between the epithelium and the mesenchyme. Some of these materials are synthesized by the epithelium, others by the mesenchyme.

In the well-studied mouse salivary gland, the extracellular material between the mesenchyme and the epithelium is composed of two layers. One consists of collagen fibers together with glycoprotein and proteoglycan. The other layer, the *basal lamina,* is rich in collagens, glycoproteins, and polysaccharides such as glycosaminoglycans. It adheres tightly to the epithelium adjacent to the plasma membrane of the epithelial cells and is important in maintaining the branched gland morphology. Bernfield and colleagues showed tat if the basal lamina is removed, the epithelium making up the gland rounds up. In one experiment the epithelium of the gland was isolated and its basal lamina was removed enzymatically. When the epithelium was recombined with mesenchyme, the epithelium lost its lobular shape and rounded up. When the epithelium resynthesized basal lamina, the branching began again. Thus the epithelium produces the basal lamina, and the basal lamina maintains the branched morphology of the epithelium. The basal lamina acts like a tight-fitting glove, preventing the epithelium from losing its form. The basal lamina is thickest in the clefts and thinnest at the rounded lobule tips. This arrangement should keep the clefts stable white allowing new clefts to form at the tips, Collagen fibers are deposited outside the basal lamina; these are also most densely packed in the clefts. Collagen fibers apparently help stabilize the indentations.

A major role of the mesenchyme is apparently to degrade the basal lamina near the rounded lobule tips so that clefts can form in these regions. Bernfield and colleagues labeled the glycosaminoglycans of the basal lamina with ^{3}H glucosamine. When mesenchyme was present, the label was lost from the basal lamina outside the rounded lobule tips. Very little label was lost from the basal lamina in the clefts. When mesenchyme was absent, no significant label was lost in any region of the basal lamina. These experiments suggest that the mesenchyme selectively degrades the basal lamina in the morphogenetically active regions of the branching epithelium (that is, the rounded tips). Once the basal lamina has been degraded, some sort of signal might stimulate epithelial cleft formation in that region. For instance, Ca^{2+} ions released during glycosaminoglycan degradation can trigger microfilament contraction in the adjoining epithelium. Alternatively, basal lamina degradation might stimulate cell division in the epithelium, which could also lead to cleft formation. The exact mechanism by which mesenchyme influences epithelial branching is not yet clear. Because of the active investigations in this area, however, more of the answers to these question should soon be in hand. In summary, some of the factors involved in morphogenesis are cytoskeletal elements (microtubules and microfilaments), localized cell division, and extracellular substances such as collagen and glycosaminoglycans.

Mechanisms of Neural Crest Migration

We noted earlier that neural crest cells differentiate into many structures some far from the original neural crest ectoderm. These cells must travel long distances to their final locations. Work from many laboratories, indicates that neural crest cells migrate along well-defined pathways. Neural crest cells of the trunk migrate in two streams. Crest cells in one stream move ventrally into the mesenchyme located between the somites and the nerve tube. Some of the cells migrate into the somites, forming dorsal root ganglia; others from sympathetic ganglia and the adrenal medulla. Crest cells in the other stream move into the surface layers of the embryo, forming pigment cell precursors. The environment in which neural crest cells find themselves seems to play a major role in controlling their differentiation. This conclusion is based on experiments in which neural crest cells transplanted from one trunk region could substitute for those of other trunk regions, and neural crest cells transplanted from one cranial region could substitute for those of other cranial regions. In one experiment, neural crest cells from the mesencephalon area, which normally give rise to the rostral parts of the skull, were transplanted to the more posterior metencephalon region. The resulting cells differentiated into the lower jaw, which is normally formed from crest cells in the metencephalon region. Thus the final environment of the crest cells can change the fate of the cells to produce the structure that normally develop at that location. These transplantation experiments used chick and quail embryos, so that the donor cells in specific structures could be distinguished from the recipient cells.

Although trunk crest can be interchanged with other levels of trunk crest and cranial crest can be interchanged with other levels of cranial crest, trunk crest cannot be completely interchanged with cranial crest, and vice versa. In other words, cranial crest cells are somewhat restricted or predetermined to form cartilage, whereas trunk crest cells cannot form cartilage no matter what their final environment is. Finally we come to one of the most intriguing questions: what are the mechanisms that control the migration f neural crest cells to distant sites in the body? The work of Marianne Bronner-Fraser has led to a better understanding of this problem. We noted above that neural crest cells migrate along well-defined pathways. Bronner-Fraser injected neural crest cells, retinal pigment cells (non-neural crest cells), or latex polystyrene beads (uncoated or coated with bovine serum albumin) onto the ventral neural crest pathway of chick embryos. All of these, including the beads, translocated ventrally along the pathway in the same way as normal neural crest cells. When the latex beads were coated with the glycoprotein fibronectin, however, the beads remained near the site of implantation and did not move ventrally. These results suggest that neural crest cell migration is influenced by some driving force imparted by the embryonic environment. Also, molecules such as fibronectin on the cell surface may serve as a recognition mechanism that prevents non-neural crest cells from entering the ventral neural crest migratory pathway. Cells of the somite, fibroblasts, and beads coated with fibronectin—all of which have fibronectin on their surfaces—cannot move onto the ventral pathway. Neural crest cells, retinal pigment epithelial cells, and latex beads without fibronectin—all of which lack surface fibronectin—can move along the ventral pathway.

Exactly what forces control the translocation of cells along neural crest pathways is unclear. These experiments show, however, that the entrance to the ventral pathway is controlled by the presence of specific surface substances. Once cells enter the pathway, cellular forces (not necessarily originating in the translocating cells) may move them to their final location. These forces may involve movements in cells surrounding the translocating particles that allow the particles to move in response to, for example, adhesiveness toward substances such as fibronectin. Fibronectin on the cells apparently prevents cells from entering the ventral pathway, whereas fibronectin concentrated at specific points along the pathway stops translocating cells at these points. This concepts is supported by the observation that both beads and cells locate in areas where fibronectin is present. If fibronectin disappears from the substratum, cells adhering to the substratum may come loose and form clusters with one another. This could account for the aggregation of some neural crest cells into ganglia. This is only speculation, but such speculations form good working hypotheses that surely will lead to additional discoveries in this exciting area.

Cell Death in Morphogenesis

Before concluding our discussion of the mechanisms that

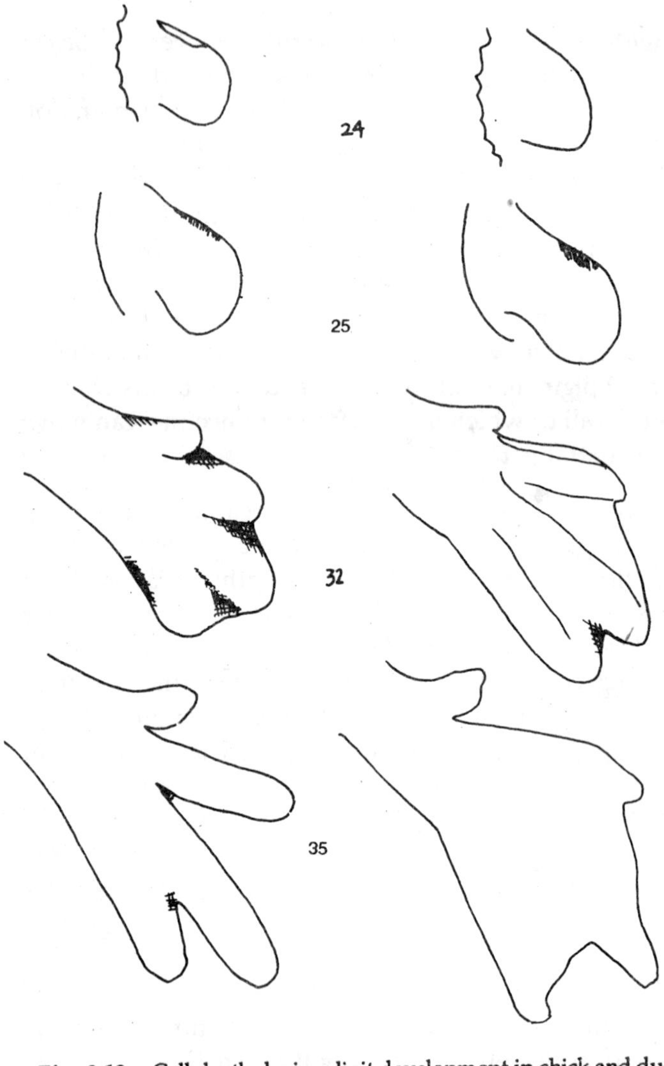

Fig. 8.13. Cell death during digit development in chick and duck embryo hindlimb. Shaded areas die in forming limb.

influence the development of form in embryos, let us briefly mention another important factor that shapes certain structures. Some structures are molded by cell death that occurs in specific regions of the structures. For example, early in embryonic development our hands and feet resemble paddles or flippers. Our fingers and toes form as the result of cell death in the regions between the prospective digits. If cell death fails to take place, certain abnormalities result, such as webbed hands or feet. Cell death occurs in many developing embryonic structures. We saw, for example, that certain structures in each sex degenerate after the indifferent stage of sexual development. The tadpole tail regresses during metamorphosis as the result of cell death. Increased levels of lysosomal enzymes (for example, *acid phosphatase, reibonuclease,* and *cathepsins*) are observed in many embryonic regions that are "programmed to die." In the urogenital system, sex hormones control death in certain structures. For example, androgens trigger cell death in Mullerian ducts in males by influencing lysosomal enzymes. In amphibians, the regression of the tail and gills is controlled by thyroid hormone. If the thyroid gland is removed from tadpoles, does not become a frog. It thyroid hormone is given to the "thyroidless" tadpoles, metamorphosis proceeds. Levels of lysosomal enzymes increase in the areas where cell death occurs. It should be stressed that specific hormones such as thyroid hormone initiate cell death only in certain amphibian cells (tail, gills, and horny teeth); in others (for example, the limb bud) thyroid hormone causes cell proliferation. The hormone may turn on certain genes in some cells, but other genes in other cells.

Pattern Morphogenesis

Pattern refers to the orderly arrangement of parts of the organism. It results from many types of morphogenetic and differentiative events. In other portions of this text, we discuss pattern formation in the context of such topics as nerve growth to end organs, limb morphogenesis and regeneration, finger and toe development, plant embryo polarity, mosaic cleavage patterns, and cytoplasmic segregations. In this section, we shall deal with a few simple systems from which some fundamental findings have come. How does an ordered arrangement of parts come about? When we think about pattern formation, we usually think of several sorts of mechanisms. For instance, there must be gene action. That is, genes that control specific differentiations (hair bristle, pigment formation) must be turned on so that the relevant parts become differentiated. The second important factor is the environment in which the differentiated cells reside. We have seen that cells often interact with other cells or with noncellular environments. This interaction initiates selective gene activation or aids functioning in responding cells. a third set of events is cell migration and selective adhesion. In many systems these events get the relevant cells to the future sites of pattern development.

In the fruit fly *Drosophila,* the first mechanism mentioned above comes into play in forming bristle patterns in the epidermis. The presence or absence of a bristle at a specific site in the epidermis depends on the genotype of that portion of the epidermis. If the tissue is mutant at a given locus, no bristle forms even if this section of tissue is surrounded by nonmutant, bristle-forming epidermis. If the tissue is wild type at that locus, a bristle forms even if the tissue is surrounded by mutant tissues. Such results indicate that specific genes control bristle formation in tissue. In other cases, however, surrounding tissue also plays a role.

Melanocytes are pigment-forming cells, most of which are derived from the neural crest. Only the melanocytes in the pigmented retina of the eye are derived from the neural tube itself. The neural crest melanocyte precursors migrate from the neural crest to a variety of sites in the body, such as the skin and hair follicles. Pigment pattern in mammals can be studied in depth because the controlling genes in such mammals as mice are known. One can transplant a piece of skin from a black mouse onto the back of a yellow mouse, or vice versa. When yellow skin is placed on the back of a black mouse, the melanocytes from this skin move out into adjacent hair follicles in the host tissue and begin to produce black pigment despite their own genotype. In the reciprocal experiment, the melanocytes from the black transplant migrate into the hair follicles of the yellow host and produce yellow pigment. Thus the formation of yellow or black pigment appears to be controlled by the hair follicle cells, not by the melanocytes. Sulfhydryl compounds, such as glutathione, may be the hair follicle factors that stimulate the formation of yellow pigment in the melanocytes. That is, the glutathione present in yellow hair follicles may inhibit black pigment synthesis and stimulate yellow pigment synthesis. This suggestion is supported by the finding that isolated, *in vitro* cultured yellow melanocytes form black pigment. If sulfhydryl compounds are added to the culture medium, however, these melanocytes revert to synthesizing yellow pigment. The most dramatic examples of pigment patterns are the spotted coats of many mammals and birds. What about an animal with white spots on a dark background? How does such a pattern form? Cytological examination of hair follicles in the white spots of such an animal indicates that differentiated melanocytes are absent in these areas. Can the hair follicles in the white spots permit melanocytes to form pigment? Yes. Melanocytes from a black skin transplant can migrate into a white spot region, enter the hair follicles, and make pigment. Thus the white-spotted areas can sustain the differentiation of melanocytes that are already mature. The white-spotting pattern is probably caused by some factors present in prewhite spots that prevent melanocyte differentiation. Even if melanocyte precursors enter the white spot areas, they do not differentiate.

Pigment pattern formation, therefore, involves all three of the mechanisms mentioned earlier:

1. Migration of premelanocytes from the neural crest to the skin and hair follicles;
2. Interaction of the melanocytes with the environment (that is, with products of the hair follicle cells); and
3. Specific gene activation in the melanocytes that results in the synthesis of pigment.

Pattern formation, therefore, is not unique. It results from mechanisms that operate in many embryonic processes. We shall completely understand the formation of organized arrangements in organisms—that is, patterns—only when we understand the molecular nature of cell-cell interactions and differentiation.

Morphogenesis In a Simple System: Cellular Slime Molds

Morphogenesis in vertebrates is complex and difficult to study. As yet, little is known about the mechanisms that control morphogenesis in vertebrate embryos. Progress is being made slowly; the major problem is that such studies must usually be performed with isolated embryo cells *in vitro*. Thus, for example, studies of cell-cell interactions are often done in culture, an environment that is not the same as that in the embryo. *In vivo* studies of a few simple systems, however, have increased our understanding of the forces that shape the organism. One of these systems is the cellular slime mold.

What are cellular slime molds? At one stage in their life cycle, these organisms are amoebae. At another stage, the amoebae aggregate to form a multicellular fruiting body composed of a stalk and a mass of spores at the tip of the stalk. The spores give rise to amoebae. This fairly simple system goes through morphogenetic processes similar to some of those occurring in higher organisms. There is an aggregation phase in which the free amoebae congregate together to form in integrated migratory cellular mass (slug). This process resembles some of the cellular rearrangements that occur in vertebrate embryos. The cells in the slug differentiate, and morphogenesis occurs in which the slug is transformed into the fruiting body (sporocarp). What is known about the mechanisms of slime mold morphogenesis? The aggregation phase in several slime molds, such as *Dictyostelium discoideum*, has been examined by several groups. The free amoebae continue to divide and remain single as long as there is an adequate supply of the bacteria they feed on. When the bacteria are exhausted, the amoebae in dense cultures begin to aggregate, forming the multicellular

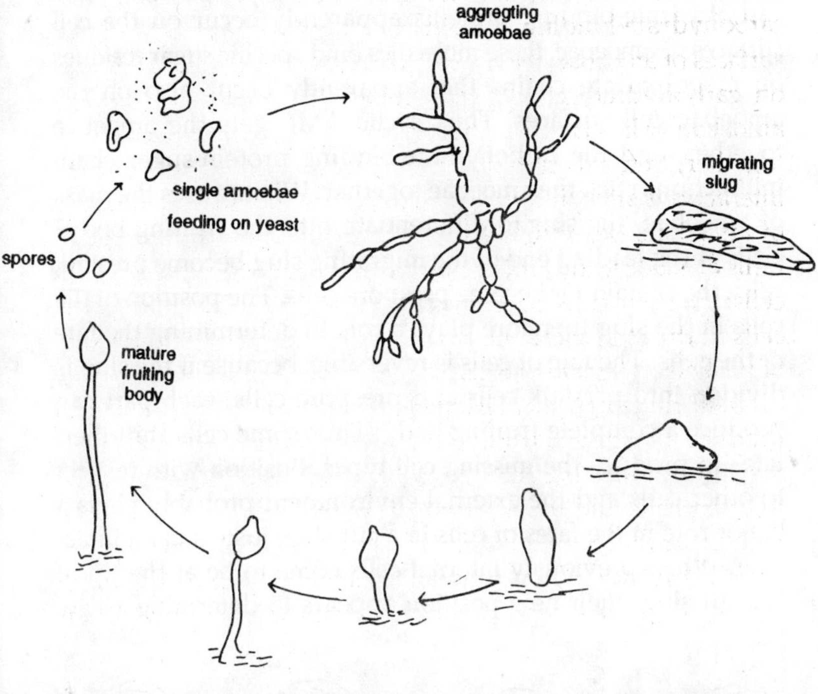

Fig. 8.14. Vegetative growth and fruiting body formation in *Dictyostelium discoideum*.

slug. What causes this aggregation to occur? Certain cells secrete the nucleotide cyclic AMP (cyclic adenosine monophosphate). The amoebae move in the direction of increasing concentration of cyclic AMP. such movement of cells is called chemotaxis. Thus specific chemotaxis controls one phase in slime mold development, the aggregation phase.

Cyclic AMP caused amoebae to move together. Other work by Rosen's and Barondes's groups suggests that the amoebae stick together by protein-carbohydrate interaction. These workers isolated carbohydrate-binding proteins (lectins) from slime mold cells that were competent to aggregate. The carbohydrate-binding proteins apparently occur on the cell surfaces of amoebae. these molecules bind specific sugar residues on carbohydrate chains that apparently occur also on the amoebae cell surfaces. Thus cyclic AMP gets the amoebae together, and the carbohydrate-binding protein-sugar chain interactions stick the amoebae together. What causes the mass of amoebae, the slug,to differentiate into the fruiting body? Cells at the leading end of the migrating slug become prestalk cells; the remainder become prespore cells. The position of the cells in the slug therefore plays a role in determining the fate of the cells. The fate of cells is reversible, because if the slug is divided into prestalk cells and prespore cells, each part can produce a complete fruiting body. Thus some cells shift their fates to produce the missing cell types. Position with respect to other cells and the external environment probably plays a major role in the fates of cells in a cut slug, just as in an intact slug. When previously internal cells come to be at the tip of the cut slug, their new position appears to determine a new fate.

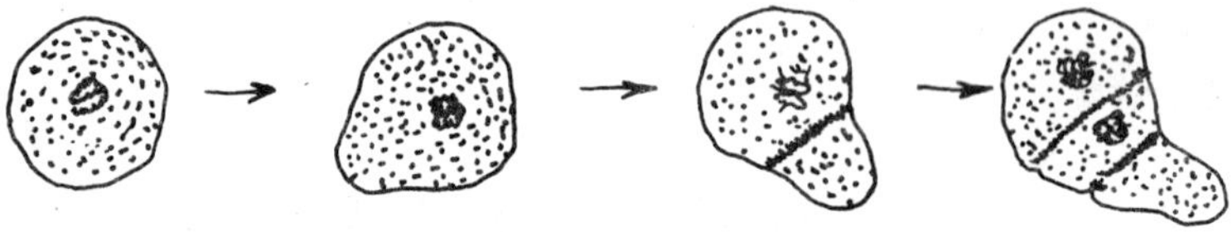

Fig. 8.15. Early development of Focus embryo.

In summary, scientists have investigated specific factors that control development in the cellular slime molds, such as cyclic AMP. Other factors such as carbohydrate-binding proteins also play roles, but these roles are somewhat less well understood. Cell position also governs differentiation in slime molds, but no one yet knows how position *"works"* at the molecular level. Still, simple systems such as this one help as understand the mechanisms that operate in the morphogenesis of higher organisms.

TOOLS FOR STUDYING THE ROLE OF FHE CELL SURFACE IN MORPHOGENESIS AND MALIGNANCY

We mentioned that lectins, carbohydrate-binding proteins or glyco-proteins, have helped us understand certain cell-surface phenomena. These molecules have provided major tools for cell biologists in their study of the cell surface. We shall now discuss these widely used molecules in more detail. *Lectins* have been isolated from the seeds of plants and from extracts of sponges, snails, crabs, fish, slime molds, and vertebrate cells. What is the function of lectins *in vivo*? We already mentioned that lectins are important in sperm-egg recognition. The sea urchin sperm binding isolated by Vacquier is a lectin that recognition. The sea urchin sperm bindin isolated by Vacquier is a lectin that recognizes carbohydrate groups on the cell surface of eggs. Lectins isolated from slime molds by Rosen's and Barondes's groups may play a role in the cell recognition and adhesiveness of slime mold amoebae as they aggregate to form a multicellular slug. Lectins, therefore, are important mediators of the cell recognition aspects of morphogenesis.

What about the function of lectins in plants, which are a major source of lectins that have been studied to date? Lectins may function as plant antibodies, protecting the plant against microbial attack. Lectins can agglutinate harmful microbes and can inhibit the hydrolyzing enzymes used by fungi to attack the plant cell wall. Lectins may help plants store and transport sugars; they may also bind plant enzymes into multienzyme systems that function together in metabolic reactions. Lectins may control cell differentiation and mitosis, especially during seed germination. Lectins can induce animal cells, such as lymphocytes, to differentiate. The lectins bind and cross-link specific cell-surface receptors that contain carbohydrate. Such cross-linking sets into motion a series of reactions that lead to differentiation and cell division.

An important function of lectins in legumes such as clover

and beans is to bind nitrogen-fixing bacteria to the roots of the plant. Legumes possess nodules of the nitrogen-fixing bacterium *Rhizobium* on their roots. These bacteria process gaseous nitrogen into a form usable by the plants and thus are of major importance to the plants. Lectins are secreted on the root surfaces of such plants. Specific lectins have been shown to bind only the strains of *Rhizobium* found on the roots of the plant from which they were extracted. For example, soybean agglutinin, a lectin isolated from soybean plants, binds only strains of *Rhizobium* found on the roots of soybeans. It does not bind other *Rhizobium* strains that form nodules on the roots of other plants. Dazzo and his colleagues have done some elegant work on the function of lectins in plants. Dazzo's group isolated a lectin from clover roots. They named the lectin *trifoliin* because it specifically binds to sugars on the nitrogen-fixing bacterium *Rhizobium trifolli*, which adheres to clover root hairs. We can speculate that evolution produced a pair of organisms, clover and *Rhizobium trifolli*, that live together symbiotically, each deriving benefit from the association. The means by which they bind together probably evolved as a lectin-carbohydrate interaction at the surfaces of clover root hair and *Rhizobium* cells.

Before we move on to discuss how lectins are used to study the role of the cell surface in morphogenesis and malignancy, a brief description of the different lectins is in order. Lectins bind terminal sugars of the cell-surface oligasoccharide chains or groups of sugars in the chains. The following is a list of lectins and the sugar(s) they preferentially bind.

Concanavalin A (Con A), isolated from the jackbean, binds b-D-mannose, D-glucose, and ß-D-fructose groups.

What germ agglutinin binds *N*-acetyl-D-glucosamine-like residues.

Ricinus communis agglutinin, isolated from the castor bean, binds D-galactose- and *N*-acetyle-D-galactosamine-like groups.

Soybean agglutinin binds *N*-acetyl-D-galactosamine- and D-galactose-like residues.

Garden pea lectin binds D-mannose groups.

Lentil lectin binds D-mannose and *N*-acetyl-D-glucosamine-like groups.

Lotus lectin binds L-fucose groups.

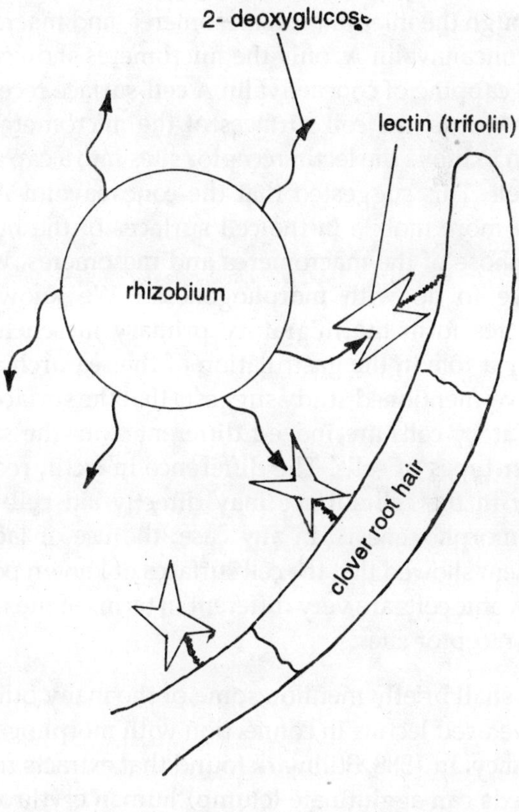

Fig. 8.16. Lectin-mediated binding of nitrogen-fixing bacterium (*Rhizobium*) to clover root hair.

These are just a few of the many lectins that have been isolated. We should stress that although these lectins preferentially bind the given sugars, the sugar specificity is seldom absolute. Most lectins will bind many different sugars, though perhaps with less tenacity than the preferred sugars. The sugar specificity of lectin binding, however, has been a key reason for its use in the study of the cell surface.

How have lectins been used to study the role of the cell surface in morphogenesis and malignancy? Oppenheimer's group found that sea urchin embryo micromeres and mesomere-macromeres had different reactions to the lectin concanavalin A. Although the micromeres, mesomeres, and macromeres all bound concanavalin A, only the micromeres showed a lectin-induced capping of concanavalin A cell-surface receptor sites. In other words, the cell surfaces of the micromeres allowed the lectin to move the lectin receptor sites into a cap at one end of the cell. This suggested that the concanavalin A receptor sites are more mobile in the cell surfaces of the micromeres than in those of the macromeres and mesomeres. What does this have to do with morphogenesis? We know that the micromeres form the migratory primary mesenchyme cells that play a role in the gastrulation of the sea urchin embryo. The above-mentioned study suggests that the surfaces of these premigratory cells are indeed different from the surfaces of the other types of cells. The difference in lectin receptor site mobility in the cell surface may directly aid cell migration during morphogenesis. In any case, the use of lacin in this experiment showed that the cell surfaces of known populations of embryonic cells are very different in terms of the mobility of specific receptor sites.

We shall briefly mention some of the many other studies that have used lectins in connection with morphogenesis and malignancy. In 1888, Stillmark found that extracts from castor been seeds can agglutinate (clump) human erythrocytes (red blood cells). Since then lectins have been extracted from the seeds of over 800 species of flowering plants, especially the legumes. Lectins have been found not only in seeds but in the leaves, roots, and bark of plants and also in extracts from many animal cell types. Numerous studies have shown that, in general, malignant tumor cells, transformed cells, and early embryo cells are agglutinated at rather low concentrations of many lectins, whereas normal adult cells are seldom agglutinated at these concentrations. Why are embryonic cells and tumor cells agglutinated when normal cells are not?

Studies using lectins labeled with the fluorescent dye ferritin or with a radioactive isotope have shown that, in general, most cells bind lectin. That is, normal adult cells, tumor cells, and embryonic cells have lectin receptor sites in their cell surfaces. It has also been shown, however, that cells that are agglutinated with lectins (tumor and embryo cells, for example) tend to have mobile lectin receptor sites in their cell surfaces. Thus capping and clustering of these sites will occur in the presence of lectins, because lectins are multivalent molecules that can cross-link several surface receptor sites. This clustering or capping of the lectin receptor sites may explain how lectins agglutinate cells. A buildup of lectin-bound receptor sites in specific regions of the cell surface may facilitate lectin-bound cell agglutination. It has also been proposed that early embryo cells and malignant cells have altered cytoskeletal elements (microtubules and microfilaments) that fail to restrict the movement of cell-surface lectin receptor sites. In normal adult cells, these cytoskeletal elements are attached to the inner membrane surface, restricting the movement of surface receptor sites. These suggestions are supported by the finding that drugs that disrupt cytoskeletal elements, such as colchicine, local anesthetics, and cytochalasin B, dramatically alter the mobility and distribution of cell-surface lectin receptor sites in a variety of cell types. Lectins, therefore, have been useful in identifying specific differences in the surfaces of tumor cells, normal cells, and embryonic cells. Other extensive studies with lectins have led to additional conclusions. Moscona's group showed that young chick embryo cells are agglutinated by concanavalin A, whereas older cells are not. Such studies were extended to other systems by several investigators. The findings in general suggest that young, motile embryonic cells are agglutinable with lectins, whereas older embryonic cells and those that do not move much are less so. When follow-up studies with labeled lectin were carried out, the results showed that agglutinable embryo cells display mobile surface lectin receptor sites like those in the sea urchin micromeres.

In summary, we can say that lectins have helped us identify cell-surface molecules containing sugar and have led to the conclusion that active cells in embryos and tumors possess mobile lectin receptor sites on their surfaces. It remains to be seen whether this characteristic plays a key role in facilitating the movements and interactions that occur during morphogenesis and the spread of cancer cells.

FILES AND WORMS

The Puzzle of Differentiation: Differential Gene Activation

Every cell of an embryo has exactly the same gene content, and genes control, directly or indirectly, all of the processes of development. How then can cells differentiate into the different body structures? One possibility is that as development proceeds, particular genes are lost from particular embryonic cells during mitosis. At the same time, other genes are lost from other cells. In that case, if gene loss could be controlled so that the same set of genes were always lost from the same cells at the same stage of development, an explanation for cellular differentiation would be possible. Gene loss, however, does not seem to lie at the bottom of the problem of how differentiation occurs, as the following two experiments illustrate. If a small piece of carrot is placed in a culture dish and then dissociated in to individual cells, one such cell, when properly stimulated, is fully capable of developing into a complete carrot plant. Similarly, if a nucleus is taken from an intestinal cell of a white tadpole and injected into an enucleated ovum of a green frog, this artificial, uniparental zygote can develop into a normal white frog. These experiments show clearly that, in two entirely different organisms, no genes required to direct the full course of development are lost during the process of cell differentiation; all genes are still present, ready to be activated in the right cells at the right times to assure normal development.

Polytene Chromosome Puffs

Directly observable evidence of gene activation and deactivation is provided by the giant polytene chromosomes of *Drosophila* and other Dipteran insects such as *Chironomus.* At regular times during larval development, "puffs"—local, fuzzy enlargements of the chromosome—appear at and later disappear from specific places along the chromosomes. A puff, which forms as a result of the loosening of the tightly coiled DNA around a gene, provides visible evidence that a gene at the location of the puff is in the process of synthesizing mRNA (that is, is turned on). When the puff disappears, we know that the DNA has returned to its tightly coiled condition and the gene is turned off. The reason that the DNA must uncoil (producing the visible puff) before the gene can be turned on is that RNA polymerase molecules needed for mRNA synthesis cannot bind to the gene and initiate synthesis when the DNA is tightly coiled. When mRNA synthesis is finished, the DNA becomes tightly coiled again and the puff disappears. The location of the puff and the timing of their appearance vary from tissue to tissue, because the particular genes that must be activated at a given time in one tissue are often different from those whose products are needed at that same time in some other tissue.

The Control of Puffing

Intact salivary glands can be removed from *Drosophila* larva and put in insect Ringer's solution, where they can survive in tissue culture for several hours. When chromosome preparations are made from these cultured glands at successive intervals, the specific puffs that are produced can be correlated with the age of the larva at the time of its dissection. In *Drosophila,* for example series of puffs arise at specific chromosome locations just before the onset of puparium formation. If a larva is

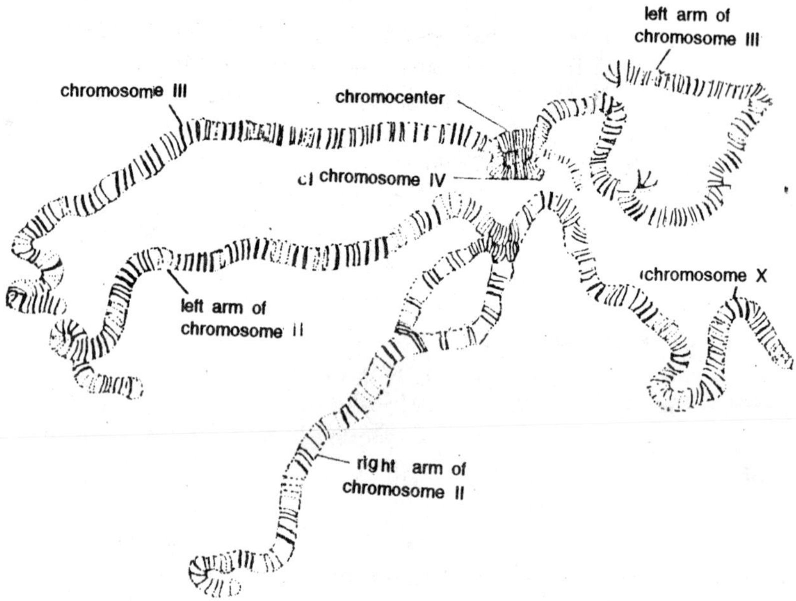

Fig. 8.17. Polytene chromosome.

dissected before that time, relatively few puffs will be seen, and the locations of most of these puffs will be different from those of the puffs that appear just prior to pupation. Some years ago attempts were made to find out whether any specific chemicals could artificially induce the set of pupation-related puffs in glands taken from larva too young to have begun to express them. In due course success was achieved: the insect hormone *ecdysone* was indeed able to induce the formation of those puffs. This finding was not altogether surprising, because ecdysone was known from previous studies to be needed for developing insects to undergo molts, including pupation. Further study showed that ecdysone induced only three or for "primary" puffs; the remainder were secondarily induced by products of the few primary puffs. Other work showed that another insect hormone, *juvenile hormone*, antagonizes the activity of ecdysone. In fact, adding juvenile hormone after the ecdysone-induced puffs have formed results in a recession of the puffs. If the appearance of a puff marks the turning on of a gene and its recession indicates that the gene has been turned off, as we believe to be the case, then we must conclude that these hormones, ecdysone and juvenile hormone, are intimately involved in the regulation of gene activity in Dipteran insects. Unfortunately, very few other organisms have polytene chromosomes in which the control of puffing can be studied so directly. Thus more indirect kinds of evidence relating to gene-control mechanisms must be sought in the case of most plants and animals.

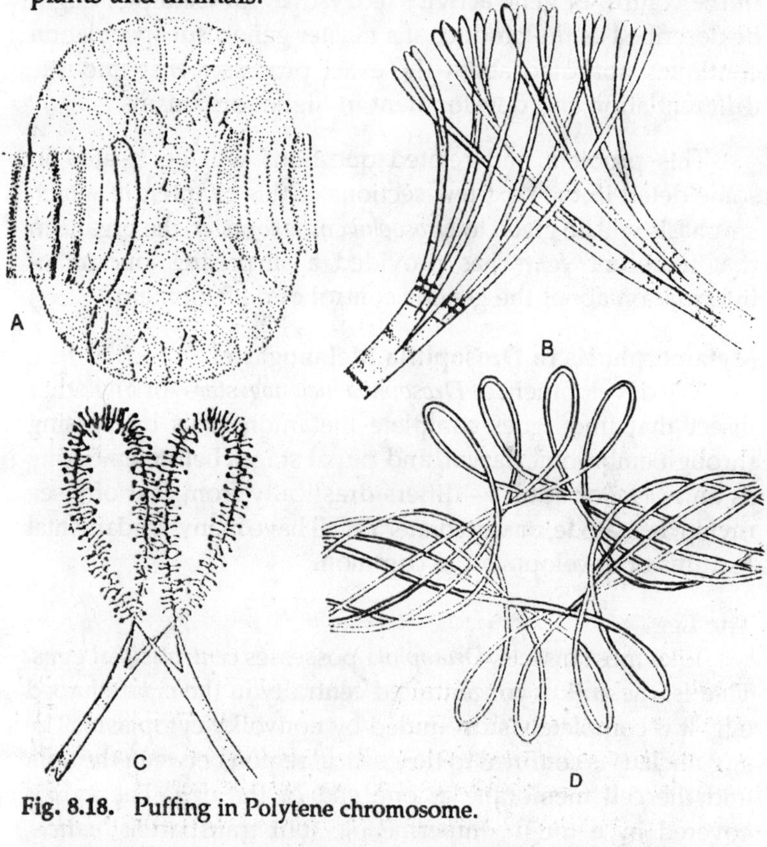

Fig. 8.18. Puffing in Polytene chromosome.

Genetic Control of Development in Higher Organisms

Much information has been gained from the analysis of *E. Coli* and other bacteria in which the operon system of gene organization provides for co-ordinate regulation of gene activity.

However, analysis of the genetic control of development is an order of magnitude more difficult in multicellular organisms than in unicellular bacteria. As operons do not seems to be present in *Drosophila* and other eukaryotic animals and plants, we must search for "master" genes that can turn other genes on an off, no matter where they may be located in the genome. We have some specific information about the control of development in eukaryotes, such as the implication of hormones in the control of gene activity (ecdysone, for example, might be described as the product of a master gene), but speculation continues unabated about the exact processes that underlie differentiation and development in higher organisms.

This problem and related questions will be treated in some detail in the next few sections of this chapter, in which emphasis will be given to *Drosophila melanogaster*, the organism that in recent years has provided a surprising amount of information about the genetic control of development.

Metamorphosis in Drosophila Melanogaster

The development of *Drosophila melanogaster*—or any other insect that undergoes complete metamorphosis by passing through embryonic, larval, and pupal stages before emerging as an *imago*, or adult — differs drastically from that of a sea urchin, nematode, or vertebrate. Yet all have many fundamental features of development in common.

The Egg

Like most insects, *Drosophila* possesses *centrolecithal* eggs. That is, the yolk is concentrated centrally in the cigar-shaped egg; it is completely surrounded by nonyolky cytoplasm. The egg nucleus is confined to the cortical region between the yolk and the cell membrane at one end of the egg. The egg is covered by a tough, impermeable (but transparent) *cuticle*, which cannot be penetrated by sperm. However, at the posterior end of the egg is a small conical *micropyle* containing a hole that allows sperm to pass through the cuticle and penetrate the egg.

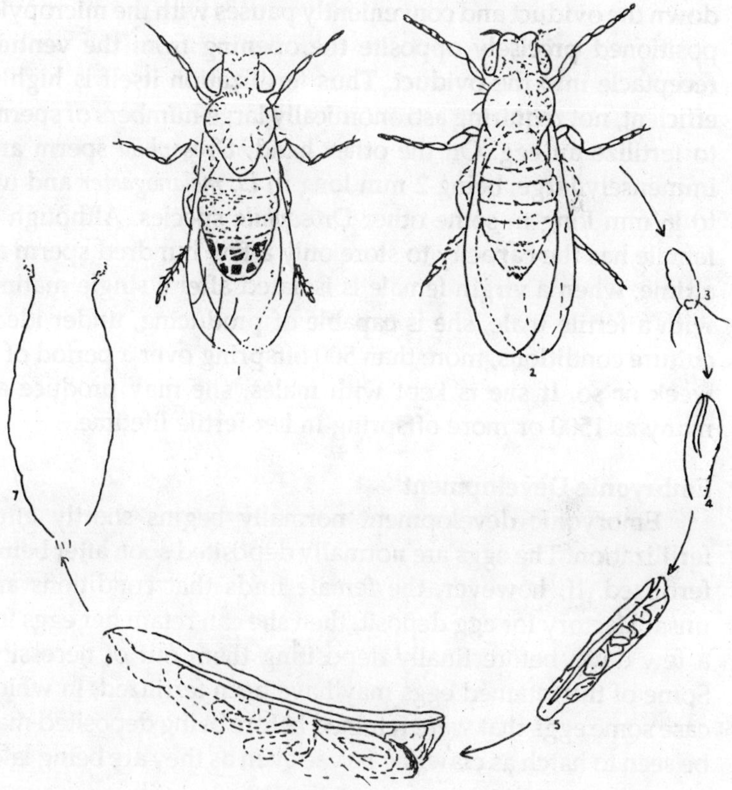

Fig. 8.19. Drosophila life cycle. (1) Adult male. (2) Adult female. (3) Egg. (4) Larva, first instar. (5) Larva, second instar. (6) Larva, third instar. (7) Pupa.

Fertilization

Fertilization is internal. Following mating, sperm from the male are deposited in one of the female's three sperm-storage organs: the larger *ventral receptacle* and two smaller *spermathecae*. Sperm can survive in these female organs for ten days or more, during which time a stored sperm is capable of entering an egg through its micropyle when the egg passes down the oviduct and conveniently pauses with the micropyle positioned precisely opposite the opening from the ventral receptacle into the oviduct. Thus fertilization itself is highly efficient, not requiring astronomically large numbers of sperm to fertilize an egg. On the other hand, *Drosophila* sperm are immensely large, being 2 mm long in *D. melanogaster* and up to 16 mm long in some other *Drosophila* species. Although a female has the capacity to store only a few hundred sperm at a time, when a virgin female is isolated after a single mating with a fertile male, she is capable of producing, under ideal culture conditions, more than 500 offspring over a period of a week or so. If she is kept with males, she may produce as many as 1500 or more offspring in her fertile lifetime.

Embryonic Development

Embryonic development normally begins shortly after fertilization. The eggs are normally deposited soon after being fertilized. If, however, the female finds that conditions are unsatisfactory for egg deposit, then she can retain her eggs for a few days, before finally depositing them out of necessity. Some of the retained eggs may have been fertilized, in which case some eggs that were retained before being deposited may be seen to hatch as crawling larvae even as they are being laid.

The earliest period of embryonic development within the egg cuticle begins after the sperm and egg pronuclei fuse in the clear cytoplasm of the cortical region of the embryo. An extremely rapid series of 12 or 13 mitoses occurs, giving rise within about two hours to a *synkaryon* of over 4000 nuclei, which migrate peripherally all around the central yolk mass to

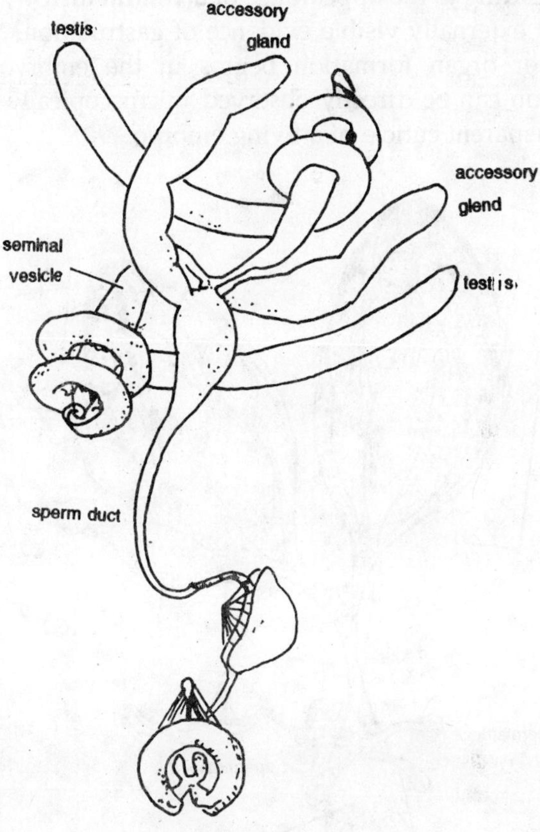

Fig. 8.20. Male *Drosophila* reproductive system.

form a monolayer of nuclei just under the egg surface. Only then do cell membranes appear that separate individual cells from one another. This signals the blastula stage. A cluster of *polar granules* is present at the posterior end of the egg cytoplasm. At the time of cellularization, the cells that come to enclose these granules, the *polar cells,* are predetermined to become gametes if they are later included in the gonads, which most are. Mutant females that produce defective polar granules, or none at all, are sterile. After about three hours of embryonic development at 25°C, a series of cellular movements takes place, leading to the appearance of a ventral furrow, which is the first externally visible evidence of gastrulation. Not long thereafter organ formation begins in the embryo. Organ formation can be directly observed microscopically through the transparent cuticle of a living embryo.

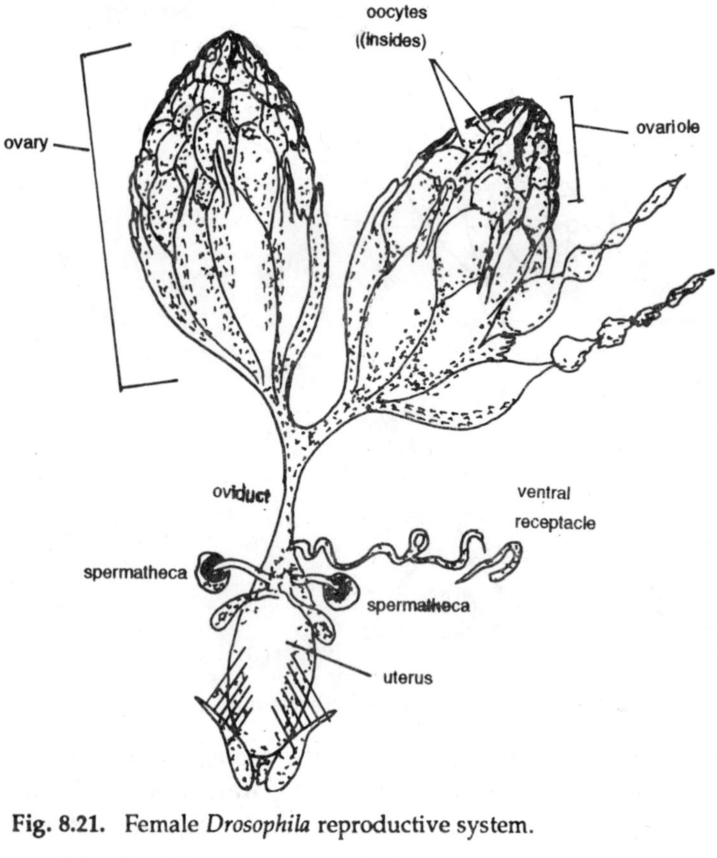

Fig. 8.21. Female *Drosophila* reproductive system.

Larval Development

Embryonic development continues until about 24 hours after the eggs are deposited, at which time the embryo hatches from the egg as a crawling, segmented larva. During the next 96 hours or so, the larva develops through three successive *instars*, or larval molts, crawling through the culture medium and eating voraciously all the while until a fully developed, third-instar larva crawls out of the culture medium, everts its *spiracles*, becomes inactive, and *"glues"* itself to a dry surface as a prepupa. At that time *puparium* formation begins, to be followed some hours later by another molt and the appearance of the true pupa. Larvae of the higher Diptera, of which *Drosophila* is one, are described as being headless: there are no signs of eyes, antennae, or mouth parts. On the other hand, the thoracic region can be distinguished from the abdomen, even in the very early larva. In a sense, adult structures such as swings, halteres, and legs, as well as eyes, antennae, and genitalia, are present in larvae at an early stage of development, but only in the form of undifferentiated *imaginal discs* (two for eyes, two for wings, six for legs, and so on). (Obviously, these discs are not "imaginary;" the term pertains to *imago*, or adult.) The discs will not begin to undergo metamorphosis and differentiate into their appropriate adult structures until the pupal stage. In contrast, most of the gut, the central nervous system, the Malpighian tubules (which function in excretion), and the gonads are present as organ systems that persist and grow directly through both the larval and pupal developmental periods.

Pupal Development

During the pupal period, purely larval structures such as the salivary glands, together with their polytene chromosomes, are broken down and disappear. At the same time, all of the adult structures arise, as cells of the various imaginal discs differentiate and grow. After a four-day pupal period, the imago emerges from its pupal case as a result of gene activity initiated by the hormone ecdysone (which also is required for the transformation of eggs into larvae and larvae into pupae). Mutants that cannot synthesize ecdysone will not develop very far. Some mutants with reduced ecdysone activity become larvae, but they eventually die as *"giant"* larvae, unable to pupate. It is interesting to note that because of its ability to antagonize the activity of ecdysone, juvenile hormone, when overproduced, also prevents development (or *"growing older"*)

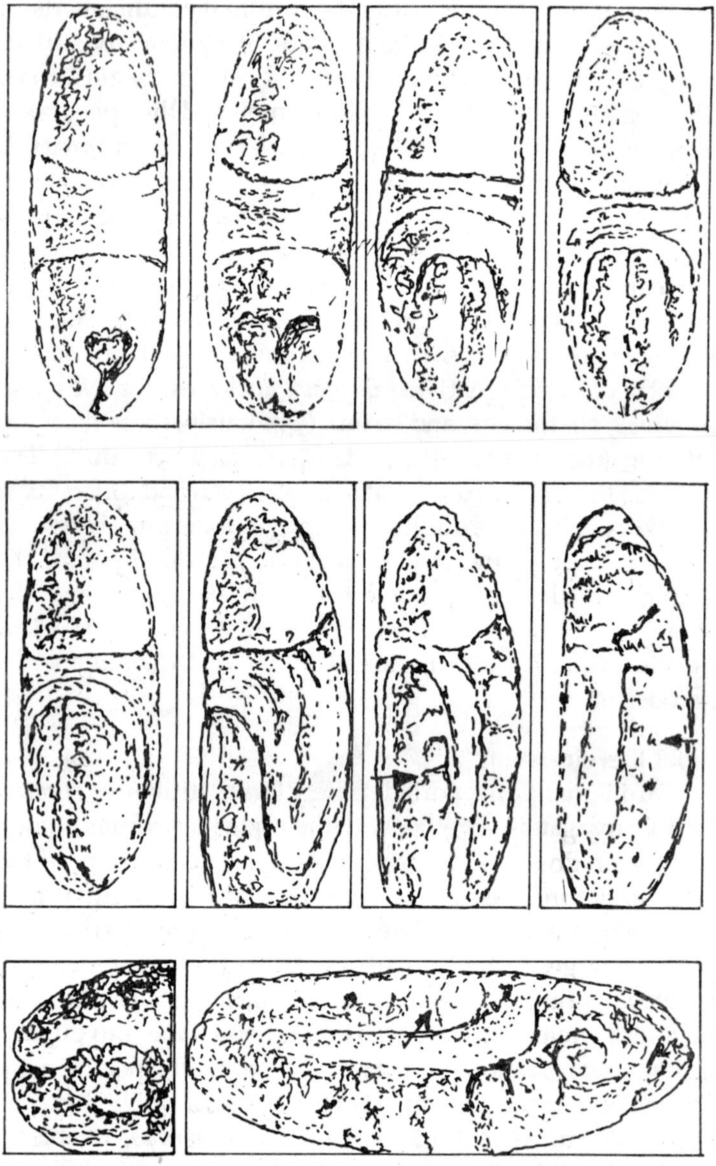

Fig. 8.22. Early development of *Drosophila*.

from proceeding beyond an early stage. Many plants produce chemical substances that process juvenile hormone-like activity, thereby protecting themselves against the ravages of parasitic insects that develop from eggs deposited in or on them.

Development of the Segmented Body

In many kinds of organisms, ranging from tapeworms to mammals, a segmented body plan underlies the greater part of the anatomy. In the most primitive organisms, a typical segment includes a full complement of virtually all of the structures required for living, including respiration, locomotion, sensation, excretion, and even reproduction. Except for a mouth in the first segment and a gut running all the way through, the organism, for all practical purposes, is formed by a series of virtually identical segments, giving an impression of development by Xerox. This condition is closely approached by the tapeworm. Most higher organisms, however, through the course of evolution, have diverged from that plan. Some, such as roundworms and echinoderms, show no segmentation; others have varied segmental expressions, with some segments fused together or retaining only one or two of the different organ systems. Instill higher forms, segmentation is seen not in every system, but only in a few, such as the nervous and circulatory systems. In fact, often the different segments no longer much resemble each other, so the underlying segmental pattern is not readily apparent. In the following material we shall examine segmentation in *Drosophila melanogaster*, which is commonly used in genetic studies. (*D. melanogaster*, though called a *"fruit fly,"* is a harmless insect, not the highly destructive Mediterranean fruit fly.)"

The Adult

Although the segmental body plan is first discernible about halfway through the period of embryonic development, not until the adult stage does the segmental body plan become obvious. The adult *Drosophila* has an abdomen with seven easily counted, similar segments (A final posterior segments

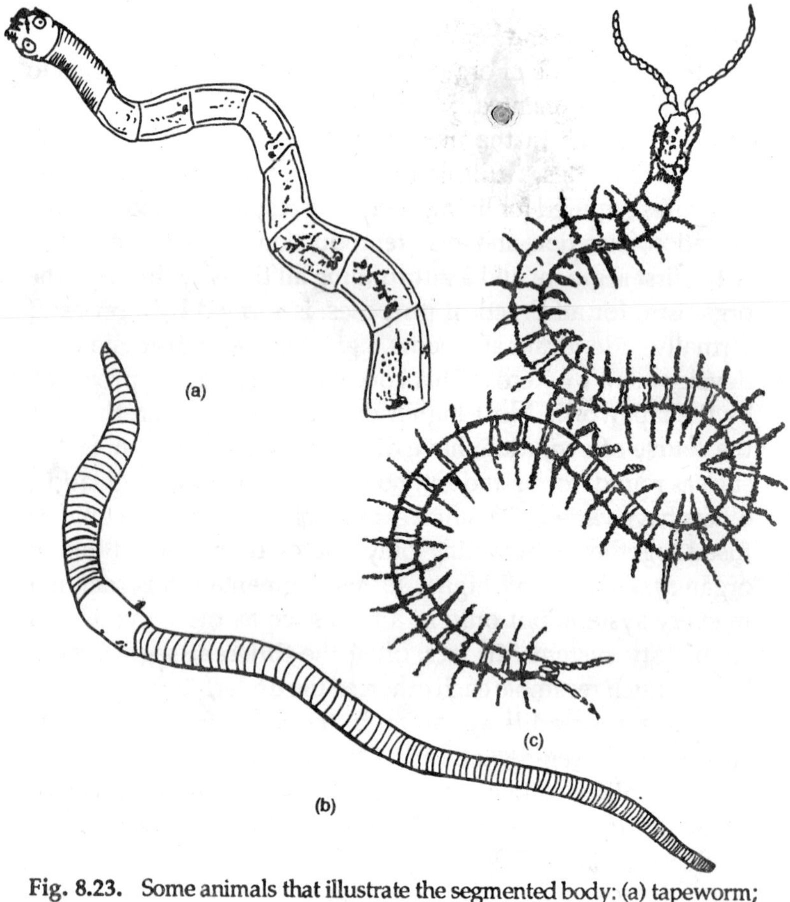

Fig. 8.23. Some animals that illustrate the segmented body: (a) tapeworm; (b) earthworm; (c) centipede.

obscured and distorted by the genitalia, and it probably is a product of the fusion of three terminal abdominal segments.) Three pairs of legs reflect the much less obvious fact that the thorax contains three segments: the *prothorax*, the *mesothorax*, and the *metathorax*, each of which bears a pair of legs, or ventral appendages. Although primitive Diptera also had a pair of dorsal appendages, or wings, on each thorcic segment, no modern insect develops prothoracic wings. Almost all insects have both mesothoracic and metathoracic wings, but the Diptera (true flies) have lost their metathoracic wings, which leaves only the two mesothoracic wings for flying. Dipteran metathoracic wings have been reduced to small knob-like *halteres*, the so-called balancers.

The head shows no obvious segmentation, but the antennae, the eyes, and the mouthparts are, developmentally speaking, appendages. The suggests that the head is a completely fused product of what, in evolution, was once at least three, but more likely five, or even more, head segments.

Genetic Control of Segmentation in Drosophila

The genetic control of the segmental body plan of *D. melanogaster* has been most thoroughly analyzed by E. B. Lewis and his colleagues at Caltech. They have identified a cluster of ten or more genes lined up near the center of the left arm of chromosome 3 at salivary chromosome position 89E. This group of genes, now known as the *bithorax complex* (BX-C for short), controls the differentiation of the thoracic and abdominal segments. Lewis has shown that a gene at one end of the BX-C becomes activated first; it is involved in the control of development of the prothoracic segment. The next gene in line is then activated to control development of the mesothoracic segment; the third, the metathorax; the next, the first abdominal segment; and so on down to the last gene and the last abdominal segment. Through observations of the phenotypic effects of mutations of these genes and molecular analyses of most of them, Lewis and his collaborators, particularly W. Bender, have provided remarkable insights into the genetic control of segmentation.

Historically, the first mutation in a BX-C gene was called *bithorax (bx)*, because at first glance the thorax seemed to be partly duplicated and the halteres were enlarged into rudimentary wing-like structures. In time, other BX-C mutations were discovered that produced similar, but different, modifications of the thoracic and abdominal phenotypes. For example, *post-bithorax (pbx)* is much like *bx* in a appearance, but it can be separated from *bx* by crossing over. The homozygous double mutation, *pbx bx/pbx bx*, has a strong, four-winged phenotype. At least 12, and by now surely more, different mutations of BX-C genes have been found and analyzed by Lewis and others. As we said above, BX-C genes control the development and differentiation of the thoracic and abdominal segments. At first it appeared that the most common effect of a mutation in a BX-C gene was the transformation of a structure characteristic of one segment into an inappropriate structure characteristic of a different segment, such as that of a haltere (a metathoracic structure) into a wing (a mesothoracic structure). Closer study showed, however, that the *bx* transformation is not simply a conversion of haltere to wing, but rather a partial or complete transformation of about half the metathorax into mesothorax, after which the transformed mesothorax, like the normal mesothorax, develops a wing. Changed of this sort are thought to result from a mutational inactivation of the *bx* locus. A different mutation is able to convert the first abdominal segment, which has no appendages, into a metathorax, which has halteres. Inactivation (or loss) of one of the other BX-C genes generally transforms the segment it controls into the next anterior one. There is some evidence to suggest that duplication of a BX-C gene transforms the segment it controls into the next posterior one. If the entire BX-C is deleted, however, all of the segments develop as a mesothorax, which appears to be the basic segment from which the others were derived. Mutations like these, which replace a given structure with an inappropriate one, such as a leg with an antenna, are now called *homeotic mutations*.

Transmission of Developmental Signals

Gerald Rubin and his colleagues successfully cloned a gene needed for the proper development of one of eight specific

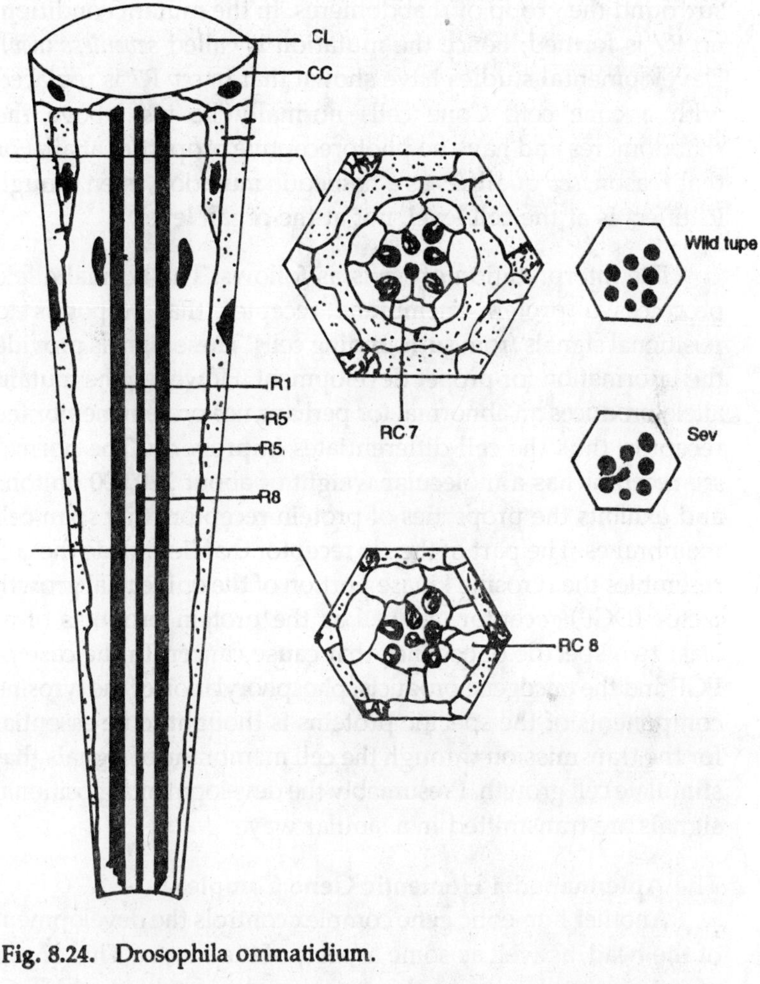

Fig. 8.24. Drosophila ommatidium.

photoreceptor cells present in each of the 750 to 800 *ommatidia* (facets) of the fly compound eye. Normally each of these eight cells forms a structure called a *rhabdomere,* which contains the visual pigments. Six of the eight rhabdomeres (R1-R6) form a trapezoidal arrangement around the other two (R7-R8), which are shorter and are stacked on top of each other. A cross section of an ommatidium cannot cut through both R7 and R8 simultaneously. The cell bodies of the photoreceptor cells surround the group of rhabdomeres. In the mutant condition, no R7 is formed; hence the mutation is called *sevenless (sev)*. Developmental studies have shown that in *sev* R7 is replaced with a cone cell. Cone cells normally lie just above the rhabdomeres and have no photoreceptive properties at all. For that reason, *sev* qualifies as a homeotic mutation, even though its effect is at the cell level, not at the organ level.

The interpretation of *sev* is as follows. The normal allele produces a protein membrane receptor that responds to positional signals from surrounding cells. These signals provide the information for proper development. However, the mutant allele produces an abnormal (or perhaps no) protein membrane receptor; thus the cell differentiates improperly. The normal *sev* receptor has a molecular weight of about 220,000 daltons and exhibits the properties of protein receptors that span cell membranes. The part of the *sev* receptor that lies inside the cell resembles the tyrosine kinase portion of the epidermal growth factor (DGF) receptor, as well as the protein products of at least two specific oncogenes that cause cancer. In the case of EGF and the oncogene products, phosphorylation of the tyrosine components of the specific proteins is thought to be essential for the transmission through the cell membrane of signals that stimulate cell growth. Presumably the developmental positional signals are transmitted in a similar way.

The Antennapedia Homeotic Gene Complex

Another homeotic gene complex controls the development of the head, as well as some aspects of the thorax. This group of genes, now known as the *Antennapedia complex*, or ANT-C, has been thoroughly studied by T. Kaufman at the University of Indiana. It was first recognized through the dominant mutation called *Antennapedia*, which produces an identifiable leg on the head in place of the antenna. Whether by coincidence or not, the ANT-C is located near the BX-C at position 84A on the left arm of chromosome 3. Several ANT-C mutations are known including one called *proboscipedia (Pb)*, in which a leg replaces the proboscis. Another ANT-C mutation has provided great insight into the function of genes that control segmentation. In the next section we shall discuss the significance of this mutation, which has been given the name *fushi tarazu (ftz)*, a Japanese word meaning *"not enough segments"*; *ftz* embryos have only half the normal number of segments.

The Homeobox and its Potential Significance

Among ANT-C mutations, *fushi tarazu* is most interesting because it does not give rise to segmental transformation; rather, it appears to control the segmentation process itself. In addition, a thorough molecular analysis of *ftz* led to the identification of a special nucleotide sequence, long enough to encode 60 amino acids, that is exactly the same as (or almost identical to) sequences found not only in many of the homeotic BX-C and ANT-C genes of *Drosophila* but also in genes of several other organisms ranging from yeast to humans. Not all of these genes give rise to homeotic mutants; several, like *ftz*, appear to play a regulatory role during development. Because the significance of this DNA sequence was first recognized in a gene of the ANT-C homeotic group and because a majority of the genes known to have this particular nucleotide sequence can produce homeotic mutants, the sequence is now called the homeobox. The sequence of the 60 amino acids encoded by the homeobox is now known. The question is, what does this small protein do? Work with yeast has provided some insight. To a significant degree, the protein produced by the homeobox sequence is homologous to the yeast mating-type proteins MAT a1 and MAT α2. Because the mating-type genes control fundamental aspects of cell differentiation in yeast, they are considered to be master control genes. The protein is believed to be able to bind specifically to DNA so as to turn certain genes on or off. Thus in all likelihood the homeobox protein in organisms ranging from fungi to primates plays a major role in controlling gene activity during development. If so, the homeobox sequence of nucleotides must have been highly conserved throughout a long period of organic evolution.

As provocative as this concept may seem, it is too early to be sure of it; much more research is needed to reach a clear-cut decision about the fundamental significance of the homeobox. For the moment, however, it is quite inviting to speculate that the homeobox protein can regulate the activity of the appropriate genes to ensure not only that the proper number of segments will develop in their proper sequence in *Drosophila,* but also that the various segments will become equipped with the appendages normally associated with that segment. When a homeobox-containing gene is in the mutant condition, the genes it controls do not turn on or off at the correct times; as a result, some structure such as a segment does not differentiate correctly.

Other Genes Regulating Early Development in Drosophila

Several genes that are not known to contain the homeobox sequence have an effect on early development in *Drosophila.* One, called *dorsal,* is involved in establishing dorsal-ventral polarity. Females homozygous for an extreme allele of *dorsal* produce dorsalized embryos that lack all lateral and ventral structures; however, dorsal structures are found throughout the body. Various other nonallelic genes besides *dorsal* have also been found to affect dorsal-ventral polarity. One of these other genes is *toll.* In the homozygous recessive condition, *toll* dorsalizes the embryo, but if cytoplasm from an embryo with a normal genotype is carefully injected into a *toll* embryo, the embryo can be "rescued" from the dorsalizing effects of the *toll* gene. The ability to rescue organisms from the effects of these genes appears to reside in a fraction of the poly-A mRNA that is stored in the egg during oogenesis. The *dorsal* gene has been cloned; it codes for a 2.8-kilobase length of mRNA that can be found both in the ovaries and in embryos up to 2.5 hours old. A remarkable number of genes are now known to control one or another specific aspect of *Drosophila* development, although the protein products that they encode and the mechanisms by which these products produce their effects remain largely unknown. As we shall describe next, a nematode worm has provided the material needed for progress in this exciting area of study.

Caenorhabditis Elegans: The Embryologist's Choice

Let us begin consideration of the small, free-living nematode worm with the long name *Caenorhabditis elegans* (*C. elegans* for short). Some years ago S. Brenner, a well-known British molecular geneticist, recognized the potential of *C. elegans* for the analysis of the genetic control of development. Overnight he became an embryologist and set up the first laboratory for the study of the developmental genetics of this free-living worm, so small that the adult is only one millimeter in length and has less than 1000 cells altogether.

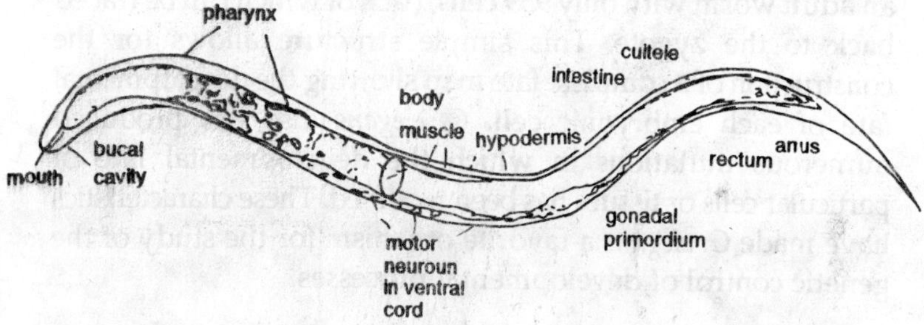

Fig. 8.25. Larva (L1) of *C. elegans*. The nematode worm at this stage is little more than a tube from mouth to anus. Every cell is in exactly the same position in every normal individual.

Because *C. elegans* has so few cells, it is possible to map and follow all of the cell lineages throughout the embryonic period of development. In other words, cells making up a give structure can be traced back to the single cell from which they all originated in the early embryo. This kind of information is vital to gaining insights into the mechanisms involved in development. Moreover, a very large number of mutations are known to affect specific facets of development in *C. elegans*. These features have persuaded many embryologists to join Brenner in the study of the development of *C. elegans* A huge literature devoted to the genetics and development of *C. elegans* has now accumulated, far out of proportion to the size of the worm.

The Life Cycle

C. elegans passes through its short life in the fast lane. After only six hours of embryonic development, its full schedule of cleavage divisions is finished. Eight hours later, the embryo hatches as an active larva. In other day and a half, the larva becomes an adult hermaphroditic worm that can produce about 250 progeny in the next four days or so. When it first emerges from the egg, the larva is composed of precisely 558 cells. After passing through four larval molts, the larva becomes an adult worm with only 959 cells, each of which can be traced back to the zygote. This simple structure allows for the construction of a complete fate map showing the developmental fate of each embryonic cell. *C. elegans* also has produced numerous mutations in which the developmental fate of particular cells or tissues has been modified. These characteristics have made *C. elegans* a favorite organism for the study of the genetic control of developmental processes.

Experimental Studies on Development

Many of the adult structures of *C. elegans* have been traced back to one or another single cell of the early embryo. By using a microlaser beam, an experimental embryologist can kill one such progenitor cell. Then, if the related adult structure fails to develop, the embryologist knows that the developmental potentialities of the destroyed cell cannot be assumed by other cells of the embryo (that is, the structure is completely determined early in development). On the other hand, if the structure does appear as it would normally, the embryologist can conclude that determination of the structure is liable (that is, other cells can assume the role óf the destroyed cell). Many experiments of this sort have made it clear that the majority of structures in *C. elegans* are fully determined very early in development, though some are not.

Specific Genes That Control Development in C. elegans

The search for specific genes that control important developmental mechanisms has led to some interesting discoveries. In certain mutants, some cells continue dividing rather than differentiating into functional cells when they normally should. This result suggests that particular genes control the precise number of cell divisions that a given cell must undergo before a specific adult structure will differentiate. Other genes have been identified that seem to have a major role in controlling development. For example, *lineage (lin)* genes play a key role in determining whether a cell will have one developmental fate or another. A specific example of how a *lin* gene works is provided by *lin-14*, which is active in determining the cuticle. R. Horwitz and his colleagues at MIT have shown that an active *lin-14* allele assures that so-called seam cells secrete larval cuticle. When *lin-14* is turned off, the seam cells produce adult cuticle. Thus development of a specific structure is dependent on whether a specific gene is turned on or off. Think of a train coming to a switch in the track. Which way the train will go thereafter depends entirely on whether the switch is turned to the right or to the left. The train does not control its own destiny (or destination); the switch does.

What happens when *lin-14* undergo mutation? If the mutant allele cannot turn off in the proper cells at the proper time, those cells retain larval cuticle, even through the remainder of the worm has adult cuticle. In the case of other *lin-14* mutants, the gene is licked off and cannot turn on. Cells in this situation never express larval cuticle. Another *lin* gene, *lin-5*, provides a different example of the genetic control of fundamental developmental mechanisms. When *lin-5* is in the mutant condition, no postembryonic cell division occurs, yet chromosome replication continues much as in the larval tissues of *Drosophila*. (Unfortunately, the replicated chromatids do not stay together so as to produce analyzable polytene chromosomes.) Note that *lin-5* affects only larval cell division, not embryonic cell division. Further, the consequences of mutation are restricted to cytokinesis, not mitosis. This demonstrates how finely tuned genetic control is, even in simple organisms. *C. elegans* may well turn out to be the organism of choice in the continuing search for an understanding of development. Many genes, not only in *C. elegans*, are known to play major roles in controlling the structure of organisms. Such genes may also play a fundamental role in evolution of the species because their mutants fail to survive the change in timing of various basic developmental processes. Often the consequences are lethal, but one some occasions such mutational changes might lead to a novel adaptive improvement that could start a new line of descent.

9

The Mechanism of Transdifferentiation in Animals

Transdifferentiation may be generally defined as the change of one recognizable cell type to another different cell type. The term was first used by Selman and Kafatos (1974) to denote the change of the cuticular cells of the moth larval silk gland to those producing HCO_3 during metamorphosis/development and has since been used in many different contexts. So as not to produce a welter of semantics to replace the term already in use. Transdifferentiation can further be categorized into primary (or true) transdifferentiation would include the cell-type conversion or cell metaplasia that is so well documented to occur in some amphibian eye tissues *in vitro* and in amphibian (newt) eye tissues in *situ*. This level is characterized by verifiably postmitotic cells, terminally differentiated and producing a specific cell product, transforming into a completely different cell type with differing cell product(s). Secondary transdifferentiation is marked by the conversion of those cells or tissues not definitely demonstrable as terminally differentiated, i.e., from an embryonic or possible stem-cell source. Also included is the concept of transdetermination (Hadorn, 1965), in which certain groups of cells in *Drosophila* occasionally become determined or committed to a developmental fate different from that expected. Tertiary transdifferentiation would encompass other purported/reported changes of tissue types, e.g., that of muscle to cartilage, and of striated to smooth muscle in Anthomedusa as reported by Schmid and Alder (1984) and Weber *et al.* (1987). The well-known plasticity of plant tissues, especially *in vitro*, is the rule, rather than the exception, and as a topic of transdifferentiation is beyond the scope of this chapter.

These three levels of transdifferentiation are reviewed and the relevant experiments critically examined. Because of greater relevance to problems of cell differentiation, with the least equivocal cell source, most of the discussion centers upon primary (true) transdifferentiation. Possible mechanisms permitting this reprogramming of differentiated cells are explored, with emphasis on the changes in gene expression that accompany, or are causal to, this phenomenon.

Primary Transdifferentiation

Ocular tissue, for unknown reasons, is the site of most documented cases of transdifferentiation. *In vitro*, this capability is shared by neural retina cells, pigmented retina cells, and pigmented epithelia of other eye tissues, all derived from adult newt, and embryonic/fetal tissue from other vertebrates. Primary transdifferentiation, involving pigmented iris epithelial cells *in vitro*, has been documented, as well as *in vivo/in situ* primary transdifferentiation. The latter is restricted in adult vertebrates to the transdifferentiation of (1) pigmented epithelium of the dorsal iris to lens, after removal of the eye lens; and (2) pigmented epithelia of the retina and ciliary body epithelia to neural retina, after removal of neural retina.

Recent research has centered chiefly on the regeneration of the eye lens after lentectomy. Because of its primacy in the field of transdifferentiation research, some background follows. This phenomenon, in which complete removal of the lens is followed by formation of a new lens at the dorsal pupillary margin of the iris epithelium, was first reported by Colucci (1891) and later by Wolff (1895), for whom this (Wolffian) regeneration is named. The capacity for lens regeneration from iris is restricted to a small number of genera, all urodeles, e.g., *Triturus, Diemyctilus, Cynopus*, and *Notophthalmus* (synonymous genera); also in *Taricha granulosus, Pleurodeles waltlii, Typhlotriton spelaeus, Eurycea lucifuga, Eurycea bislineata, Salamandra s. salamandra,* and *Salamandra perspicillata.* Lens regeneration has never been confirmed to occur in birds or mammals, despite many attempts to demonstrate this phenomenon in higher organisms.

The classic idea that the source of the regenerated lens is the dorsal iris epithelial cells has been confirmed experimentally. Organ culture of explanted pure populations of dorsal iris epithelial cells has demonstrated that these cells are transformed into lens cells now possessing the lens-specific structural proteins, the crystallins. Immunofluorescence of antibodies, specific for total crystallins and γ-crystallins, applied to sections through such cultures were positive, confirming longstanding microscopic evidence.

The morphogenesis of the regenerating and regenerated lens rudiment is superficially similar to that of the embryonic lens rudiment in normal lens development, despite the fact that it is produced by adult tissue cells. Some differences, however, should be noted between embryonic and regenerative formation of the lens. In normal development, a single layer of head ectoderm gives rise to the lens, whereas in regeneration, the double-layered iris, derived from neural ectoderm, is the source of the lens. Also, in normal development, the lens rudiment separates early from surrounding tissue and becomes an independent system, in the sense of no further cell contribution by other tissues, whereas in regeneration, the lens rudiment remains connected with the iris epithelium via a "stalk" and continues to receive cells from it until late in lens development. It has been well documented for several decades that lens regeneration from the dorsal iris *in vivo* is dependent on a stimulus coming from the neural retina. Despite much research, such a "factor(s)" has not been identified or characterized directly. Iris epithelial cells in cell culture, however, can be found to transform into lens cells without the presence of retina; the effect of neural retina thus appears to be permissive, rather than instructive, and its presence is not an absolute condition for transformation of iris epithelial cells into lens cells. Yamada and Beauchamp (1978) suggest that the "factor" involved with lens regeneration *in situ* acts indirectly, to preferentially shorten the cell cycle time in the dorsal iris epithelial cells *in vivo* as compared with ventral iris, permitting the dorsal iris epithelial cells to go through the requisite number of cell cycles for conversion into lens cells before mitosis ceases once again in the iris epithelium.

Concerning the events of lens regeneration per se, the first cellular alterations after lentectomy, in condensation of chromatin and in ultrastructure of the nucleoli of iris epithelial cells, can first be recognized on day 2 after lentectomy [i.e., lens regeneration stages]. The process continues to days 4-5, with the condensed chromatin characteristic of these cells progressively decreasing and nuclear volume increasing and nucleoli becoming larger.

These structural alterations of nuclei and nucleoli are associated with enhanced incorporation of labeled precursors into ribosomal and heterogeneous nuclear RNA. These events are followed immediately by replication of DNA, starting on days 3-4.

Although the iris epithelium of adult newts is normally a non-dividing tissue, mitotic activity can be demonstrated in the tissue from day 4 postlentectomy, soon after the start of DNA replication. A high level of mitotic activity and DNA replication can be seen throughout the iris epithelium from day 5 to approximately day 20. When iris epithelial cells are in the cell cycle, the number of melanosomes per cell decreases. During the phase of depigmentation, the volume of cytoplasm increases, extensive intercellular space appears, and cytoplasmic processes are produced in iris epithelial cells. Melanosomes are extruded via exocytosis as a membrane-bound complex or individually. Macrophages invade the iris epithelium and take up the discharged melanosomes. Depigmentation is most rapid at the dorsal margin of the epithelium. The subsequent increase in non-pigmented iris epithelial cells is due to replication of non-pigmented cells as well as completion of depigmentation in other cells. Only those iris epithelial cells that become completely depigmented participate in formation of the lens rudiment (vesicle, i.e., with a central cavity), which appears around days 14-16 at the mid-dorsal iris margin. Those proliferating iris epithelial cells located outside the dorsal marginal zone only partially depigment, withdraw from the cell cycle, and resynthesize melanosomes. With iris epithelial cells that have not replicated, they reconstitute the iris epithelium.

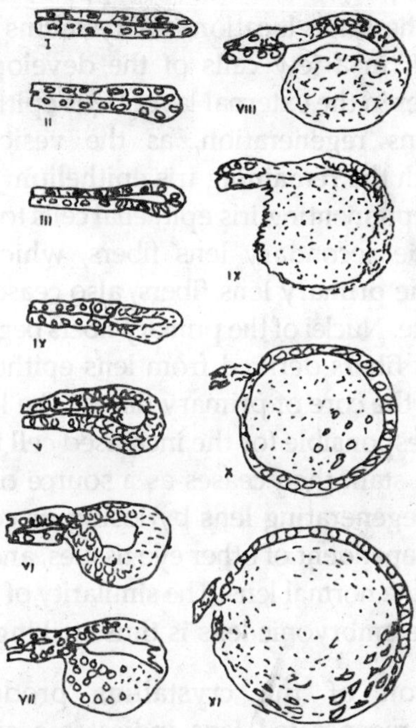

Fig. 9.1. Diagrams showing morphological stages of wolffian lens regeneration (from dorsal iris) in adult *Notophthalmus viridescens*. Sections are through the mid-dorsal pupillary margin of the iris, oriented perpendicular to the main body axis. The cornea is above, the retina below, and dorsal is to the left. The pigmented iris cells are indicated by black cytoplasm, and depigmented regenerate cells by white cytoplasm. Incomplete depigmentation is shown as dotted areas. Iris stroma cells and ameboid cells are not pictured.

In the internal wall of the newly formed lens vesicle, replication of cells soon stops, and they enter the terminal, non-dividing phase. The internal wall thickness by elongation of individual cells, now called primary fibers. The begin to synthesize crystallins. The external wall of the lens vesicle expands and thins, forming the lens epithelium. In contrast to lens fibers, the lens epithelium retains its proliferative capacity. The ontogeny and localization of specific crystallin classes in lens regeneration have now been determined. β-Crystallins are first detectable in the thickening internal layer of the stage V lens (vesicle); γ-crystallins also first appear, albeit erratically, at stage V in the same location, α-Crystallins do not appear until stage VII, in a few cells of the developing lens fiber region, and later in the external layer/lens epithelium. During subsequent lens regeneration, as the vesicle grows, the connection with the remaining iris epithelium (stalk) forms a pathway for depigmenting iris epithelial cells to join the vesicle epithelium. The secondary lens fibers, which form to the periphery of the primary lens fibers, also cease to divide and begin to elongate. Nuclei of the primary fibers begin to disappear, and secondary fibers derived from lens epithelium continue to be added to the core of primary fibers. The lens epithelium is thus solely responsible for the increased cell numbers in the fiber area. The stalk then ceases as a source of iris epithelial cells into the regenerating lens by losing its connection. The lens is now independent of other eye tissues, and subsequently grows as does the normal lens. The similarity of developmental events with the embryonic lens is thus striking.

Examination of the crystallins produced by the regenerating/ regenerated lens indicates a similar, yet not identical, profile. The ontogeny and localization of the crystallins in regenerating lens differs from that seen in the normal development of the newt lens. Chromatographic and isoelectric focusing analysis of fully regenerated lenses (6 months to 1 year postlentectomy) demonstrates a decrease in high-molecular-weight components (α-and β_{HMW}-crystllins) and an increase in low-molecular weight γ-crystallins). This tendency becomes more marked upon successive or serial lentectomies—up to six times—as shown by thin-layer isoelectric focusing and sensitive high-performance liquid chromatography (HPLC). These results suggest a reactivation of the dormant crystallin genes in the newt dorsal iris (pigmented) epithelium, with varying gene expression. Implicit in the relentectomies described is the interesting developmental question of why some dorsal iris epithelial cells will not participate in the formation of an initially regenerated lens, yet upon a successive lentectomy will contribute to another, new lens regenerate—certainly diametric differentiative decisions.

Research on the molecular events of *in situ* lens regeneration in the newt has been sparse, probably because of logistical problems due to the small amounts of available tissue present. In fact, no molecular biology/nucleic acids studies have been reported since before the advent of recombinant DNA technology. The last reported in an analysis of DNA per se from regenerating newt irises (up to day 7), a change in DNA configuration from duplex to single-stranded or nicked, and an increase in template activity. Collins himself notes the difficulty in interpreting these experiments. Earlier reports on ribosomal RNA enhancement in regenerating newt iris constitute the meager background on newt lens regeneration in this area. The prospects for the use of recombinant techniques in this system are explored at the end of this chapter.

Assuredly, *in vivo* lens regeneration in the newt has the major relevance to problems of health, i.e., replacement of lenses in mammals caused by cataract, as well as to the control of differentiation. The system, however, is complex, and several laboratories have therefore examined transdifferentiation in vertebrate ocular tissues *in vitro*. This has enabled the use of modern research techniques, including recombinant DNA, and, albeit with some provisos discussed below has provided us with a clearer picture of crystallin gene expression in the transdifferentiative process in cultured cells. This approach began during the 1970s with the tissue, primary, and clonal culture of newt dorsal iris epithelial cells for the first time. The conversion of these cells into lens cells (lentoid bodies) involved depigmentation as *in situ*; however, several important differences became apparent. Unlike the *in situ* situation, in which only dorsal iris epithelial cells (from the dorsal pupillary margin) participate in lens regeneration, cells derived from both dorsal and ventral iris will transdifferentiate into lens cells with comparable ability. In addition, the time interval needed for appearance of lens phenotype is almost twice as long as that needed *in situ* (28 versus 15 days). Also, a factor elicited by the neural retina has long been known to be necessary for lens regeneration *in situ* iris epithelial cells *in vitro* have no such requirement.

Yamada (1977, 1982) used these and other observations concerning cell cycle times in newt lens regeneration to hypothesize that the number of cell cycles undergone is a primary determinant of lens regenerative ability in iris epithelial cells. *In situ*, the dorsal iris epithelium undergoes $\geq$ 6 cell cycles during the same period that the ventral iris cells traverse $\leq$ 4 cell cycles. At the end of this period, lens phenotype (detection of crystallins) appears in the dorsal iris, but the ventral iris cells stop dividing and repigment. *In vitro*, there is no such block to ventral iris cells, and both dorsal and ventral iris epithelial cells have similar cell cycle times, both can undergo at least 6 cell cycles, and thus ventral iris cells are capable of transdifferentiating into lens cells. The neural retina *in vivo* supplies a mitogenic factor that enables the population of dorsal iris to progress through at least 6 cell cycles; *in vitro*, both populations can traverse the critical number of cell cycles, obviating the need for the neural retina. Crigoryan *et al.* (1987) recently confirmed this putative relationship between cell cycle number (6-7) and ability for transdifferentiation, with newt ventral iris in organ culture with a mitogenic factor. Although support for this hypothesis is largely circumstantial, it provides a framework for discussion and future research heretofore unavailable.

The retina of the adult newt closely resembles that of birds and mammals, with a single outer layer of pigmented epithelium closely adherent to an inner multilayer neural or sensory retina. If these layers are merely separated or the neural retina is surgically removed or caused to degenerate the pigmented epithelium of the retina (pigmented retina) will undero *in situ* transdifferentiation into neural retina. Although not as extensively investigated as transdifferentiation of iris epithelial cells into lens in the adult newt, there is good autoradiographic and functional evidence for this phenomenon. Restricted in adults to newts (although negative in other adult urodeles and anurans, some evidence may exist for tadpoles), the source of the complete new neural retina has been found by autoradiography to be not only from centrally located pigmented retinal epithelial cells but also partially from the *"anterior complex"* (inner epithelial layer at the junction of the *pars ciliaris retina* and *ora serrate*). The latter component apparently contributes to the peripheral part of the regenerated neural retina. Analogous to regeneration of lens from iris epithelial cells, the pigmented retinal epithelial cells will, during the course of transdifferentiation, depigment, extrude melanosomes, and re-enter the cell cycle (from which they had withdrawn), with an outer layer then redifferentiating into pigmented epithelium and an inner layer becoming increasingly stratified and histologically differentiated as neural retina. This is reminiscent of the divergent fate of the dorsal iris epithelial cells upon lentectomy. Also strikingly similar is the reported acceleration of the cell cycle in the regenerating neural retina compared with the pigmented epithelium, creating two populations of cells with differing proliferative capabilities. After several months, the resulting neural retina exhibits a normal electroretinogram and vision.

Thus, in many ways transdifferentiation *in situ* of the histologically simple pigmented retinal epithelial cells into the complex neural retina resembles that of iris epithelial cells into lens. That the resultant tissue exhibits even greater diversity of cell type than does lens recommends this rather neglected system for further study.

Secondary Transdifferentiation

Included in secondary transdifferentiation are reported cases of transdifferentiation that (1) are not unequivocal, as is *in situ* regeneration of neural retina from retinal pigmented epithelium or lens from dorsal iris epithelium, or *in vitro* conversion of iris epithelial cells to lens; (2) cannot document/demonstrate restricted cell lineage; or (3) arise from cells not withdrawn from the cell cycle. It is clear, however, that the following cell types would never normally express the resulting altered differentiated state. The conversion from one phenotype to another is striking and will provide us with insight into the molecular events of transdifferentiation. The system most commonly used and most valuable is that involving *in vitro* culture of embryonic chick neural retina and retinal pigmented epithelium (usually 3.5-9 days of incubation age). Coulombre *et al.* (1963) reported for neural retina, and Stroeva and Mitashov (1983) for pigmented retina, that mitosis is still present in some sectors of these tissues of the embryonic chick eye until 10-12 days; thus, the source of cells in most reports is not totally withdrawn from the cell cycle. In addition, when embryonic chick neural retina cells are clonally cultured, two sets of cells are eventually produced: those expressing the lens phenotype, and pigmented cells (i.e., the initial cell population exhibits heterogeneity). Thus, in the transdifferentiation of embryonic chick neural retina glial (Muller) cells into lens cells lentoid bodies, it is not possible to demonstrate unambiguously single Muller precursor cells because of their lack of a definitive marker.

In a series of elegant experiment, the laboratories of Clayton in Scotland and of Okada in Japan have investigated the molecular biology of the *in vitro* transdifferentiation of embryonic neural retina and pigmented retina cells into lens. Most commonly, δ-crystallin mRNA accumulation (is the most abundant embryonic chick crystallin) had been assayed: specific cDNA probes have also been used. Apparently, the crystallin genes, especially δ-crystallin, are leaky and extremely low levels of transcripts can be found in non-lenticular tissue, although translation cannot be detected. *In situ* hybridization analysis confirms the massive increase in -transcripts reported in transdifferentiating cultures of neural retina and retinal pigmented epithelium in 3.5- and 8-day chick embryos. Localized almost exclusively in the nucleus, these δ-transcripts are high molecular weight, indicating that they are unprocessed. Not until 14 days of neural retina culture (7 days for αA) and even later in pigmented epithelium cultures, are these mRNAs detectable. Errington *et al* (1985) have never detected any αA mRNA other than mature, however. It is obvious that there are varying transcriptional controls in transdifferentiating embryonic chick eye tissues.

Another example of secondary transdifferentiation is the phenomenon of transdetermination in *Drosophila*. When the imaginal discs of *Drosophila* are fragmented or dissociated and reaggregated, followed by culture in the abdomens of adult females, the cells proliferate without differentiating (hemolymph serves as the culture media). Upon serial culturing and reimplantation into a host larva, the development state of the disc, elicited by ecdysone of the host, may be assayed. Such analyses demonstrate that discs are highly determined with disc-specific development states reproducible for as much as 55 generations, e.g., genital disc structures. Despite this stability, changes in groups of cells from one disc-specific state of determination to another may occasionally occur, hence transdetermination. These switches are generally to recognizable alternative disc types, in frequencies and directions characteristic for different discs (e.g., leg disc to wing) and *Drosophila* species. Occurring in groups of cells, these are epigenetic rather than somatic mutational changes, and transdetermination involves a switch from one determined state to another. There is multiple cell co-operativity in transdetermination, resembling that seen in homeotic mutants (in which one body structure is replaced by another): minimally, polyclones are involved in these switches. The transdeterminating cells are thus committed, yet undifferentiated, in the sense that their original cell-type specificity has not been expressed at the time the cells are shifted from their original pathway.

Tertiary Transdifferentiation

Tertiary transdifferentiation includes those examples of a switch in phenotype from tissue composed of progenitor cells without overt terminal differentiation to one exhibiting properties of the differentiated state. These are frequently the least well documented of cases involving transdifferentiation, without the assurance of known cell lineages *in vivo/in situ* and without clonal culture of the progenitor cells *in vitro*. That "stem cells" are the origin of the transdifferentiation observed in these systems is usually clear. Further research, as seen in examples of primary and secondary transdifferentiation, to quantify and delineate is necessary in most cases, so as to attain possible resolution of the cellular pedigree and elucidation of the cell/molecular biology of the transformation. A discussion of several representative cases follows.

When *Xenopus laevis* larvae are lentectomized, a new lens regenerates from the inner cell layer of the outer cornea. Regeneration of the lens, with respect to histology and crystallin content, is reminiscent of normal lens development in *Xenopus*, albeit itself somewhat aberrant of other anurans. The necessity of a factor emanating from the neural retina, as in Wolffian lens regeneration, for successful transplant or *in vivo* corneal regeneration has also been demonstrated. Unlike the dorsal iris epithelium in Wolffian lens regeneration, however, the transdifferentiating source is undifferentiated and still in the cell cycle. It is not possible to elicit lens regeneration from the cornea in *adult X. laevis*. It is known that skeletal muscle (embryonic rat) can form hyaline cartilage *in vitro* and bone *in vivo*, when demineralized bone in some form is present. This can be considered as differentiated cells capable of altering their phenotype with the proper stimulus and is an inherently interesting phenomenon Nathanson and Hay (1980) present evidence, however, that suggest both skeletal muscle and cartilage arise from a common cell precursor. Even though the cells would not have formed cartilage if the extract were not present, the transformation is viewed as a continued differentiation, elicited by a signal resulting in a phenotypic change.

Other examples of tertiary transdifferentiation are as follows:

1. Transformation of rat pancreatic cells (of questionable origin) into hepatocytes through the use of drugs (ciprofibrate) or by Cu^{2+} depletion. [It is important to note the embryonic origin part of the pancreas from liver and the possibility of trapped liver cells in the rudiment].
2. Progressive restriction of developmental fates of the neural crest, wandering cells that will eventually form a plethora of structures, including pigment cells, neurons, and glial cells of the peripheral nervous system, the adrenal medulla, and head and face structures [This varied phenotypic expression is apparently brought about by the local milieu in which the neural crest cells, or a subpopulation, find themselves. The provenance of this signal is unknown and acts on what can be considered originally as stem cells; it is unclear when neural crest derivatives are cultured (e.g., non-neuronal cells of sensory ganglia that will melanize), whether we are dealing with a homogeneous stem cell population or a determined subpopulation of these cells.]
3. Changes in their differentiated state of the epithelial cells (and some nerve cells) in Hydra during their constant mitotic division and change in location during growth.
4. Limb regeneration *in vivo* in newts, in which pluripotential fibroblasts of the dermis are converted into chondrocytes.

Also included in this category of transdifferentiation is the elegant research of Schmid and Alder (1984) and Weber *et al.* (1987) on the transdifferentiation of striated muscle cells to smooth muscle cells in anthomedusae, marine jellyfish. In an earlier *in vitro* system, they demonstrated the only case in which cultured animal cells can transform into a functional organ (striated muscle into the sexual and feeding organ). In additional, fragments containing 20-50 striated muscle cells, after collagenase treatment to *"destabilize,"* are cultured in artificial seawater. Transformation (only with this treatment) to new cell types (including smooth muscle cells) occurs. According to the authors, the cultured "piece" is not contaminated with other cells, and the striated muscle has an appropriate ultrastructural profile. The resulting smooth muscle cells exhibit *de novo* flagella and non-striated contractile myofilaments. Inhibition of transcription and translation prevents this transdifferentiation. This work is by far the *"cleanest"* demonstration of transdifferentiation in invertebrates (and many vertebrates), but the phylogenetic niche of this organism must be considered. In contrast to what we know of vertebrate skeletal muscle, composed of multinucleated myotubes, terminally differentiated and stable in culture, medusa striated muscle cells are mononucleated and, after treatment, capable of change *in vitro*. Although such coelenterates exhibit specialized cell type, they generally have prodigious powers of regeneration. The spectrum of these types is limited, however (mesoderm is not present), and the versatility and scope of expression of a smaller number of individual cell types than that of vertebrates necessitate an inherently greater differentiative plasticity per cell type. Because of this less restrictive developmental birthright, the eventual information concerning gene expression may not be directly applicable. The system does provide us with a unique opportunity for elucidation of evolutionary links in metazoan differentiation mechanisms.

It can be seen that an obvious hierarchy exists in the expression of transdifferentiation in animals. The levels described in an attempt to discriminate among these examples is admittedly arbitrary, and overlap is possible. These inherent differences are valuable, however, in an effort to highlight those most stringent (primary) cases of transdifferentiation. An understanding of the experimental and biological parameters surrounding this spectrum will lead to the formulation of new and more rewarding interpretations and approaches to the problem and transdifferentiation, and ultimately phenotype expression.

POSSIBLE MECHANISMS IN TRANSDIFFERENTIATION

The most likely general candidâtes for roles in effecting transdifferentiation now appear to be alterations in the cell surface, growth factors, and elements affecting gene expression. A discussion of, and evidence for, these mechanisms, probably acting in concert rather than individually, follows. Most of the recent research that implicates them have used transdifferentiating eye tissues (usually of embryonic chick) as an experimental system.

Alterations in the Cell Surface

Transdifferentiation studies of embryonic chick, adult newt, and fetal human eye tissues have uniformly concluded that disruption of the tissue in some way is necessary for transdifferentiation to take place. The intact basement membrane of these tissues is thought to act as a stabilizer, to reduce the frequency of possible transformation *in situ*. Thus *in vitro* culture is usually necessary to elicit their transdifferentiation capability, and only under specified conditions, e.g., aggregate, rather than spreading, cultures will not transdifferentiate nor will cultures with low-serum concentrations, adult serum, supplementary glucose, or F-12/M-199 media. Only newt eye tissue among adult vertebrates will undergo primary transdifferentiation *in situ*. Even then, removal of the normal lens or neural retinas to initiate regeneration from dorsal iris epithelium and pigmented retina, respectively, disrupts the intimate contact of the pupillary margin of the eccentrically placed lens with the dorsal iris or the contract of original neural retina with adjacent pigmented retinal epithelium. Yasuda (1979) reported that collagen substrates inhibit transdifferentiation *in vitro* of pigmented retina cells into lentoids. Indeed, Reh *et al.* (1987) demonstrated that laminin promotes *in vitro* transdifferentiation of pigmented retina cells into cells that bind a neuron-specific monoclonal antibody — apparently transdifferentiation to lens cells is suppressed. Lentectomy in adult newts stimulates $^{35}SO_4$ incorporation into newly synthesized glycosaminoglycans in newt iris, associated with de-differentiation (with loss of pigment) of iris epithelial cells prior to lens regeneration. Hyaluronidase activity is also increased in the newt dorsal iris (the only lens regenerating sector *in situ*) compared with ventral iris after lentectomy. Again, upon lentectomy in the next, changes occur in the proliferating iris epithelial cells, resulting in discharge of melanosomes by exocytosis via elevation of intracellular [Ca^{2+}] (Patmore and Yamada, 1982); this elevation is mediated by cyclic adenosine monophosphate (cAMP), whose level is highest at this time, and the cell configuration also becomes altered. Zalik and Scott (1972, 1973) had earlier reported changes in electrophoretic mobilites of newt iris epithelial cells after lentectomy, indicating alterations in cell surface net charge. So, too, in the transdifferentiation of striated to smooth muscle in medusae discussed previously, collagenase treatment of the striated muscle cell fragment is necessary to "destabilize" the cells and permit formation of smooth muscle. Thus, we have increasing direct and indirect evidence of the importance of the cell surface in transdifferentiative events.

Growth Factors

The necessity of a neural retina factor necessary for newt lens regeneration *in situ* from dorsal iris has been noted. Earlier, lens regeneration was found to occur in intact irises *in vitro* if neural retina was provided or if the iris was cultured in contact with pituitary glands. A mitogenic factor has been isolated from bovine retina and looms as a candidate for the as yet uncharacterized neural retina factor operative in newt lens regeneration. The extract has been fractionated into eye-derived growth factors (EDGF) I, II and III. EDGF I and II have now been reported to be the retinal forms of basic and acidic Fibroblast Growth Factor from pituitary gland. Preliminary observations with an EDGF preparation injected into the eye of lentectomized newts whose dorsal iris had been removed, suggest a role in stimulating lens regeneration. Ventral iris, normally non-lens regenerating *in situ*, subsequently will frequently regenerate a lens. Using purified EDGF fractions, two laboratories recently reported success in eliciting lens regeneration from cultured newt irises. Cuny *et al.* (1986) report a stimulation of lens regeneration with EDGF II and III, while Connelly and Green (1987) find that only EDGF II has lens-regeneration stimulating activity. The discrepancy may be due to differing interpretations of baseline morphological lens regeneration. In any case, a growth factor is now available whose major characteristics (of retinal origin, and mitogenic, so as to increase the number of cell cycles the iris epithelial cells must undergo to form lens) fit the mold of the putative neural retina factor. Further reconciliation of the *in vitro* results, together with extension to the all-important *in vivo* situation (since transdifferentiation can occur so readily in eye tissues *in vivo* under varying conditions without exogenous growth factors) is eagerly awaited. Certainly a recurring problem will be the logistics of the tissue, because of the small amounts available (a single dorsal iris contains approximately 2000 epithelial cells), classic growth factor experiments (receptor affinity, with Scatchard plots, etc.) will always be difficult.

Elements Affecting Gene Expression

Transdifferentiation *in vitro* or *in vivo* indicates that previously inactive genes become activated, either by explanation to culture or, for example, by lentectomy. What permits these genes to express themselves? Knowledge of gene structure for the differentiated cell product may provide insight as to possible gene structural bases for the plasticity of gene expression. Some of the generalized features of the eukaryotic gene (and its flanking regions) and the processing of its mRNA transcript suggest themselves. The RNA polymerase (II) molecule probably recognizes and binds to certain regions of DNA that are upstream (toward the 5' ends) from the specific initiation site where transcription of DNA into RNA will begin. Certain consensus DNA sequences that have now been found appear to serve as promoter regions for RNA polymerase binding. One of these sites is centered 25-30 bases upstream from the initiation site and has the consensus sequence TATA (the TATA box, or Goldberg - Hogness box). Another region has the consensus sequence CCAAT and is located 70-80 nucleotides upstream from the initiation site. The deletion or alteration of these putative promoter regions is usually known to result in varying degrees of suppression of the levels of transcription, indicating that they are somehow involved in controlling the initiation of transcription. It is now believed that the TATA box is important in determining the precise initiation site and that the other regions upstream are involved in the efficiency of the transcription process. Enhancer elements (whose nucleotide sequence differs in different sources but seem to have a core sequence GTGGXXG,X=A, or T) were first reported in mammalian viruses. Enhancer sequences can stimulate transcription when placed either upstream or downstream of the 5' end of the gene, even when the orientation is inverted, and even from a considerable distance (several kilobases) from the beginning of the gene. They have been proposed as tissue-specific modulators of transcription in some mammalian genes.

There is also growing evidence for the importance of downstream control elements in eukaryotic genes. The results of Charnay *et al.* (1984) and Wright *et al.* (1984), using hybrid gene techniques, suggest that the β-globin gene, although it has functionally important upstream promoter elements, also contains an element within the structural gene necessary for expression in erythroid cells. These findings, together with other evidence of downstream regulatory sequences, e.g., *Xenopus* 5S ribosomal RNA gene and some mouse immunoglobulin (Ig) genes, with an enhancer element within an intron, illustrate the necessity to consider sequences both upstream and downstream of the transcription start site. It would be remiss not to mention, however, that alterations in the consensus CCAAT box sequence do not always result in reduced transcriptional efficiency e.g., the herpes simplex virus TK gene injected into *Xenopus* oocytes, and that some eukaryotic genes do not contain a consensus CCAAT sequence, e.g., insulin and ovalbumin. It appears that these gene sequences are major pieces, but not the sole ones, in the eukaryotic gene regulatory puzzle.

It is only with the transdifferentiating systems (eye tissues) that produce newly synthesized crystallins as their end-cell product that relevant current research can be found. Evidence is beginning to accumulate that such gene structures as above may be associated with crystallin gene expression. Kondoh *et al.* (1983) reported that cloned δ-crystallin gene 1 injected into the nuclei of various mouse somatic cells was expressed poorly in all but lens epithelial cells, indicating that transcription of the genes is tissue specific (even though mouse lens would not normally produce δ-crystallin!). Of special interest is the expression of this gene when the chick DNA segment 5' of the δ-gene (including the promoter region) was replaced with a virus putative promoter sequence; the gene was no longer expressed preferentially in lens epithelial cells. This indicates the importance of the native promoter region in gene expression. A similar reciprocal experiment has demonstrated that a cloned mouse γ-crystallin promoter is active in lens explants derived from 14-day chick embryos but inactive in cells of non-lens origin. γ-Crystallins are not normally present in the bird lens. Using the pSVO - CAT expression vector in transfected chick embryonic lens epithelia, Borras *et al.* (1985) have found evidence for differential promoter activities for the two δ-crystallin genes. The δ1 gene contained a consensus CCAAT sequence and enhancer-like sequences. Previous reports have indicated little, if any, δ2-crystallin gene activity. Although post-transcriptional events that limit δ2 mRNA formation cannot be ruled out, these results suggest a promoter role in δ-crystallin gene expression. In addition, in a similar expression vector system, a DNA region upstream of the TATA box of the mouse αA crystallin gene has recently been implicated in *tissue-specific* expression. These regulatory sequences promoted CAT activity only in lens cells, while other, non-crystallin, promoter sequences were effective in both lens and non-lens cells. Similar results on the mouse αA-crystallin promoter have been reported in the developing lens of transgenic mice containing the recombinant αA-crystallin-CAT gene.

The interaction of these regions of the gene locus with "factors" has been increasingly suggested as a means of modulating gene expression. *Treisman* (1986), has demonstrated in the *c-fos* promoter region an element responding to serum growth factors; cell fusion experiments with heterokaryons have implicated positive and negative regulatory factors operating upon the rat prolactin promoter region. Artishevsky *et al.* (1987) report sequences in the 5'- and 3'-termini of a hamster histone gene involved in cell-cycle-related transcriptional regulation; and (more relevantly) Borras *et al.* (1987), using deletion experiments, suggest the presence of both positive and negative regulatory domains in the promoter region of the δ-crystallin gene. The DNA footprinting technique has been increasingly used to detect cellular factors that bind with eukaryotic genes, especially to the regulatory sequences previously describes. Thus, Zinn and Maniatis (1986) employed it to demonstrate the binding of repressor molecules at two different sites, and a transcription factor at another, in the regulatory region of the human β-interferon (IFN_β) gene.

Such experiments suggest that this promising approach to the problem of transdifferentiation should have as longer-term objectives (using newt crystallin genes but applicable to other gene products of transdifferentiated cells):

1. Expression of isolated and characterized crystallin genes in *in vitro* systems.
2. Determination of the role, if any of DNA methylation in the *in vivo* lens regeneration process [Hypomethylation of DNA is frequently associated with regulation of gene expression during development and cell phenotypic change.
3. Investigation into the existence of and possible role(s) for a "transcription complex", as well as possible factors interacting with the gene regulatory regions, as a basis for the differential crystallin gene activity seen in newt iris, before and after lentectomy.
4. The use of newt crystallin cDNA probes as markers for differentiation in single and multiple (sequential) lens regeneration and for chromosomal localization of the new crystallin genes using *in situ* hybridization.

Only with such a multifaceted attack upon the cell and gene control mechanisms operative in the varying levels of transdifferentiation can we progress beyond the descriptive stage. Untouched upon in this treatment is the always unstated paradox or dilemma of explaining and reproductive the necessary integration of cells, resulting from transdifferentiation, into well-ordered tissue. In the transdifferentiation of neural retina, pigmented retina, and dorsal iris epithelium into "lens" *in vitro*, for example, the "lens" (actually lentoid) formed has no external layer/epithelium surrounding a fiber mass, as occurs with *in situ* lens regeneration. The tissue, cell, and molecular bases of transdifferentiation would provide a necessary framework for understanding the larger problem of differentiation in animals, per se.

10

Neoplastic Cells

For at least a century and a half, biologists have been aware that, relative to normal cells, many cancer cell appear morphologically undifferentiated. While some have concluded that cancer cell must arise by the differentiation of normal cell, others consider it more likely that many cancer cells develop from normal cells that never fully differentiated. Since these undifferentiated cancer cells can be shown to share many fundamental characteristics with normal cells, it is not altogether surprising that some neoplasms can be induced to differentiate and become postmitotic benign cell. If only one or a few malignant tumors had the potentiality for differentiation, the phenomenon would be of interest primarily because of its rarity. However, many types of malignant neoplasms, such as plant teratomas, fish tumors, amphibian cancers, and malignant neoplasms of all three embryonic germ layers of mammals, can be induced to differentiate. This indicates that the phenomenon is nor rare, and it suggests the possibility that the control of differentiation could be exploited in the treatment of cancer.

It is the purpose of this chapter to consider the diversity of cancers that differentiate and the multiplicity of inducers that elicit these differentiations. It would be impractical to list all the literally hundreds of agents known to induce differentiation of cancer cells, but a sufficient number will be mentioned to indicate the heterogeneous nature of agents with this property. Moreover, inasmuch as induced embryonic differentiation and induced cancer cell differentiation share some commonalities, both phenomena are considered. Although research into the control of cancer cell differentiation is a timely endeavour, it is interesting to note that speculation and experimentation in this area have their roots in nineteenth and early twentieth century biology. A brief chronicle is included in this chapter to trace the evolution of current notions concerning cancer cell differentiation.

There is ample evidence that terminal differentiation abrogates malignancy. That abrogation of malignancy is a principal rationale for continued study of cancer cell differentiation.

CANCER AND STEM CELLS

It is not inappropriate to compare normal cells and cancer cells. This is because studies of cell anatomy and cell function show that cancer cells could not survive *in vivo* or *in vitro* if they were not provided with the usual array of nucleic acids, enzymes, and physiologically essential molecules in an architectural order that allows for survival and proliferation. Thus, it may be stated that cancer cells share many attributes with normal cells. However, compared with normal differentiated adult cells, malignant-neoplastic cells appear incompletely or improperly differentiated. It will be argued here that the appearance of neoplastic cells is due to the fact that many of them are transformed stem cells frozen in that state by an oncogenic event.

This is not to suggest that the only difference between a cancer cell and its normal cell of origin is that the former is a mitotically active stem cell precluded from maturation because of transformation. Carcinogenesis by chemicals, viruses, and radiation obviously causes subtle changes in cancer cells that ultimately are manifested by altered cell function. However, even with significant changes in DNA, genetic defects can be bypassed in some cancer cells, and the cells can bee induced to differentiate with appropriate treatment despite the genetic lesion.

Stem cells in normal adult tissue are cells that divide and give rise to daughter cells with "unlimited growth potentiality" in a process of sell-renewal. Adult tissues that depend on a stem cell population include the lining of the gut, skin, and blood-forming cells in the bone marrow. The stem cells also provide mitotic progeny, which differentiate and replenish cells lost to wear or senescence. Differentiated cells generally do not transform into cells of another type, and — when terminally differentiated — they are usually postmitotic.

It the notion that cancer cells are similar to normal cells in many respects is accepted, it becomes useful to consider the possibility that the undifferentiated appearance of many cancer cells is not due to dedifferentiation (an uncommon event in animals) but is due to the overproduction of mitotically active transformed stem cells that retain morphological similarity, not to fully differentiated postmitotic cells, but to normally regulated stem cells. Evidence will be presented that at least some cancer cells not only resemble normal stem cells morphologically, but (as their normal stem cell compatriots) also retain the capacity for differentiation. If enough is learned about the phenomenon of cancer cell differentiation, it may be possible to induce the less than fully differentiated neoplastic cells to differentiate and become postmitotic without the usual need to kill either the cancer cell intentionally or normal cells inadvertently.

NEOPLASMS THAT DIFFERENTIATE

Crown Gall Tumor of Tobacco

All multicellular organism that undergo differentiation may become afflicted with neoplasms. This generalization applies as much to plants as it does to animals. A non-self-limiting neoplastic disease of dicotyledonous plants known ass crown gall is initiated when the T_i plasmid is transmitted to a wound on a susceptible plant by a specific bacterium. *Agrobacterium tumefaciens*. Normal cells and tissues from many mature plant can be cultured *in vitro*. Some somatic cells and protoplasts, when cultured *in vitro,* appear totipotent by virtue of their capacity to grow and differentiate all the cell types of an entire plant. Tobacco (*Nicotiana tabacum*) cells exhibit such totipotency. The crown gall with a known etiology and well-characterized mode of transfection, coupled with advanced plant cell culture technique, was an ideal subject for characterization of tumor different ative potential.

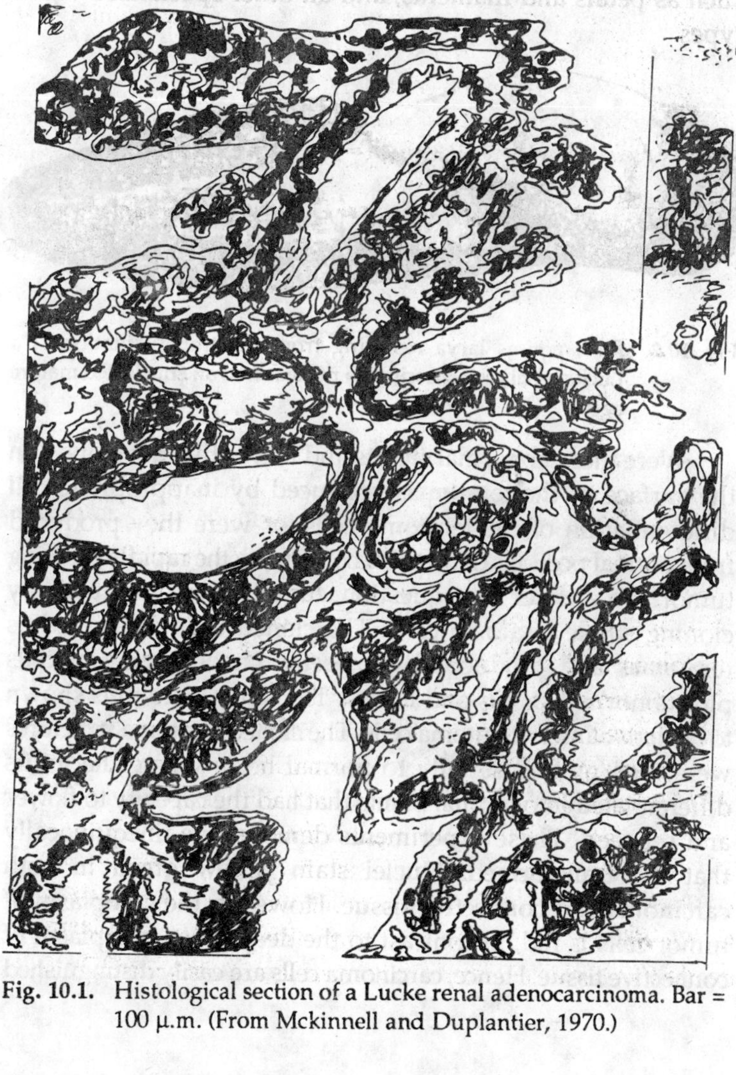

Fig. 10.1. Histological section of a Lucke renal adenocarcinoma. Bar = 100 μ.m. (From Mckinnell and Duplantier, 1970.)

Tumors in the tobacco plant induced by the T_i plasmid may grow as typical crown gall or as teratomas, depending on culture conditions. Teratomas grow as a chaotic arrangement of tissues and organs, including abnormal shoots and leaves. The normal shoots and leaves may be grafted serially (i.e., tumor tissue may be grafted to several hosts successively) giving rise to functional leaves, stems, diploid flower parts such as petals and filaments, and all other specialized tissue types.

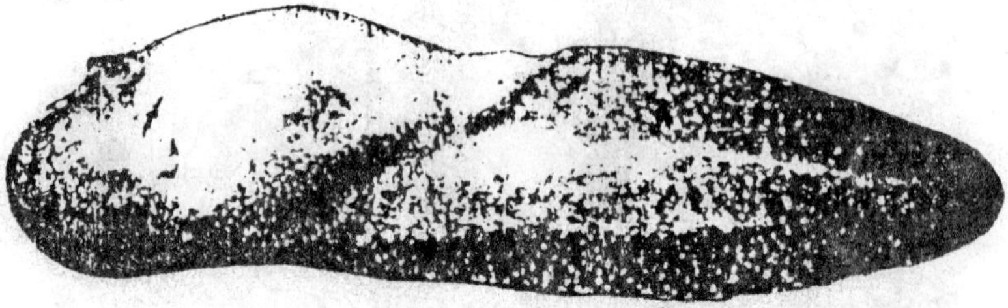

Fig. 10.2. *Rana pipiens* larva resulting from the transplantation of a Lucke renal adenocarcinoma nucleus into an enucleated mature egg.

Were these abnormal shoots and leaves that developed on the surface of the teratomas produced by inappropriate cell differentiation of the teratoma cells, or were they produced from normal portions of the tobacco plant by the rapidly growing tumor? To answer that question, tumors were produced by cloning single teratoma cells. The cloned cells developed as teratomas and produced by expected chaotic array of cells plus abnormal shoots and leaves. The latter were thus shown to be derived from teratoma cells. The cloned teratoma structures were then grafted serially to normal hosts where the grafts differentiated into normal plants that had the capacity to flower and set seed. These experiments demonstrate unequivocally that the transformed. Nuclei stain yellow green in both carcinoma and connective tissue. However, the cytoplasm of tumor cells is red, in contrast to the deep green cytoplasm of connective tissue. Hence, carcinoma cells are easily distinguished from connective tissue cells by their fluorescence in the UV range. Most (98.5%) of the dissociated cells were carcinoma, strongly supporting the view that the larvae previously reported could only have developed in response to the insertion of a renal adenocarcinoma nucleus into an enucleated ovum.

Additional evidence supporting the view that malignant carcinoma nuclei programme for the nuclear transplant larvae was obtained by transplantation of ranal adenocarcinoma nuclei whose cells were cultured in *vitro*. The cell cultures were epithelial in appearance, as would be expected from cultures of a carcinoma. Larvae were produced from transplantation of the cultured nuclei, corroborating and strengthening the earlier studies. The inescapable conclusion is that mitotic descendants of the Lucke renal adenocarcinoma nuclei are competent to differentiate into all the tissues characteristic of a swimming larva. The molecular mechanisms controlling differentiation are yet to be explained.

As is true of most tumors, the cell type of origin of the Lucke renal adenocarcinoma is unknown. However, several lines of evidence argue compellingly that the target cell for neoplastic change is either a stem cell or a less than fully differentiated precursor cell of the frog kidney. One line of evidence concerns the developmental stages that are vulnerable to Lucke tumor herpesvirus (LTHV) transformation. Only eggs, embryos, and young larvae are susceptible to the oncogenic action of LTHV. A pronephric cell (or perhaps an antecedent stem cell of the pronephros) can be transformed by the virus, resulting in pronephric carcinomas at metamorphosis. Since the pronephros exists only in immature stages, the pronephric renal tumors support the view that an embryonic cell, not a fully differentiated adult cell, is the target for LTHV transformation. In addition to the relatively rare induced pronephric tumors, many mesonephric tumors are induced by injection of LTHV. Nephric blastemas become separated from lateral plate mesoderm in recently hatched frog larvae and differentiation of mesonephric nephrons from the nephric blastema does not begin in larvae until the operculum is completed. Since eggs, embryos, and young larvae, bereft of differentiated renal cells, are highly vulnerable to the oncogenic effect of LTHV, it is reasonable to suspect that the target of LTHV is not a mature renal cell but a less than fully differentiated nephric precursor cell. Induction of tumors in adult frogs by exposure of early embryonic frogs to LTHV has interesting counter-parts in mammalian tumor induction studies where viruses and chemical carcinogens have greatest efficacy when administered to neonatal subjects.

Other evidence relating to the embryonic origin of the Lucke renal adenocarcinoma is derived from the transplantation experiments. Normal development, or extensive embryonic development in nuclear transfer embryos, is characteristic primarily of transplanted young embryonic nuclei. Nuclei from older embryos and adults generally require pretreatment either with low temperature and a polycationic amine or by pretreatment by exposure of nuclei to oocyte cytoplasm. Yet no pretreatment of any kind is required to condition renal adenocarcinoma nuclei to participate in embryonic development after nuclear transplantation. In other words, the Lucke renal adenocarcinoma nuclei behave as if they were embryonic nuclei. That might be, as is suggest, because the Lucke renal adenocarcinoma cells are embryonic stem cells frozen in that stage by exposure to LTHV.

An additional line of evidence suggests that an immature cell is target for transformation by LTHV is temperature sensitivity of the cytoplasmic microtubule complex. Cells cultured from metamorphic larvae (i.e., larvae with large posterior limbs and fully formed anterior limb enclosed within the gill cavity just before transformation into juvenile frogs) display cytoplasmic microtubules at both 7^{0}C and 28^{0}C, in contrast it cultured Lucke renal adenocarcinoma cells, which have well-developed cymplasmic microtubules at 28^{0}C but have profoundly disorganized microtubules at 7^{0}C. The sensitivity to loss of organization of the cymplasmic microtubules complex at the reduced temperature of these tumor cells is similar to, and characteristic of, early embryonic cells. These lines of evidence support the view that the Lucke renal adenocarcinoma may have an origin from less than fully differentiated precursor cells as other tumors are thought to have.

One goal of differentiation therapy is to induce the genome of neoplastic cells to yield cell progeny that are terminally differentiated and postmitotic. It may be asked if cell differentiation can suppress expression of the neoplastic phenotype without concomitant loss of proliferative potential. Differentiation of the full spectrum of larval tissues has been demonstrated in these studies of a neoplastic genome. Cytogenetic analysis of metaphase chromosomes of the swimming tadpoles is reported above. Thus, it is clear because of the presence of the metaphase chromosomes that neoplastic suppression is compatible with retained mitotic potential.

The experiments resulting in differentiated mitotic progeny of the Lucke renal adenocarcinoma would be of limited value if they had no relevance to tumors of higher organisms. However, it has been reported that a human congenital nephroma, which grew *in vitro* as undifferentiated cells totally devoid of renal structures, could be cocultured with fetal mouse dorsal spinal cord with the result that the nephroma differentiated fetal nephrons complete with Bowman's capsules, glomeruli, and proximal tubule. The human nephroma was thus shown to differentiate when confronted with a normal kidney tubule inducer. Therefore, both the frog and the human renal neoplasms retain genetic potentiality, as do many other malignant neoplasms, for differentiation of cell types not expressed in the neoplastic condition.

Epithelial Tumor of Triturus Cristatus

Skin tumors have been induced in the European newt, *Triturus cristatus,* by subcutaneous injection of a mixture of dibenzanthracene and benzpyrene. The tumors became invasive and in a number of instances metastatic nodules were detected at autopsy. Without any treatment, the tumors underwent what was termed spontaneous regression but was in fact differentiation of the neoplastic cells into keratinized skin cells and epithelial pearls, pigment cells, and mucous cells.

ECTODERMAL NEOPLASMS OF MAMMALS

Squamous Cell Carcinoma

Skin neoplasms of rats include a transplantable squamous cell carcinoma with islands of well differentiated epithelial pearls. The pearls are composed of differentiated squamous cells lacking mitoses. Rats with such tumors were pulse-labeled with [^{3}H] thymidine in order to detect the origin of the epithelial pearls. While undifferentiated squamous cell carcinoma nuclei were labeled after 2 hrs with this treatment, epithelial pearls remained unlabeled. Tumors examined at 96 hrs postlabeling, however, contained epithelial pearls that were labeled. This study indicates that the well-differentiated pearls, comprised of postmitotic cells, were derived from the malignant squamous cell carcinoma. The designations squamous cell carcinoma and epithelial pearls are morphological and do not indicate the capacity of either to form malignant growths. Therefore, fragments of carcinoma and dissected pearls were grafted to appropriate rat hosts. The epithelial pearls failed to form tumors, in contrast to many of the carcinoma fragments, which did. It was concluded that the squamous cell carcinoma stem cells differentiated into the non-malignant squamous cell pearls which were incompetent to form neoplastic growths.

Frequently, the concept of differentiation therapy is used in juxtaposition to cytotoxic therapy. This is in contrast to a therapy that seeks control of neoplasia without the need for cell death with another therapy, which by design, is lethal to cells. Squamous cells normally differentiate and die. Thus, in the special case of squamous cell carcinoma, differentiation therapy may lead to cell death. Cell death was indeed reported after squamous cell carcinoma, cultured as three-dimensional tumoroids, was treated with IFNγ. The treated cells became near-normal morphologically and contained more of a 67-kDa keratin (the 67 kDa keratin is thought to be differentiation specific because of its presence in epidermis and because it is found in reduced amounts in cell cultures and in poorly differentiated neoplasms, than the control cells.

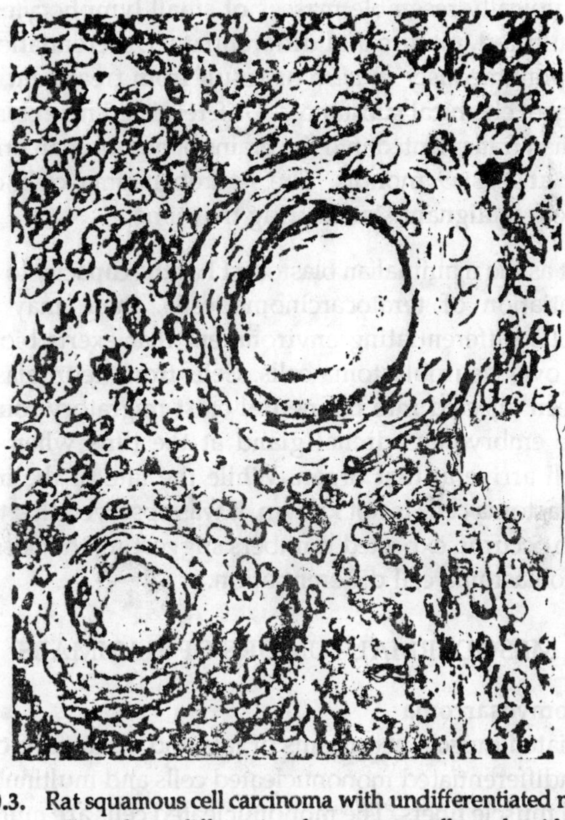

Fig. 10.3. Rat squamous cell carcinoma with undifferentiated neoplastic cells and two differentiated squamous cell pearls with keratotic centers.

Neuroblastoma

The capacity of human neuroblastomas to differentiate was reported more than 40 years ago, but the significance of the observations to the idea of cancer cell differentiation may have escaped the investigators. In an attempt to develop a procedure that would distinguish neuroblastomas from lymphosarcomas and small cell carcinomas, eight human neuroblastomas were cultured *in vitro*. Dendritic processes and axons (neurites) were a conspicuous trait of the cultured neuroblastoma cells. Since that pioneering study, other investigators have reported similar results, i.e., neuroblastomas, which typically resemble masses of small lymphocytes, will, when cultured, mature and differentiate. Differentiation may occur spontaneously *in vitro* or result from treatment with a diversity of chemical substances. Differentiated neuroblastoma cells, when transplanted into the brains of non-human primates, survive at least 9 months. The neuroblastoma cells do not revert to a malignant state during that time.

Just as the mammalian blastocyst has the capacity to induce differentiation of teratocarcinoma cells, there may be an analogous differentiating environment that exerts biological control over neuroblastoma cells. Cells from neuroblastomas have been injected into the neural crest migratory route and into the embryonic adrenal gland at the time when neural crest cell arrive at that organ. While the fate of the injected neuroblastoma cells is not known, it was observed that tumors failed to occur in expected numbers suggesting the possibility of neuroblastoma cell differentiation.

MESODERMAL TUMORS OF MAMMALS

Rhabdomyosarcoma

Striated muscle neoplasms of rats and of humans contain both undifferentiated mononucleated cells and multinucleated striated muscle fibers. The mononucleated cells are mitotically active, incorporate [^{3}H] thymidine, and have DNA values typical of dividing cells. By contrast, the postmitotic multinucleated striated fibers fail to be labeled with [^{3}H]thymidine after a 2-hr exposure. If the cells are incubated an additional 16-hr after unincorporated radioactive precursors have been removed, the differentiated cells become labeled, indicating that the origin of the muscle fibers is from the mononucleated stem cells. Undifferentiated mouse striated muscle tumor cells become differentiated when treated with sodium butyrate (Dexter *et al.,* 1981) or with *N.N*-dimethyl-formamide. Animals injected with tumor cells that have differentiated develop fewer tumors than do animals receiving undifferentiated rhabdomyosarcoma cells. Rat rhabdomyosarcoma cells will differentiate in response to hydrocortisone, which in turn inhibits metastases to other tissues. Metastasis is a fundamental characteristic of malignancy. Differentiated postmitotic cells should, and apparently do, fail to metastasize.

Myeloid Leukemia

Factors that control growth, viability, and differentiation of granulocyte and macrophage stem cells have been purified and characterized as glycoproteins produced by normal tissues. Differentiation factors induce mouse and rat myeloid leukemic cells to differentiate *in vitro*. The human myeloid leukemic cell lines ML-1 and HL-60 can be induced to differentiate *in vitro* by treatment with a number of different agents. Terminal differentiation of murine and human leukemic cell lines has been induced with human colony-stimulating factors produced with recombinant DNA techniques. A combination of interferon- and retinoic acid-induced differentiation *in vitro* of leukemic cells obtained from patients with acute myelogenous leukemia. Curiously, HL-60 cells differentiate in culture medium deprived of a single essential amino acid, despite the failure of protein synthesis inhibitors to induce differentiation.

Some myeloid tumor cell lines have the competence to grow but fail to differentiate properly. Some of these cells designated D^+ for differentiation positive, respond to an exogenous supply of differentiation inducer by producing cells that are mitotically inactive and that are morphologically and physiologically indistinguishable from normal macrophages and granulocytes. The D^+ cells respond not only to the natural inducer but by differentiation to a relatively long list of chemical agents as well. Other leukemic cells, designated D⁻ for differentiation negative, fail to respond to natural differentiation inducers and represent a different stage of neoplastic progression than do the D⁺ type cells. The D⁻ myeloid leukemic cells may respond to other inducers, or combinations of inducers, together with normal differentiation factors, to produce differentiated, postmitotic myeloid blood cells. The D⁻ cells may respond to steroid hormones, X-rays, some vitamins, insulin, lectins, some phorbol esters, and how doses of conventional chemotherapeutic agents. These studies demonstrate that both the D⁺ and the D⁻ cells have retained the genes necessary for differentiation, even though their chromosomes may vary from the normal euploid karyotype. It is interesting to note that while there may be a genetic basis for the malignant behaviour of leukemic cells, the genetic alteration can be bypassed with appropriate treatment.

Murine Erythroleukemia

Murine erythroleukemia cell (MELC) lines, derived from mice with Friend virus-induced leukemia, grow *in vitro* as pleomorphic populations of undifferentiated cells and cells at various levels of erythroid differentiation. Changes in MELC that accompany differentiation include greatly increased levels of globin mRNA, α- and β-globin synthesis, appearance of erythrocyte membrane antigens, synthesis of heme and haemoglobin, decreased cell volume and cell water, and a limited capacity for cell renewal.

Chemicals known to induce differentiation of MELC *in vitro* include dimethylsulfoxide (DMSO), hexamethylene bisacetamide (HMBA), purines and purine analogues, short-chain fatty acids, hemin, cardiac glycosides, and metabolic inhibitors. Recently, the antibiotic trichostatins and a protein differentiation factor were added to those chemicals that have the capacity to induce differentiation of MELC and thus may be added to the more than 300 known inducers of differentiation.

ENDODERMAL TUMORS OF MAMMALS

Human Colon Cancer

Human colon carcinoma cell lines grow in soft agar and form tumors when injected subcutaneously in mice. Treatment of those cell lines with the polar solvent *N, N*-dimethylformamide induces differentiation of the cells and the loss of clonogenicity in semisolid medium as well as a major reduction in the tumor forming capabilities of the cells. Similarly, the growth of grafted colon carcinoma cell lines was inhibited in nude mice treated with polar solvents. It is not known whether the inhibition of growth in treated nude mice is caused by differentiation of the grafted colon carcinoma cells or by some effect of the host against the inoculum. Be that as it may, an induction of differentiation with a concurrent reduction of tumorigenicity were demonstrated with the human colon carcinoma cells. More recently, a human colon carcinoma cell line (HT-29), cultured in the absence of glucose, was shown to exhibit differentiation characterized by gut cells that are polarized, have brush borders, and have enhanced glycosaminoglycans synthesis as well as cells which secrete mucous . Another colon carcinoma cell line (LIM 1863) similarly differentiates columnar cells and goblet cells *in vitro*. Colon cancer cell differentiation in response to chemicals, culture conditions (confluent or postconfluent), and other factors was recently reviewed.

Teratocarcinoma

It may be appropriate to place teratocarcinoma under the caption endodermal tumors, but the rationale for placing them here relates to the fact that they are considered germ cell tumors and primordial germ cells in the mammal have an origin on the posterior yolk sac of the embryo.

Teratocarcinomas occur spontaneously in the gonads of some strains of mice or are produced by grafting embryos to ectopic sites. Mouse teratocarcinomas are composed of a wide diversity of tissues in various stages of differentiation, including

Central Nervous System (CNS), cartilage, bone, stratified squamous epithelium, striated muscle fibers, respiratory and alimentary mucosae, and glands in a hodgepodge arrangement described by Needham (1942) as an approximation to chaos. Sometimes structures appear that resemble early mammalian embryos; these are called embryoid bodies. The internal cells of the embryoid bodies were thought to be multipotential and were believed to give rise to the diversity of cell types by differentiation. Grafts to the anterior eye chamber or subcutaneously in mice established experimentally the multipotentiality of the cells. Kleinsmith and Pierce (1964) transplanted single teratocarcinoma cells isolated in capillary tubes into mice and obtained experimental tumors composed of embryonal carcinoma and many well-differentiated somatic tissues.

Brinster (1974) first demonstrated that teratocarcinoma cells, transferred to a mouse blastocyst, will participate in normal development. He illustrated his study with a random bred Swiss albino mouse displaying a stripe of agouti fur. The teratocarcinoma cells, carrying the genotype for agouti colour, had differentiated into fur-producing cells in the presence of the normally differentiating embryonic host cells. Brinster's studies were extended by others, who injected genetically marked teratocarcinoma cells into blastocysts that were subsequently placed in foster female mothers (the methodology for this procedure is described by Bradley, 1987). Normal live mosaic mice were born. The teratocarcinoma cells differentiated all tissue types found in-normal mouse development.

A teratocarcinoma cell line, E Ca 247, has the competence to differentiate trophectoderm after treatment in the blastocyst. If inner cell mass cells are the normal counterpart of malignant teratocarcinoma, one might suspect that some inner cell mass cells could differentiate trophectoderm also. Some do. In this interesting study, knowledge of cancer cell biology has led to a new appreciation of the differentiative potential of normal embryonic cells.

Teratocarcinoma-derived cell lines differentiate *in vitro*. Retinoic acid will induce such differentiation, and varying the concentration of retinoic acid used will induce different cell types. A subclone of P19 teratocarcinoma cells exposed to retinoic acid at 10^{-9} M differentiates into cardiac cells, whereas retinoic acid at 10^{-8} M induces skeletal cells, and retinoic acid at 10^{-5} to 10^{-7} M causes the differentiation of neurons and astroglia. Retinoic acid treatment of teratocarcinoma cells affects DNA methylation, expression of homeobox genes and X-chromosome inactivation and will decrease poly (ADP-ribose) synthesis. Cell-Surface glycolipid antigens change as an embryonal carcinoma cell line (NTERA-2) differentiates in response to retinoic acid.

Not only will murine teratocarcinoma cells differentiate but so too will the human counterpart, both *in vitro* as well as *in vivo*.

CANCER CELL DIFFERENTIATION

The possibility of exploiting induced cancer cell differentiation as a new mode of cancer therapy is becoming increasingly attractive. One might view the biological research leading to contemporary clinical trials as existing only in the recent literature. It may therefore come as a mild surprise that notions concerning cancer cell differentiation have their roots in early nineteenth century biology.

As early as 1829, Recamier envisioned neoplastic cells as embryonic cells that persisted during development and into adulthood. The embryonic cells were thought to become neoplastic because of differentiation in an inappropriate anatomical site or at an improper time in ontogeny. This view was shared by Remak in 1854. Cohnheim (1875, 1889) and Beard (1902). These early workers invoked the need to consider the origin of cancer cells as it relates to morphology (i.e., some tumors appear to be undifferentiated and embryonic compared with adult tissue) and mitotic potential (i.e., embryos and neoplasms frequently share similar growth rates that contrast with the reduced or absent growth of many normal adult tissues).

The maturation and differentiation of leukemic cells were reported early in this century by Awrorow and Timofejewsky (1914) and Timofejewsky and Benewolenskaya (1929), followed by Bessis (1954), FRiend *et al.* (1966), and Ichikawa (1969).

Needham (1936, 1942) was aware of the Lucke renal adenocarcinoma and proposed an experiment that he hoped would "master" the *"wildly growing material."* Needham suggested grafting Lucke renal adenocarcinoma fragments to the limbs of urodeles (which regenerate). After the graft had gown, he proposed that the limb be amputated with the thought that the intense biological activity of the regeneration blastema might force the frog neoplasm to differentiate. Rose and Wallingford (1948) and Rose (1949) attempted the Needham experiment and reported preliminary success; other failed in their attempt to repeat the Needham experiment.

It is worth reconsidering the Needham/Rose experiment in light of current concepts. One would speculate that, in view of the stability of cell-type specialization of most tissues, it would be unlikely that renal adenocarcinoma would differentiate limb tissues. If grafted Lucke renal adenocarcinoma in the blastema differentiated normal cells, it would probably differentiate normal kidney cells. However, limb blastema would seem an inappropriate environment for the induction of kidney cell differentiation. Consider the fact that teratocarcinoma cells differentiate in the blastocyst and that neuroblastoma cells may differentiate in the neurula. These examples would appear to suggest that cancer cells differentiate in an environment appropriate for the differentiation of normal stem cell counterparts, pointing to the embryonic kidney as an appropriate site for differentiation of the Lucke renal adenocarcinoma. No one, thus far, has studied the capability of differentiating embryonic kidney to induce differentiation of the grafted Lucke renal adenocarcinoma. The Needham/ Rose experiment is considered here because in the evolution of ideas it proposed a mode of inducing differentiation, it was ahead of its time, and contemplation of the experiment helps clarify the need for special environments for induced cancer cell differentiation.

CONTROL OF DIFFERENTIATION OF EMBRYO CELLS AND CANCER CELLS: ARE THERE COMMONALITIES?

Earlier in this chapter it was stated that the anatomy and function of normal cells and cancer cells have much in common. Following that discussion, a number of cancers that undergo maturation and differentiation were reviewed. Inasmuch as embryo cells also undergo maturation and differentiation, one could ask if there are common mechanisms that lead these two categories of cells to the differentiated state. The question is unanswered because primary embryonic induction remains an unsolved problem after over a half century of study. In fact, Gurdon (1987) stated that the "molecular basis of induction and the inductive response remains almost totally obscure." That statement is as applicable to cancer cell differentiation as it is to embryo cell differentiation.

While the molecular mechanisms of embryonic induction are admittedly obscure, much is known about the operational aspects. For example, it is recognized that primary induction of the embryonic axis and neural plate occurs through an interaction of the ectoderm with underlying chordamesoderm. Induction of neural tissue can occur experimentally in heterotopic sites *in vitro* or as a result of the interaction of inducer and responding tissue in sandwiches *in vitro*. Curiously, living inducer is not required for induction, making a forceful argument that the reaction is a chemical, not biological, event. Following the realization that a chemical substance may trigger the events leading to differentiation, there was frenzied activity to identify chemicals critical to this event. Within a short time, many unrelated and frequently bizarre chemicals were identified that have the capacity to induce neural differentiation of ectoderm fragments *in vitro*. The understanding of induced differentiation was further complicated when it was found that neural differentiation *in vitro* would occur with no inducer at all. It would seem that there are too many inducers and the inducers appear to be unrelated chemically. This observation, coupled with the reports that induction occurs with no inducer at all, led to the notion that perhaps some (many?) of the chemicals were not inducers themselves but were substances that act to liberate inducers intrinsic to the reacting cells. It was postulated that they did this by injury to the plasma membrane resulting in a sublethal cytolysis that in turn activated substances leading to neural differentiation. Curiously, the concept of cell injury, which provided a common denominator linking diverse inductive reactions, reappeared recently in a discussion of induced cancer cell differentiation. While prospective neural plate seems poised and ready to differentiate in response to a diversity of seemingly unrelated stimuli, it should be remembered that in the intact embryo embryonic ectoderm differentiates as epidermis in the absence of chordamesoderm.

There are operational similarities in the differentiation of cancer cells to that described for the differentiation of normal embryonic cells. Spontaneously, squamous cell carcinoma differentiates in intact animals and humans. So too does neuroblastoma. Thus, cancer cell differentiation occurs *in vivo*, as it is known to occur in intact embryos. Tumor cells such as teratocarcinoma and neuroblastoma can be placed in a new embryonic environment in which differentiation occurs, perhaps analogous to heterotopic differentiation in experimental embryos. And cancer cells will differentiate in response to a diversity of substances (which might be described as unrelated and even bizarre) *in vitro* much as normal embryonic differentiation can be induced *in vitro*.

Biological agents that induce cancer cell differentiation encompass a substantial literature. Biological response modifiers discussed with myeloid leukemia are the Macrophage Granulocyte Inducers (MGI), also known as granulocyte macrophage colony stimulation factor (GM-CFS). A protein, the Erythroid Differentiation Factor (EDF), has been isolated from a human leukemic cell line that induces differentiation of murine erythroleukemia cells. Glial Maturation Factor (GMF) is an acidic protein that promotes differentiation of cultured astroblasts and cultured human glioma cell lines (Lim *et al.*, 1986). Interferon alone has no effect on differentiation of HL-60 cells, but retinoic acid induced-differentiation is enhanced when combined with interferon treatment. Similarly, tumor necrosis factor and retinoic acid potentiate the differentiative capacity of each other. Other biological agents that effect differentiation of cancer cells include factors in the cytoplasm of the amphibian egg, which inhibit expression of those genes characteristic of the Lucke renal adenocarcinoma and awake from dormancy the genes associated with normal differentiation of amphibian larvae, and the interior of the murine blastocyst, which permits differentiation of teratocarcinoma cells.

Cancer cells also respond to a wide spectrum of relatively simple chemical substances. The induced differentiation of teratocarcinoma cells occurs by a variety of chemical substances: retinoic acid, retinal, retinol, 5-bromouracil 2'-deoxyribose (BUdR), 5-iodouracil 2'-deoxyribose (IUdR), hexamethylene bisacetamide (HMBA), dimethylacetamide (DMA), dimethylsulfoxide (DMSO), and ouabain. Promyelocytic leukemia cells (HL-60) are induced to differentiate by retinoic acid, 12-O-tetradecanoyl-phorbol-13-acetate (TPA), phytohemagglutinin-leukocyte-conditioned medium, cyclopentenyl and cyclopentyl analogues of cytidine, dimethylsulfoxide (DMSO), and dimethylformamide (DMF), and tunicamycin, as well as recombinant human leukocyte interferon in combination with other inducers. This is only a partial listing of differentiation agents. Hickman and Friedman (1988) report that 73 compounds are known that induce the terminal differentiation of HL-60 promyelocytic leukemia cells.

The HL-60 list seems small compared with the 300 agents known to induce differentiation in mouse erythroleukemia cells. Thus, it is beyond the scope of this chapter either to list all the agents or explain their function. Perhaps the diversity of chemical differentiation inducers is more impressive than the sheer numbers of these substances. That diversity is reminiscent of the range of chemicals that induce CNS development in vertebrate embryos.

One might observe here, as was noted in the discussion of CNS differentiation: "there are too many inducers." Hickman and Friedman (1988) suggest that an understanding of how the various agents act may be simplified a bit if it is recognized that there may be a limited repertoire of cellular responses to a multiplicity of chemically induced changes or lesions. These investigators not that many of the agents that have efficacy in inducting cancer cell differentiation are cytotoxic drugs used at "marginally toxic concentrations." More specifically, it may be noted that the concentrations of polar compounds (e.g., *N,N*-dimethylformamide, *N*-methylformamide, and related ureas and acetamides) required to induce differentiation of the promyelocytic leukemia cell line HL-60 were marginally lower than those that were cytostatic or cytotoxic, which suggests that " a toxic threat to the cells was sufficient to induce differentiation." HL-60 cells differentiate in response to an enormous diversity of chemical gents; Langdon and Hickman (1987b) note that these agents include not only the polar solvents they studied, but chemical compounds as different as anthracycline antibiotics, methotrexate, and tunicamycin as well. A property of all of these agents is that they are either cytotoxic or cytostatic, or both. It could be that the more recent studies of cancer cell differentiation will provide new insight into the old, and almost forgotten, sublethal cytolysis of differentiating embryonic CNS proposed a generation ago by Holtfreter. Furthermore, it is at least within the realm of possibility that the adjective non-cytotoxic will have to be deleted from at least some forms of differentiation therapy. A useful catalogue that sorts agents into five categories is that of

Gabrilove (1986), in which inducers of cancer cell differentiation are divided into chemotherapeutic agents, polar compounds, phorbol esters, vitamin analogs, and physiologically important cytokines. The multiplicity of chemical substances that induce cancer cell differentiation provides no insight into molecular mechanisms responsible for that differentiation. However, two areas of research are converging that have the potentiality of providing new understanding of cancer cell differentiation. Those two areas involve research into growth factor and the study of oncogenes. Growth factors are peptides that cause an increase of cell proliferation by binding to specific cell membrane receptors. Some neoplastic cells exhibit an escape from growth factor control by displaying a decreased or absent requirement for specific growth factors.

There is a direct connection between several important growth factors and cellular oncogenes. Examples of the relationship of specific peptide growth factors and specific oncogenes include platelet-derived growth factor (PDGF), which is structurally related to the protein product of the oncogene *v-sis*. Production of PDGF may lead to the expression of a second oncogene, *c-myc*, which is turn is believed to be related to cell proliferation. This series of sequentially related events (i.e., an oncogene that codes for a growth factor, the production of which in turn causes the expression of another oncogene that is related to cell proliferation) has been described as a hierarchical control of cell proliferation.

Growth factor receptors may be encoded by certain oncogenes. For example, *c-erb* B codes for a protein with close similarity to the Epidermal Growth Factor (EGF) receptor (Downward *et al.*, 1984) and *c-fms* gene product is similar to colony-stimulating factor-1 (CSF-1) receptor. The p21 gene product of *ras* has been termed an obligatory intermediate in the transduction of a growth factor signal. Further linking of growth factors with oncogenes is evidence by growth factors that increase transcription of certain oncogenes. Obviously, then, growth factors and oncogenes are intimately interconnected. That leads to the question of the relationship of oncogene expression to cancer cell differentiation.

If cellular oncogenes are involved in malignancy, and there is ample evidence that they are cancer cell differentiation described in this chapter would seem an appropriately rich arena for establishing the role of various oncogenes (and growth factors). This is because it should be possible to ascertain the level of *c-onc* mRNA in a cell population while the cells are phenotypically malignant, during the time that they are differentiating, and after they display a benign phenotype. It is difficult to imagine a direct role in oncogenes is for a particular *c-onc* if transcriptional activity of that gene remains invariant with phenotypic change. However, if the level of mRNA of the *c-onc* is directly related to the state of differentiation, a role for that gene can be envisioned. As would be expected, there is an emerging literature concerning this question. The literature remains somewhat difficult to interpret.

An example is the proto-oncogene *c-myc*. The *c-myc* mRNA level is reported to be cell proliferation dependent. It is expressed in normally growing cells as well as neoplastic cells. Ordinarily, with growth arrest, *c-myc* expression is greatly reduced (as expected). *c-myc* mRNA reappears when cells are stimulated to resume cell division. The oncogene is expressed especially high in the human promyelocytic leukemia cell line, HL-60. As anticipated, there is an almost total inhibition of *c-myc* transcription after induced differentiation of HL-60 cells with either Me_2SO or retinoic acid. Similarly, *c-myc* mRNA levels are decreased with differentiation induced by the dihydroxymetabolite of vitamin D_3. Removal of the vitamin D metabolite after the onset of differentiation results in the resumption of elevated levels of *c-myc*. Likewise, decreased levels of *c-myc* mRNA with differentiation were reported by Einat *et al.* (1985), Grosso ad Pitot (1984), and Slungaard *et al.* (1987). Contrast those studies with chronic lymphocytic leukemic cells that were induced to differentiate by the action of a phorbol ester. Transcription of *c-myc* was increased and remained high during the differentiation process. In this instance, it is clear that downregulation of *c-myc* is not a prerequisite for differentiation. In another study, HL-60 cells were treated with DMSO with or without growth inhibition. The kinetics of *c-myc* expression parallel closely the increased proportion of differentiated myeloid cells but not by changes in the proportion of proliferating cells. Finally, with Friend erythroleukemia cells treated with DMSO, there is a marked decrease in *c-myc* mRNA for 12 hr, followed by an increase in *c-myc* mRNA to the pretreatment level by 18 hrs, followed by a gradual decline. One would expect variant expression of *c-myc* in different neoplastic cell lines. Such variant expression is reported. However, it is difficult to generalize at the present time from these studies whether to expect *c-myc* expression to increase, decrease, of follow a pattern of alternating levels with cell maturation. Hickman and Friedman (1988) commented that much of the work reported concerning transcriptional activity of oncogenes has been without parallel measurements of the protein gene products. These workers suggest that post-translationally regulated changes may be as important as transcriptional modulational. they state that " at this moment in time, it cannot be claimed that a clear role for oncogene expression in differentiation has emerged."

The caption of this section asks the rhetorical question as to whether there are commonalities in the differentiation of embryonic and cancer cells. While a definitive answer to that question is lacking because of a paucity of understanding concerning the molecular mechanisms of either, it seems likely that heightened insight of the one will augment understanding of the other — with the possibility of a practical event emerging from this greater enlightenment. That event would be the continued and enhanced development of non-toxic differentiative cancer therapy. The multiplicity of recent papers concerning cancer cell differentiation, only some of which are cited here, witnesses the increasingly widely held view that differentiation therapy has potential value in the treatment of malignancy.

11

The Control of Cell Differentiation

The cell differentiation, it includes diverse cellular changes responsible for generating the many cell types that make up the adult organism. At the same time, differentiation also entails a sequence of stages: the *commitment* of cells to a particular pathway, the irreversible *determination* of that commitment, and the final stage of *terminal differentiation* as the cell begins to perform a specific function. Commitment to a specific differentiation pathway is a stepwise process; as it takes place, as series of decisions are made between alternative fates, and the cell's developmental potential is increasingly restricted. A determined cell follows its set course of differentiation regardless of the influences of external stimuli. In additional, once a cell has reached this developmental state, all of fits daughter cells are similarly determined. Thus, *the determined state is heritable.*

To discover experimentally whether a cell has become determined, the cell must be placed in a new site where it receives new external input. Such transplantation sites may be within a developing embryo, in a particular location in an adult organ, or in an artificial environment, such as a culture dish. In each of these situations the transplanted cells must also remain available for future study so that the outcome of the experiment can be evaluated.

It has been seen that cell determination can be assessed by transplantation *in vivo*. Such experiments have limited value, however, because of the complicated environment in which the transplanted cells develop. In contrast, cells maintained in tissue culture exist in an environment that can be tightly regulated. Because *in vitro* tissue culture plays a crucial role in our understanding of cellular determination and differentiation, we shall begin by outlining and discussing the various methods that experimenters employ and the advantages and disadvantages of each. Subsequent sections of the chapter will focus on selected examples as we consider *in vitro* studies of cell determination in depth.

THE ADVANTAGES OF TISSUE CULTURE

Cells can be studied *in vitro* in two basic ways: either as small fragments of isolated tissue (*explants*) or as individual cells separated from the tissue by enzymatic or mechanical dissociation (*dispersed cells*). The present discussion will be restricted to the second category. After isolation, dispersed cells are maintained in a *nutritive medium* that often contains *serum* — a cell-free fraction of blood that contains factors to promote cell proliferation. By combining cell culture with techniques to identify and isolate desired cell types, it is possible to analyze the behaviour of homogenous cell populations and to study interactions between different cell types. One of the most significant advantages of tissue culture is the fact that *in vitro* cell populations are removed from many of the complicating influences of other cells and materials present *in vivo*. This simplified situation has a number of consequences, but perhaps the most important benefit is that the cellular environment is much easier to manipulate. For example, the responses of particular cell types to drugs, hormones, or specific nutrients can be analyzed simply by adding them to the culture medium. *In vitro* culture has thus been an important tool for characterizing the basic nutrient requirements of somatic cells and defining the hormones and growth factors that regulate cell proliferation. In some cases, the resulting insights have made it possible to simplify cell culture methods, including the elimination of serum from culture media. Thus, some cell types are now grown in a *defined medium*, in which the chemical characteristics of all of the constituents are known. One important goal of such work is to develop media in which cells will proliferate and undergo the stages of differentiation so that the nature of the external factors required for these basic processes can be fully understood. Currently, however, maintaining living, proliferating embryonic cells *in vitro* generally requires either the addition of fetal serum or the supplementation of non-fetal serum with an extract from embryonic tissue.

Another feature of *in vitro* culture is that experimenters can synchronize various events in cell populations and so increase the number of cells participating in a single event. For example, in the chick limb *in vivo*, terminal differentiation of muscle cells is initiated at about 11 days of development. The subsequent recruitment of cells to form multinucleate myotubes extends over several days. *In vivo*, however, the process unfolds much more rapidly; by manipulating the medium, as much as 60 per cent to 70 per cent of an *in vitro* chick myogenic population can be persuaded to undergo cell fusion in as little as 12 hours. The cell cycle of cells *in vitro* can also be synchronized by removing essential medium components, such as serum or essential amino acids, by adding drugs that reversibly block cells at specific points in their cycles or by simply shaking off loosely adhering mitotic cells. The general advantage of synchronizing cell events is the facilitation of biochemical analysis.

A third type of *in vitro* manipulation involves individual cells. In the technique of *somatic cell fusion*, cells are treated with fusing agents, or *fusagens*, such as the Sendai virus or polyethylene glycol. The treated cells fuse together to generate *heterokaryons*, single cells that are combinations of two similar or different whole cells. Parts of cells, such as cytoplasms and nuclei, can also be fused. After cells are exposed to the fusagen, they are placed in a selective medium, which allows only the heterokaryons to proliferate. Next, by first exposing the hybrid cells to drugs that increase the mutation rate and then selecting for extensive proliferation *in vitro*, variant cells with new desired

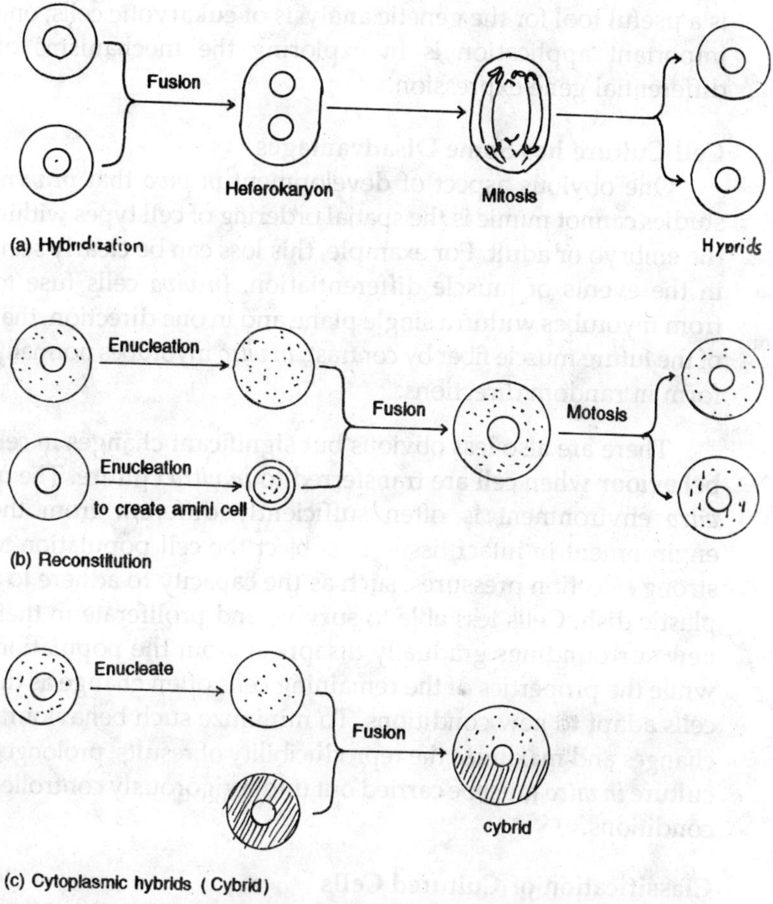

Fig. 11.1. Cellular hybridization and reconstruction techniques. (a) Two complete cells are fused to yield a heterokaryon. A single spindle forms at the first mitosis, and two cells, each containing the chromosomes from both parents, are formed. (b) In reconstitution, cytochalasin B and centrifugation are used to generate an enucleate cytoplasm (cytoplasm) and a nucleus surrounded by a thin ring of cytoplasm (minicell). Fusion of the two then results in a reconstituted cell, which has the nucleus from one cell and the cytoplasm from the other. (c) "Cybrids" are formed when an enucleated cell is fused to a normal cell.

characteristics can be identified and isolated. This technique is a useful tool for the genetic analysis of eukaryotic cells; one important application is in exploring the mechanisms of differential gene expression.

Cell Culture has Some Disadvantages

One obvious aspect of development *in vivo* that *in vitro* studies cannot mimic is the spatial ordering of cell types within the embryo or adult. For example, this loss can be clearly seen in the events of muscle differentiation. *In vivo* cells fuse to from myotubes within a single plane and in one direction, that of the future muscle fiber by contrast, *in vitro* myotubes normally form in random directions.

There are also less obvious but significant changes in cell behaviour when cell are transferred to *in vitro* culture. The *in vitro* environment is often sufficiently different from the environment in intact tissue to subject the cell population to strong selection pressures, such as the capacity to adhere to a plastic dish. Cells less able to survive and proliferate in their new surroundings gradually disappear from the population, while the properties of the remaining cells often change as the cells adapt to new conditions. To minimize such behavioural changes and maximize the reproducibility of results, prolonged culture *in vitro* must be carried out under rigorously controlled conditions.

Classification of Cultured Cells

Cell proliferation *in vitro* in induced in three distinct phases. In the first phase, the *primary culture,* cells recently transferred from a living organism adjust to their new environment. Because most cell types require a surface on which they can proliferate, they must become attached to their artificial substrate before proliferation is initiated. This phase ends when the rate of proliferation reaches a plateau, and the cells exist either as a dense suspension or as a continuous, single-layered sheet. The culture is now ready for its first subcultivation *passage,* in which a few cells are transferred to fresh medium to reduce the density of each culture. This transfer constitutes the start of phase two, the *log phase.* If the cells are subjected to frequent medium changes and subpassaging, they will undergo extensive and rapid proliferation. At the end of phase two, the rate of proliferation slows and degenerative changes begin, which mark the beginning of a third, *terminal phase.* During this phase the nuclei of the cells become atypical in both shape and size, and cells exposed to fresh medium fail to proliferate and ultimately die.

Not all cells cultured *in vitro* follow this proliferation pathway, however. Some cells, which appear in the culture spontaneously, escape the terminal phase and become *immortal.* Such immortal cells may result from selection pressures caused by extensive proliferation; these are called *established cell lines.* Alternatively, immortal cells known as *neoplastically transformed cells* may come about when the culture is exposed to a virus or carcinogen. Established and transformed cell lines have been generated from a wide range of cell types, including a number of embryonic tissues. Examples include teratocarcinoma lines, derived from germ line cells; the erythroleukemic Friend cell line, derived from a precursor of the red blood cell; and cell lines derived from myoblasts, the precursors of skeletal muscle cells. Interestingly, teratocarcinomas and the cell lines derived from them resemble the cells of early embryos. But although they can differentiate into one of several embryonic cell types— a capacity that makes them uniquely suitable for the study of cell commitment — the degree to which these cells reproduce the behaviour of their normal *in vivo* counterparts remains uncertain.

The basis of the altered behaviour seen in immortal cells is not well-understood, although such changes may depend on mutations in genes that control growth. Normal cells make the transition to immortality rapidly if they are treated with viruses or carcinogens, both of which can alter the organization and expression of genes. In addition to having a new karyotype, immortal cells possess other potentially useful features; for instance, they usually have new cell surface properties, are less fastidious about their nutritional requirements and therefore easier to culture, and can be produced in large numbers. One characteristic that facilitates the culture of immortal cells is that established and transformed cell lines are often superficially more consistent than either primary cultures or early passage cells. This trait has let to the erroneous belief that such cells are genetically homogeneous. Genetic homogeneity in cultured cell lines is not common, however. To achieve genetic homogeneity, an individual cell must be selected, isolated, and allowed to proliferate to generate a group of cells, known as a biological *clone*. Once a clone with the desired characteristics has been identified, maintaining its genetic constitution requires further culture under rigorously controlled conditions.

Despite their altered properties, established and transformed cells can differentiate. The rat myoblast cell line known as L8 is an example of an established cell line with the capacity to undergo differentiation, in this case to form muscle. Generated from embryonic rat muscle, L8 is immortal, and individual cells can fuse in culture to form myotubes in a manner closely resembling that of primary rat myoblasts. In contrast, the differentiation of another cell line, L6, is deviant because unlike L8, which arose spontaneously, L6 was generated through the use of carcinogens.

The transformed erythroleukemic Friend cell mentioned earlier offers insight into the nature of the cellular changes that signal the transformation to immortality. Friend cells have been employed extensively as substitutes for erythroblasts, the normal stem cells that give rise to red blood cells, because erythroblasts have very limited proliferative potential. However, although a Friend cell can still undergo aspects of the differentiation from erythroblast to erythrocyte, its behaviour deviates from the normal erythroblast to erythrocyte, its behaviour deviates from the normal erythroblast in two ways. First, it has no sensitivity to the hormone erythropoietin, which normally induces maturation in erythrocytes and it is incapable of undergoing terminal differentiation unless exposed to artificial inducing agents. Second, the extent to which these cells produce globin proteins, the timing of protein synthesis, and other molecular characteristic of the differentiating Friend cell are deviant, suggesting that its regulatory mechanisms are defective. Because of such differences in the performance of normal and transformed cells, investigators have recognized that there are important limits on the use of transformed cells in the study of cell differentiation.

Choosing the Correct Cell Type

To study the effects of interactions between differentiating cells and their environment, it is essential that the cell population remain sensitive to its external milieu. For the experimentalist a number of important factors may be introduced into this environment, including hormones, extracellular matrix components, or other cell types. Environmental factors are likeliest to influence primary cells are early passage cells, whose *in vivo* history is still an important factor in determining their behaviour. Cultures of these cells are susceptible to contamination by other cell types that are present when theses cells are first isolated from the animal or plant. This difficulty can be minimized by techniques that enrich or isolate the particular cells(s) required. When genetic homogeneity is a priority, it is often preferable to use a clonally derived, established cell line, preferably one generated without the use of viruses or carcinogens.

CELL DIFFERENTIATION USING TISSUE CULTURE

Having examined the methods, advantages, and disadvantages of tissue culture, we can now turn our attention to some specific examples. In this section we shall focus on the differentiation of two particular types of cells, those that flow in the blood and those that make skeletal muscle. The differentiation of each of these cell types presents specific problems, and their study in culture has helped investigators address many of the general problems of cell differentiation.

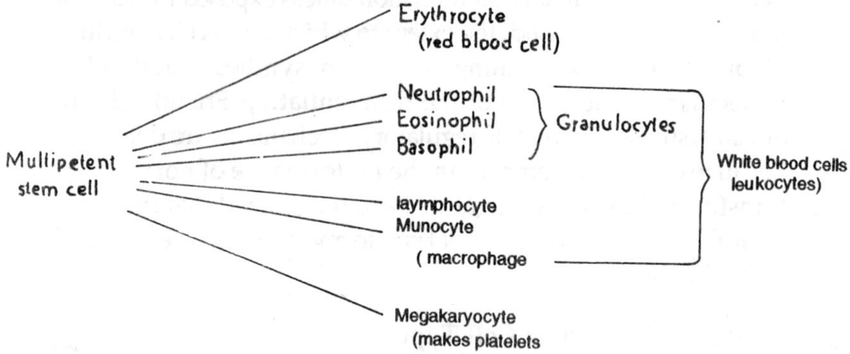

Fig. 11.2. A summary of the range of cell types produced during hemopoiesis.

Haemopoiesis

The haemopoietic tissue system generates the various differentiated cells of the blood. These cells differ from one another in form and function; they include haemoglobin-rich red blood cells, a range of white blood cells, including the T and B lymphocytes active in the immune response, granulocytes, phagocytic cells macrophages derived from monocytes, and massive multinucleate cells called megakaryocytes, which make the platelets involved in blood clotting. New cells are generated at specific locations in the body in association with a non-haemopoietic support tissue called stroma. For example, T lymphocytes are produced in the thymus, and red blood cells are produced in the bone marrow. In each location *multipotent stem cells* divide to produce two daughter cells, thus serving as the starting points in the haemopoietic production line. One daughter remains as a multipotent stem cell, which can generate all of the cell types in the hemopoietic system. The other becomes a *progenitor cell,* whose potential is restricted to a subset of the possible cell types in the system. The progenitor cell is termed a *pleuripotent stem cell,* which means that its progeny are divided into two groups: those that remain as stem cells and those whose fate is further restricted. Those in the second category are destined to differentiate terminally; thus they will choose a particular pathway and become determined. Each of terminally differentiated cell types — erythrocytes, for example — ceases division, and the cells pass into the bloodstream where they survive for a relatively short time, on the order of a few months.

Two central questions must be answered if we are to understand how different kinds of blood cells are generated: What choices do the progenitor cells have, and what are the potencies of the stem cells themselves? These questions revolve around the phenomena of cell commitment, cell determination, and terminal differentiation. In recent years numerous experiments concentrating on a particular phase of the haemopoietic system and using particular types of *in vitro* culture have been carried out. We shall begin our description of some of the key experiments by seeing how the stroma, the support tissue of the hemopoietic system, plays a crucial role in balancing the production of red and white blood cells.

Normal Cellular Interactions from Genetically Abnormal Animals

Numerous examples of genetically inherited defects affecting haemopoiesis occur in mice. For instance, the mutant SL/SL^d mouse has anemia, a blood defect in which the proportion of red cells is abnormally low. This condition cannot be cured by directly injecting normal hemopoietic stem cells but is alleviated by the transplantation of a normal spleen — a finding that strongly suggests that the lesion lies not in the stem cells but the spleen stroma. In contrast to SL/SL^d mice, W/W^v mice are afflicted with an anemia that does respond to the injection of normal stem cells; hence it is thought that the corresponding defect affects the pleuripotent stem cells. The nature of both types of lesions can be further examined in culture.

When stromal and hemopoietic tissues from either SL/SL^d and W/W^v mice are established in culture, they produce only low numbers of red blood cell progenitors. Likewise, few red blood cell precursors are made when stroma from SL/SL^d mice and haemopoietic cells from W/W^v mice are cultured together. However, when the arrangement is reversed and haemopoietic cells from SL/SL^d animals and stroma from W/W^v animals are cultured together, normal numbers of both stem and progenitor cells are produced. This finding supports the notion that SL/SL^d mice have a defective stroma and W/W^v mice defective stem cells. And because these results underscore the role of the stroma in actively controlling differentiation of the red blood cell, they illustrate the importance of appropriate cellular interactions in generating haemopoietic cells.

We shall now consider the first phases of the blood cell production line, taking into account the characteristics of multipotential and progenitor stem cells. The most recent results from experiments on long-term cultures of bone marrow have provided new details on the potency of these stem cells.

Long-Term Bone Marrow Cultures

The normal bone marrow stroma comprises a number of cell types in addition to the hemopoietic stem cells. In long-term cultures of bone marrow, the first cell types to be established include a haeterogeneous collection of phagocytic mononuclear cells, epithelioid cells, and giant fat cells, but haemopoietic stem cells do not survive. The cells that do survive adhere to the surface of the culture vessel, and only when a second inoculum of the less adherent bone marrow cells is introduced to the stem cells survive and proliferate. The stem cells then remain viable for several months, generating multipotential stem cells and their progeny. Within the initial adherent layer, extensive cellular interactions take place between the stem cells and other cells in the layer, and specific cells are restricted to particular locations. For example, the granulocyte precursors associate specifically with the giant fat cells.

In this type of experiment the adhering population of stromal cells is apparently reenacting its *in vivo* role of providing the necessary environment for the successful proliferation of haemopoietic stem cells and their progeny. One feature of this environment appears to be close association of the haemopoietic stem cells with specific non-haemopoietic types of stroma cells. The capacity of this system to sustain both the renewal and differentiation of stem cells is evident from its prolonged production of granulocytes, the precursors of erythrocytes, megakarocytes, and lymphocytes; it thus presents investigators with the opportunity to examine directly the control mechanisms that regulate stem cell activity.

Progenitor cells can be examined in isolation using a specially designed, semisolid gel technique. In a simple experiment, bone marrow cells are seeded into a semisolid agar medium; in the agar only a small number of the cells can proliferate, and those form colonies. Cells within each colony then differentiate into one or more recognizable types. The proportion of the different cell types obtained from a colony depends upon the presence of specific factors that stimulate proliferation and differentiation. For example, the macrophage and granulocyte m-g progenitors require *colony-stimulating factors* (CSF). These growth promoters are present in serum and are synthesized and secreted by several tissues, endothelial cells, fibroblasts, some macrophages, and lymphocytes. Manipulating CSF levels affects cell differentiation; thus high CSF levels result in large colonies, most of which consist entirely of granulocytes, whereas low levels yield smaller colonies consisting only of macrophages.

There are two possible reasons for this result. One hypothesis involves selection; that is, experimental manipulation creates a specific condition that favours the growth of a particular subpopulation of cells. In this case the originating cell need only be *unipotential*. A second hypothesis is interaction, in which the pattern of progenitor cell differentiation depends upon which of the two environmental conditions is present. If this is the case, then the cell is *bipotential* and its direction of commitment is influenced by its external environment.

To test these two possible explanations, several individual stem cells were carefully observed under the microscope. When they divided, the two daughter cells were separated and transferred to paired cultures containing either identical concentrations of CSF or one high and the other low. The proliferation of each of the single cells was then monitored, and after six days the number and types of cells in each colony were scored. When both daughters were placed in identical high concentrations of CSF, similar proportions of pure granulocyte colonies were produced. Exposure of one or both daughters to low CSF levels, however, resulted in these cells producing higher numbers of macrophage colonies — the same outcome obtained for cells grown in low levels of CSF from the start. This result clearly demonstrates the bipotential nature of the progenitor cells. In effect, the fate of the daughter cells is influenced by the CSF level, an external factor known to be secreted by other cells of the haemopoietic system. Little is known about the decision between alternative cell fates except that it appear to be linked in some way to the proliferative cycle. Not only are the colonies that differentiate into macrophages smaller than their granulocyte counterparts, indicating that constituent cells undergo fewer rounds of cell division, but the duration of their cell cycle is lengthened.

It is likely, however, that the control of macrophage and granulocyte numbers is more complex than this experimental result suggest, because four different macrophage-granulocyte CSFs have now been purified from mice. Although each CSF is a glycoprotein of similar molecular weight, each has a different amino acid sequence. Membrane-bound receptor molecules unique for each of these CSFs have also been identified on most granulocytes and macrophages, implying that each factor plays a role in controlling macrophage-granulocyte differentiation.

In order for the blood cell production system to work effectively, the correct numbers of each terminally differentiated cell type must be manufactured. This means that the correct balance of multipotential and progenitor stem cells must be maintained and that sufficient numbers of the differentiated derivatives of the progenitor stem cells must follow each possible path. If the process controlling the stem cell divisions or the balance of choices for terminal differentiation go wrong, the blood-stream may become flooded with one particular cell type. The results of such an aberration are likely to be serious for example, leukemias, diseases of the blood, are caused by massive increases in white cell numbers. In a normal blood count, approximately 8000 white cells are found per cubic millimeter, whereas in some forms of leukemia this number is over 300,000. Such dramatic alterations in the balance of the haemopoietic system are commonly fetal.

The Erythropoietin

Erythropoietin, a hormone produced mainly in the kidney, regulates the numbers of mature erythrocytes released into the bloodstream from the bone marrow. The hormone is a glycoprotein, and the amount released by the kidney is determined by the number of circulating red blood cells; a shortage of red blood cells stimulates the release of erythropoietin, and an excess curtails it. The mechanism of action or erythropoietin has been studied using dispersed bone marrow cells in culture. The results show that the number of red blood cells produced by the stem cells of the erythrocyte lineage is highly sensitive to the concentrations of erythropoietin in the medium. Thus, if the hormone is absent, no red cells are produced. However, if a low concentration of hormone is added to the medium, small colonies of red cells appear over a 3-day period, each colony comprising less than 100 cells. In contrast, when the concentration of erythropoietin in the medium is significantly increased, colonies of red blood cells are produced over a longer period, up to 10 days, and each colony comprises thousands of cells. The reason for this effect of concentration lies in the nature of the stem cell production line of the erythrocyte lineage. A low concentration of erythropoietin has a short-term effect on an erythrocyte stem cell that is both committed and close to terminal differentiation. This stem cell can nonetheless divide a limited number of times, giving rise to fewer than 100 red cells. The effect of a high concentration of the hormone targets a committed erythrocyte stem cell that is further from terminal differentiation and able to divide more actively, producing thousands of cells. These two types of unipotential stem cell within the erythrocyte cell lineage can be histologically identified, the one closer to the terminally differentiated erythrocyte being smaller.

DIFFERENTIATION OF SKELETAL MUSCLE

Cell Culture is also a crucial technique in the example we take up next the differentiation of skeletal muscle cells. The main difference between the differentiation of muscle cells and that of blood cells lies in the number of choices open to the precursor cells. In the haemopoietic group, at least two eventual cell types can form from the progeny of the multipotent stem cell. By contrast, myogenesis involves a single choice — whether to differentiate into a muscle cell. Little is known about the initial decision of a cell to join the myogenic cell lineage, but *in vitro* studies have provided a great deal of information about the later differentiation of the myoblast. In this section we shall discuss what is known about commitment, determination, and terminal differentiation of the skeletal muscle cell.

The Differentiation of Myoblast

The skeletal muscle of the forelimbs and hindlimbs of vertebrates is thought to originate from cells that migrate into the limb from the somites, the segmental blocks of tissue located in the mesoderm of the embryo. Once inside the limb, the myoblasts proliferate extensively and gradually align in rows along the direction of the future muscle fibers. In the chick, the rate of proliferation of these cells has decreased by the 8th or 9th day of development. By 11 days the cell membranes between neighbouring, longitudinally aligned cells begin to fuse, creating the multinucleate myotubes characteristic of skeletal muscle. DNA synthesis and cell division cease, and many muscle-specific proteins are synthesized. These include the proteins actin and myosin, which are associated with the contractile apparatus, the receptor for acetylcholine, and enzymes such as creatine phosphokinase (CPK) and acetylcholine esterase. The final maturation of each myotube requires contact with a nerve, which promotes further changes to generate a specific type of muscle fiber.

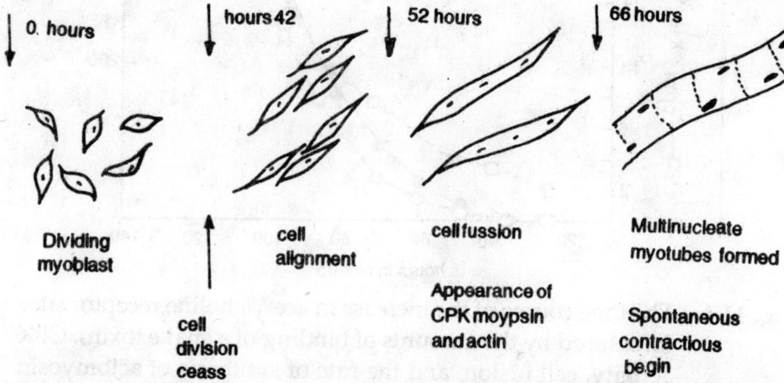

Fig. 11.3. The sequence of events associated with the differentiation of chick myoblasts in culture.

Embryonic myoblasts isolated from the myogenic regions of the limbs and breast have been studied *in vitro*. Since only muscle is generated from the cells that migrate into the limb, it is probable that the limb cells migrating from somites are already committed to the myogenic pathway. Embryonic myogenic cells cultured *in vitro* undergo changes that parallel those seen *in vivo*. Thus, after a variable phase of proliferation whose duration depends upon both the initial cell density and the concentration of growth factors, the cells begin to fuse and to accumulate muscle-specific products. The typical time course of differentiation obtained for avian myoblasts seeded at high density and exposed to fresh medium after 24 hours of culture.

The myotubes gradually mature, eventually exhibiting cross-striations, the *in vitro* muscle even twitches spontaneously.

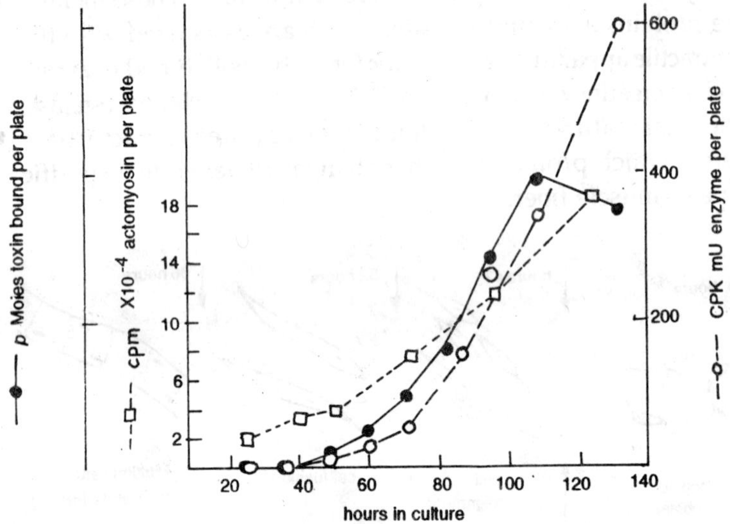

Fig. 11.4. The time course of the increase in acetylcholine receptor sites (measured by the amounts of binding of a snake toxin). CPK activity, cell fusion, and the rate of synthesis of actomyosin in chick myoblast cultures.

Molecular studies of muscle differentiation *in vitro* are possible because (1) the myoblast has extensive proliferative potential; (2) the proportion of nonmyogenic cells can be kept very low, or if desired, reduced to zero by clonal selection; and (3) a high proportion of the myoblast population can initiate differentiation with a high degree of synchrony. Differentiating myotubes produce high levels of CPK, acetylcholine receptor sites, and actomyosin. It is worth noting that similar biochemical studies cannot be carried out *in vivo* because the individual cells would undergo differentiation asynchronously and over a much longer period of time.

From *in vitro* studies such as those just cited, it is cleat that commitment to the myogenic pathway is a very early step that is independent of the myoblast's subsequent decision to undergo terminal differentiation. Cells prepared from the myogenic regions of 18-day rat embryos proliferate extensively after exposure to growth factors and with frequent passaging. Only when transferred to conditions compatible with differentiation do these cells express their myogenic nature by terminally differentiating into mature myotubes. Because this terminal differentiation is not affected by the duration of the proliferation phase, we can conclude that the committed state is very stable and is inherited by successive generations of daughter cells. Experimental results show, however, that the myogenic commitment alone is not sufficient to produce a muscle; terminal differentiation requires an appropriate environment.

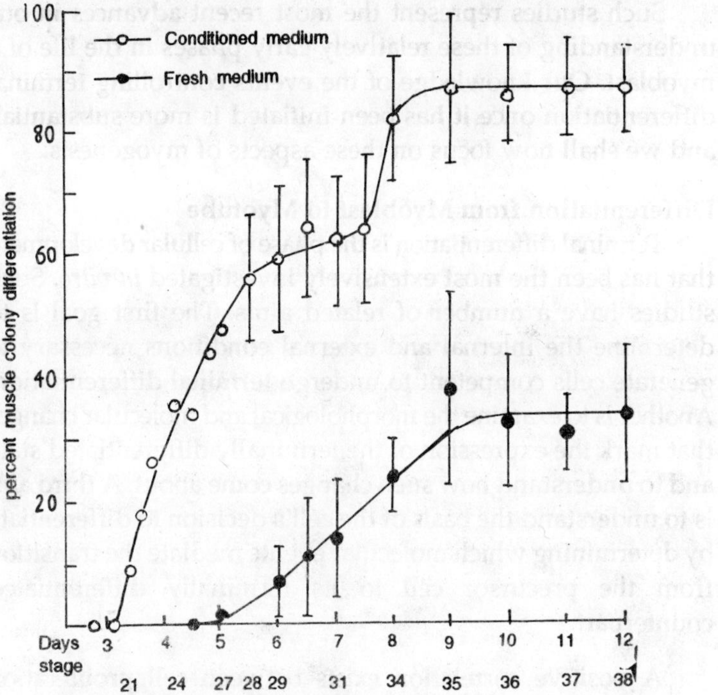

Fig. 11.5. The muscle-colony-forming cell fraction of the clonable cell population isolated from chick embryos limbs at different stages of development. Clones were grown either in medium taken from differentiating mass cultures of chick myoblasts conditioned medium or in fresh medium.

We can conclude, then, that the events of commitment and terminal differentiation are separable and can be independently regulated. In addition, it is interesting to note that the ability of cells to undergo terminal differentiation may arise gradually within a population of committed myoblasts. This possibility is suggested by a set of experiments involving cells taken from the developing chick limb bud. If cells derived from the myogenic region of a young limb bud are placed on a culture dish, only a few undergo terminal differentiation. With increasing age of the embryo, however, an increasing proportion both of clones and of cells within a clone undergo terminal differentiation.

Such studies represent the most recent advances in our understanding of these relatively early phases in the life of a myoblast. Our knowledge of the events controlling terminal differentiation once it has been initiated is more substantial, and we shall now focus on these aspects of myogenesis.

Differentiation from Myoblast to Myotube

Terminal differentiation is the phase of cellular development that has been the most extensively investigated *in vitro*. Such studies have a number of related aims. The first goal is to determine the internal and external conditions necessary to generate cells competent to undergo terminal differentiation. Another is to examine the morphological and molecular changes that mark the expression of the terminally differentiated state and to understand how such changes come about. A third aim is to understand the basis of the cell's decision to differentiate by determining which molecular events mediate the transition from the precursor cell to its terminally differentiated counterpart.

A positive correlation exists between cell proliferation and the degree of terminal differentiation in primary myoblast cultures. This finding suggests that the cell cycle must be completed for the majority of the cells to undergo terminal differentiation *in vitro*. Consistent with this idea is the observation that the drug fluorodeoxyuridine, which prevents the initiation of DNA synthesis, can also prevent the onset of differentiation. A connection between the cell cycle and myoblast fusion is further supported by the observation that fusion occurs only during the G_1 phase, which is prior to the next round of DNA synthesis.

Environmental Approach

Although these observations suggest a relationship between the cell cycle and the decision of myoblasts to fuse, number of observations indicate that environmental cues trigger the event. Once fusion is set in motion, the myoblasts can only fuse during G_1. The evidence for environmental influences controlling the fusion event comes from several different experiments. Of particular importance is the influence of the culture medium on the timing and extent of terminal differentiation in the myoblast population; for instance, exposure to fresh medium and to certain growth factors markedly delays the onset of differentiation. The reverse is also true. When medium from differentiating cultures (*condition medium*) is placed in contact with younger, rapidly proliferating cultures, the young cultures respond by undergoing terminal differentiation much sooner than normal.

The cell's substrate also plays a crucial role in terminal differentiation. Typically, the chick myoblast is grown on a collagen substrate in a culture medium that contains serum and embryo extract as growth promoters. Under these conditions, the myoblasts proliferate, take up a bipolar shape, and align along their longitudinal axes. The direction of alignment appears to be influenced by the extracellular matrix molecules, collagen and fibronectin, which form the substrate. If either collagen or fibronectin is applied to the dish in a single direction, alignment and fusion follow the direction of the underlying fibers. This influence of the underlying substrate suggests that it may also be important in determining the alignment of muscle fibers *in vivo*.

Finally, some recent evidence suggests that myoblasts begin terminal differentiation prior to fusion. If this is so, then environmental cues must be at work, with fusion being a secondary event despite its clear relation to the cell cycle. This evidence centers on the detection of myosin, a protein component of the contractile machinery of the differentiated myotube, in prefusion myoblasts. Myosin can be identified using a sensitive staining method involving antibodies raised against the myosin molecule. It is clear from these results that mitotically active, non-fused myoblasts can begin to terminally differentiate.

We have already described three of the major aspects of terminal differentiation — the production of myosin, the fusion of myoblasts, and the fact that fusing myoblasts no longer divide. Although the causal sequence of these events remains unclear, a further sequence of changes in gene expression is known to follow these early events. The differentiation of both the plasma membranes and the cytoplasm of the muscle cell has been examined. This work has focused on when the muscle-specific cellular components first appear and which points during transcription and translation regulate these components. Products such as contractile, proteins and acetylcholine receptors begin to accumulate in the muscle cell at approximately the same time as fusion begins. The amounts of these substances increase linearly for about two days after the myotube forms. After this time, the number of acetylcholine receptors that are inserted into the muscle plasma membrane levels off, while the quantities of contractile proteins and creatine phosphokinase continue to increase. It is thought that the extended production of the contractile proteins and related components reflects the gradual increase in the number of myofibrils in the myotubes, even though the size of the plasma membrane remains roughly the same.

Some clues about how protein synthesis is regulated can be gained by examining the overall pattern of protein synthesis in myoblasts. One method is to incubate the myogenic cultures with the radioactively tagged amino acid methionine, separate the resulting labeled proteins by two-dimensional gel electrophoresis, and visualize these proteins by autordiography. By using this method, it is possible to display the pattern or "fingerprint" of the proteins being actively synthesized by the cells. By comparing such patterns in myoblasts and myotubes, the number and type of synthetic changes can be monitored. The majority of proteins visualized by this method — that is, those that are coded by mRNAs belonging to the more prevalent categories — are present in both myoblasts and myotubes. However, the autoradiographs do reveal both quantitative and qualitative changes. As expected, a number of proteins - such as those associated with the contractile apparatus — appear for the first time, while others disappear. Similarly, the rate of synthesis increases for some proteins and falls for others. A final category of proteins is synthesized in an irregular pattern. Using cultures that have been highly synchronized by manipulation of the medium, researches have examined the initial increases of some of the newly appearing proteins more closely. In these synchronized cells, many new proteins begin to be synthesized simultaneously. In addition, the rates of synthesis are virtually identical for seven out of nine of the contractile proteins studied. Taken together, these experiments demonstrate co-ordinated changes in the patterns of protein synthesis that are part of the expression of the differentiated state of the muscle cell. Despite the diverse nature of such molecular changes, it is clear that they are governed by a high degree of temporal order.

The Regulation of Protein Synthesis

When protein synthesis is inhibited with cycloheximide, the production of muscle-specific proteins is blocked almost immediately. However, even after most mRNA synthesis in primary cultures of rat myoblasts has been inhibited by actinomycin D for up to six hours before the onset of cell fusion, the initial accumulation of muscle-specific proteins after fusion does not stop. This finding suggests that transcription is regulated prior to the fusion of myoblasts and that mRNAs

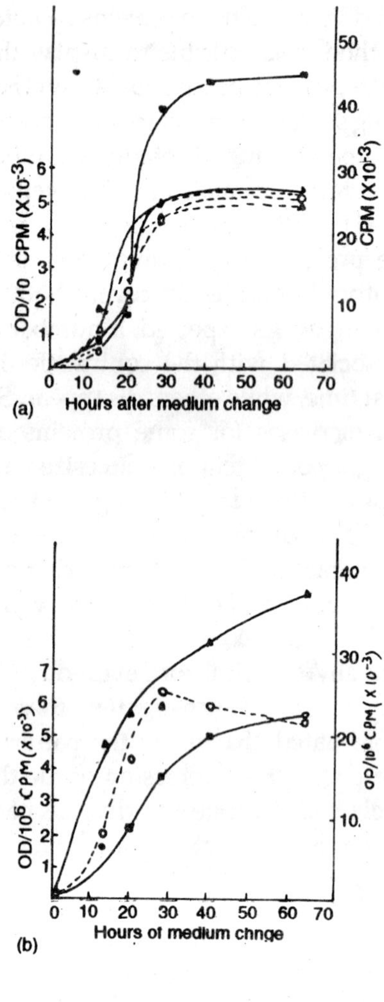

Fig. 11.6. The kinetics of the synthesis of contractile proteins during myoblast differentiation. Labeled proteins were isolated at different times of culture and separated by two-dimensional gel electrophoresis: (a) Light chains 1 (●) and 2 (O), troponin t (Δ) and c (Δ), and myosin heavy chain (■), (b) Tropomyosin 1 (O) and 2 (●), and actin (Δ).

are stored for later use in the myotube. However, a more direct examination of mRNA production, using cDNAs specific for the individual mRNAs that code for muscle-specific-proteins, has shown that for many muscle-specific proteins there is little or no lag between the production of mRNAs and the accumulation of proteins in the muscle cell. It is not yet known whether the increase in mRNA concentration results from higher transcription activity or a change in the nuclear processing. The fact that a significant increase in myosin mRNAs is correlated with a 20- to 30-fold rise in RNA polymerase molecules associated with the myosin gene is evidence that new transcription does indeed take place. A high frequency of transcription would also explain the relative insensitivity of myosin mRNA to inhibition by actinomycin D.

12

Hormones and Differentiation

It has already been seen that during embryogenesis in plants and animals many cell types differentiate. These cells are organized into tissues and organs; each organ then grows to a predetermined size to form a properly proportioned adult body. In part, each tissue and organ is responsible for controlling its own growth, especially in the early stages of development. As the adult form becomes more complex, however, some overall co-ordination of growth and tissue homeostasis is required. This co-ordination of continued growth and differentiation is provided by the chemical messengers known as *hormones*. Hormones are particularly important for timing developmental events. Examples that we shall consider in some detail include metamorphosis in amphibians an insects, in which hormones trigger large-scale remodeling of body parts, and the maturation and release of oocytes, which occur in a highly ordered sequence. The hormones that control these and other events of differentiation vary in structure and mode of action.

GENERAL CHARACTERISTICS OF HORMONES

Hormones are the chemical messengers or internal secretions. These are organic molecules synthesized and secreted in the animal body by specific cell types known as endocrine cells; after diffusing into adjacent capillaries, hormones are carried through the circulatory system to parts of the body that can be distant or nearby. The various hormones have their effects when minute quantities interact with specific target cells to produce specific responses.

Hormones work in their respective target organs and are regulatory molecules; that is, they do not participate in the evoked reaction itself, as would an enzyme or a cofactor; instead, they trigger, enhance, or inhibit preexisting capabilities or biosynthetic reactions within the target cell. For example, fat is broken down by cells in adipose tissue at a certain rate, and the hormone insulin decreases that rate; conversely, heart muscle contracts in the absence of epinephrine but beats with greater force when that hormone is present. Some developmental process are totally dependent on hormones. Thus, the maturation of mammalian oocytes requires the presence of follicle-stimulating hormone (FSH), and metamorphosis is initiated and sustained by thyroid hormones.

Chemically hormones fall under the categories : *Amino acid* derivatives live thyroxin; *catecholamine* like epinephsine; *protein* like insulin and glycogen; *glycoproteins* like FSH, LH and TSH; *fatty acid* derivatives like prostagladins; and steroids which are based on Chflestron and are lipid soluble. Some of the lipid insolubly hormones like the small peptide hormones, such as the hypothalamic peptide *thyrotropin-releasing factor* (TRF), are synthesized by non-ribosomal enzymes in the cell cytoplasm. Others, such as the neural pituitary hormones *vasopressin* and *oxytocin*, are encoded by mRNA and produced by the normal ribosomal machinery for synthesizing proteins. Most glycoprotein hormones are synthesized as pre-pro-hormones. Glycosylation and additional processing of the pre-pro-hormone complexes take place in the Golgi, and the mature hormones are packaged into membrane-bound vesicles. When the hormone-containing cell is appropriately stimulated, these vesicles move to the plasma membrane and release their contents by exocytosis into the extracellular space; the hormone then enters the capillaries by diffusion.

TABLE 12.1

The Characteristic of Hormones

	Site of Synthesis	*Principal Action*
Group 1.Steroids (Lipid-Soluble Molecules Based on Cholesterol)		
1. Estradiol	Ovary, placenta	Development and maintenance of secondary female sex characteristics
2. Testosterone	Testis	Development and maintenance of secondary male sex characteristics
Group 2. Lipid-Insoluble Molecules		
Amines (amino acid derivatives)		
1. Epinephrine	Adrenal medulla	Release of fatty acids from fat cell raises blood pressure and rate of heartbeat
2. Thyroxine	Thyroid gland	Increase in general metabolic activity in most cell types
Small Peptides		
1. Thyroid-stimulating hormone releasing factor	Hypothalamus	Stimulates pituitary to release thyroidstimulating hormone
2. Vasopressin	Posterior pituitary	Elevates blood pressure; increases water reabsorption in kidney
Proteins		
1. Insulin	Beta cells of pancreas	Utilization of carbohydrates; general stimulation of protein synthesis, lipid synthesis of fat cells.
2. Parathormone	Parathyroid glands	Bone reabsorption and consequent raising of blood Ca and phosphate.
Glycoproteins		
1. Follicle-stimulating hormone	Anterior pituitary	Stimulates egg maturation and progesterone release in ovaries; stimulates testosterone release from testes.
2. Luteinizing hormone	Anterior pituitary	Stimulates follicles to grow and release estradiol in ovaries; stimulates spermatogenesis in testes.

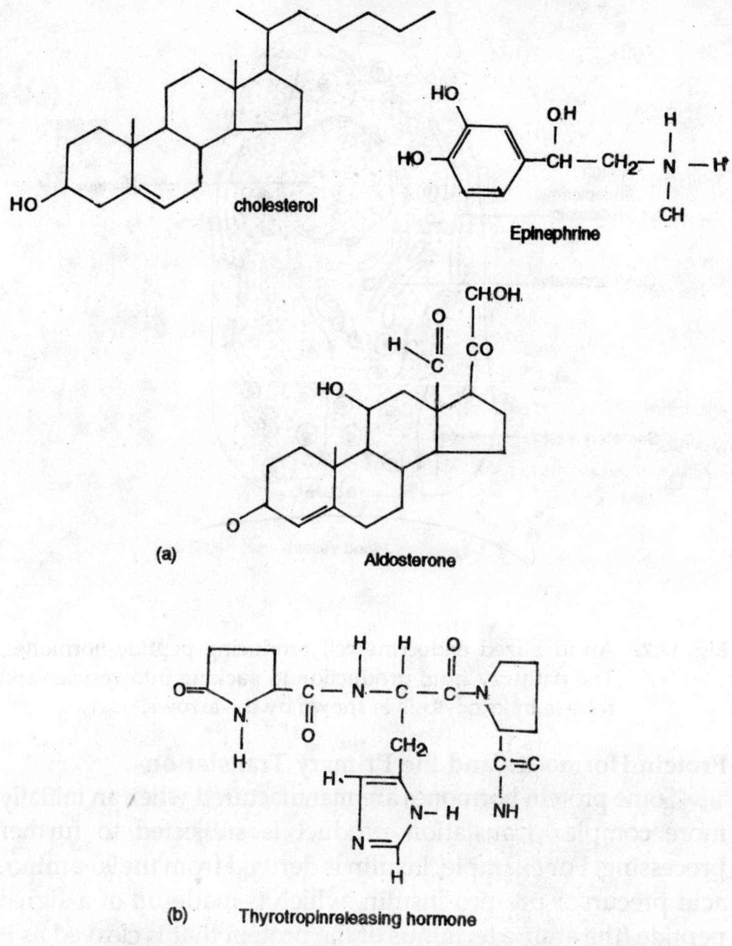

Fig. 12.1. Vertebrate hormones can be broadly categorized into two classes: lipid-soluble molecules such as the steroids and lipid-insoluble molecules. (a) Lipid-soluble. Cholesterol and aldosterone, a hormone based on cholesterol. If the COH of aldosterone were an H atom, the hormone would be corticosterone. The lipid solubility results from the extended, conjugated ring system. (b) Lipid insoluble. The hormone epinephrine and thyrotropin-releasing hormone and lipid insoluble because of the hydroxyl groups and charged quaternary amine (epinephrine) and the nitrogen-containing rigs (thyrotropin-releasing hormone).

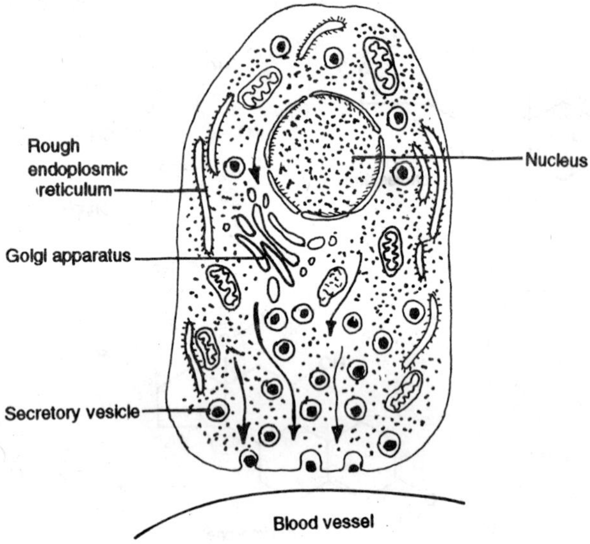

Fig. 12.2. An idealized endocrine cell producing peptide hormones. The pathway from production to packing into vesicles and release by exocytosis is shown by the arrows.

Protein Hormones and the Primary Translation

Some protein hormones are manufactured when an initially more complex translation product is subjected to further processing. For example, insulin is derived from the 96-amino-acid precursor pre-pro-insulin, which is made up of a signal peptide (the amino terminus of the protein that is cleaved as it locates the ribosome involved in the pre-pro-insulin molecules' translation on the membrane of the endoplasmic reticulum) and three regions known as A, B and C peptides. The signal peptide is removed during translation, and the 33-amino-acid internal C peptide is removed by proteolytic enzymes. This leaves the A and B chains, which form the active hormone, attached by disulfide bonds. The precursor to the hormone adrenocorticotrophic hormone (ACTH) also contains peptide sequences of at least six other hormones. In addition, some multimeric polypeptide hormones share a common chain. Thus,

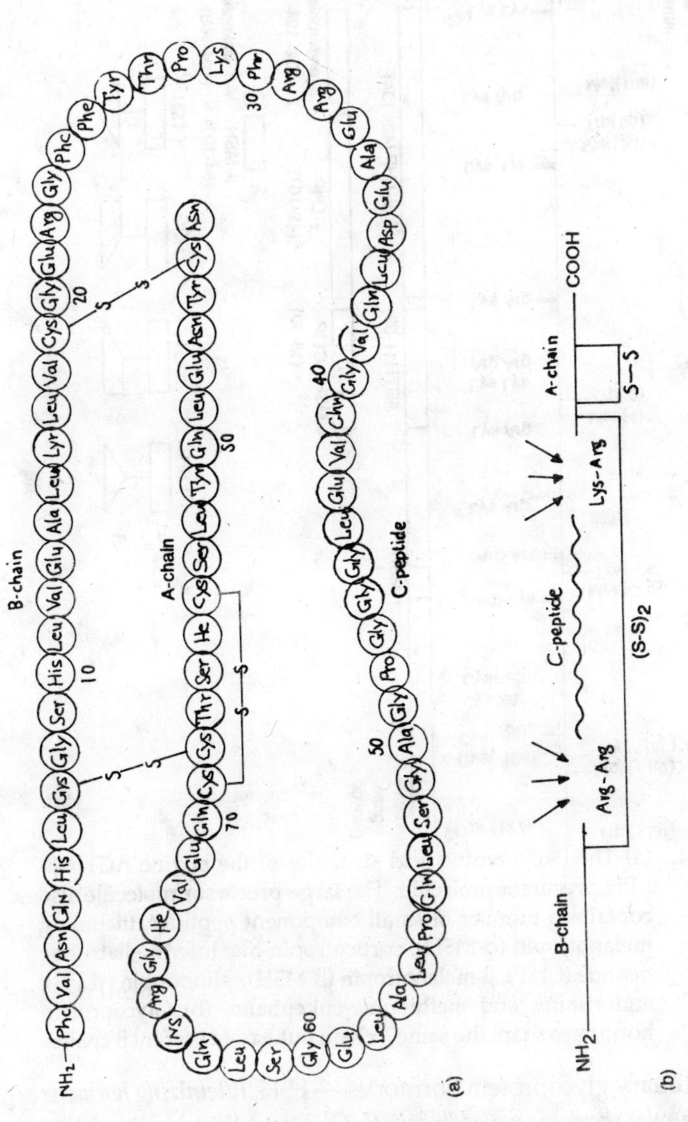

Fig. 12.3. (a) The proposed structure of human proinsulin showing A, B and C chains before cleavage. (b) The points of cleavage (arrows) by enzymes and the positions of sulfide bonds, which eventually hold the A and B chains together.

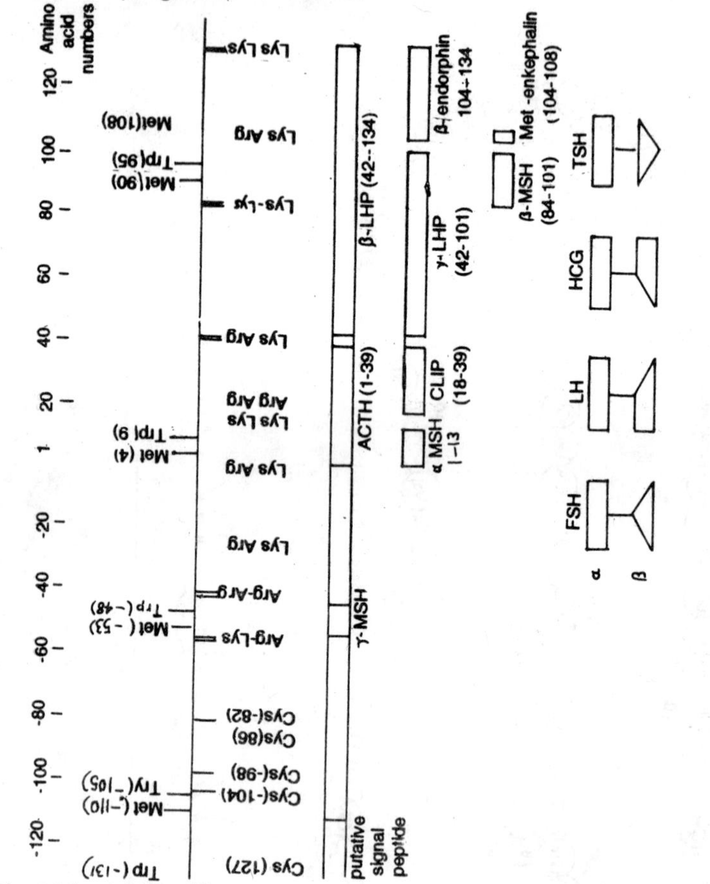

Fig. 12.4. (a) The likely amino acid structure of the bovine ACTH-β-LPH precursor molecule. The large precursor molecule also contains a number of small component peptides, including melanotropin (α-MSH), corticotropin-like intermediate lobe peptide (CLIP), β-melanotropin (β-MSH), γ-lipotropin (γLPH), endorphins, and methionine-enkephalin. (b) Glycoprotein hormones share the same _ chain but have different β chains.

four pituitary glycoprotein hormones — FSH, *luteinizing hormone* (LH), *human chorionic gonadotropin* (HCG), and *thyroid-stimulating hormone* (TSH) — are each composed of one alpha and one beta subunit; the 96-amino-acid alpha subunits of the four hormones are identical, but the beta subunits, which are the parts of the hormones that produce the functional specificity are different.

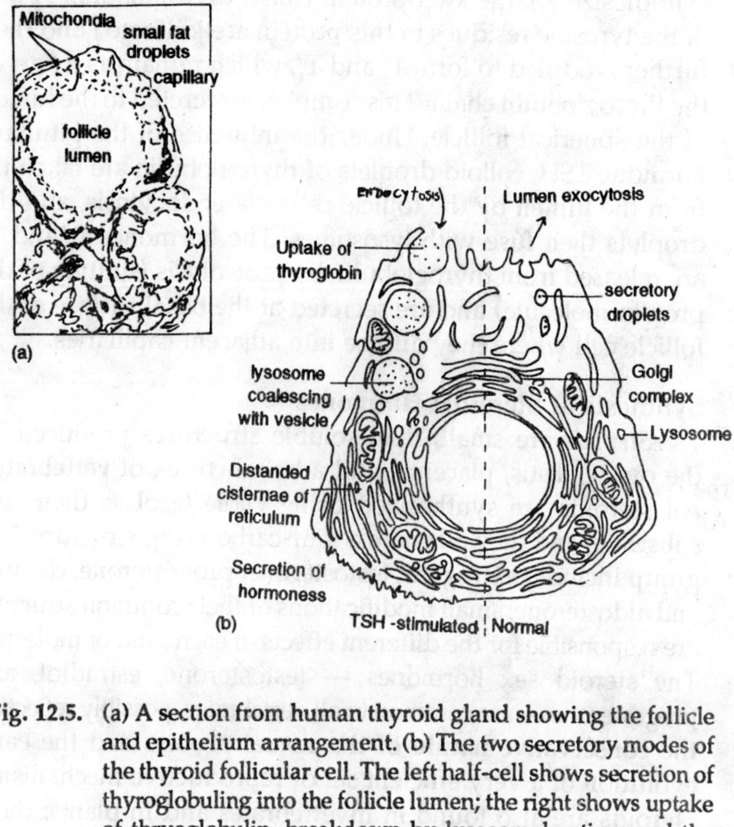

Fig. 12.5. (a) A section from human thyroid gland showing the follicle and epithelium arrangement. (b) The two secretory modes of the thyroid follicular cell. The left half-cell shows secretion of thyrogobuling into the follicle lumen; the right shows uptake of thryogobulin, breakdown by lysosomal action, and the release of thyroxine from the capillary surface of the cell.

Synthesis of Hormones

Hormones derived from amino acids can be manufactured in very different ways. For example, *epinephrine* is an amine — a modified residue of the amine acid tyrosine, and the hormone is synthesized using both mitochondrial and cytoplasmic enzymes. The molecule is then packaged as granules for secretion by exocytosis. The thyroid hormones *thyroxine* (tetra-iodothyronine, or t_4) and *tri-iodothyronine* (t_3) are also modified tyrosine residues but are derived by an entirely different pathway. In this case the follicle cells of the thyroid gland synnthesize a large swcoprotein called thyroglobulin. A few of the tyrosine residues in this protein are iodinated and then further modified to form T_3 and T_4, which remain as a part of the thyroglobulin chain. This complex is secreted to the lumen of the spherical follicle. Under the influence of the pituitary hormone TSH, colloid droplets of thyroglobulin are taken up from the lumen by the follicle cells via endocytosis, and the droplets then fuse with lysosomes. The hormones T_3 and T_4 are released from thyroglobulin by proteolysis (spitting of the protein molecule) and are secreted at the basal surface of the follicle cell where they diffuse into adjacent capillaries.

Synthesis of Steroidal Hormones

Steroids are small, lipid-soluble structures produced by the ovary, testis, placenta, and adrenal cortex of vertebrates. All steroids are synthesized using cholesterol as their first substrate, and they all share a four-carbon ring structure. The group includes estrogens, testosterone, progesterone, cortisol, and aldosterone; small modifications of their common structure are responsible for the different effects of each kind of molecule. The steroid sex hormones — testosterone, estradiol, and progesterone — are common to all vertebrates, possibly reflecting the conservative nature of the sexual process and the early evolution of a very efficient set of reproductive mechanisms. Steroids are also found in invertebrates and in plants; those obtained from plants are sometimes used as starting material for preparing vertebrate hormone analogs in the laboratory.

The Response to a Hormone

One of the most important features of hormone action is the fact that the specificity of the response lies with the target cell. For instance, the hormone prolactin is carried in the bloodstream and is "seen" by almost all the cells of the body; however, it is recognized only by certain cells in the mammary gland, which respond by producing milk proteins, lactose, and fat. Alternatively, many hormones, such as estrogen, have several different types of target cell, each showing a different, characteristic response. For example, in the chicken oviduct, cells respond to estrogen by proliferating and synthesizing large amounts of egg white proteins, whereas in the liver estrogen promotes the synthesis of the yolk protein precursor vitellogenin, which is secreted into the blood and then taken up and processed by growing oocytes.

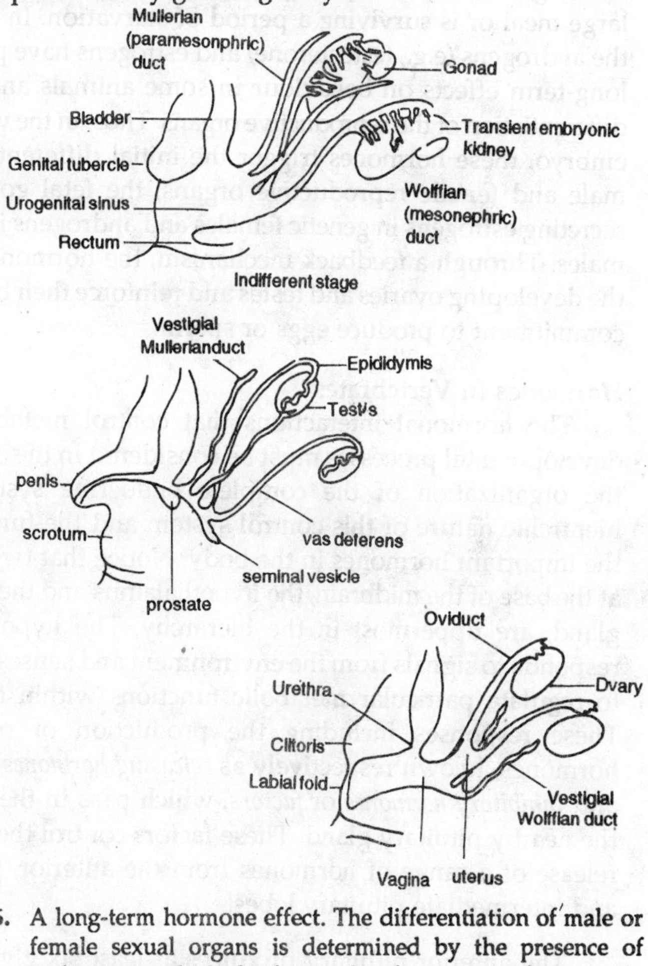

Fig. 12.6. A long-term hormone effect. The differentiation of male or female sexual organs is determined by the presence of androgens, which leads to the formation of the male reproductive structures.

Hormonal Effects

Hormones regulate two major types of activities: quick-acting, short-duration events involved in controlling metabolite levels and slower-acting, long-duration events that are characteristic of developmental processes. For example, insulin is a metabolically important hormone designed to regulate blood glucose levels quickly when the organism ingests a large meal or is surviving a period of starvation. In contrast, the androgens (e.g., testosterone) and estrogens have profound long-term effects on behaviour in some animals and on the differentiation of the reproductive organs. Thus, in the vertebrate embryo, these hormones trigger the initial differentiation of male and female reproductive organs, the fetal gonads, by secreting estrogens in genetic females and androgens in genetic males. Through a feedback mechanism, the hormones act on the developing ovaries and testes and reinforce their biological commitment to produce eggs or sperm.

Hormones in Vertebrates

The hormonal interactions that control metabolic and developmental processes must be considered in the context of the organization of the complete endocrine system. The hierarchic nature of this control system and the functions of the important hormones in the body. Notice that two regions at the base of the midbrain, the hypothalamus and the pituitary gland, are uppermost in the hierarchy. The hypothalamus responds to signals from the environment and senses the need to regulate particular metabolic functions within the body. These responses including the production of regulatory hormones, known respectively as *releasing hormones* or *factors* and *inhibitory hormones* or *factors*, which pass in the blood to the nearby pituitary gland. These factors control the levels of release of a range of hormones from the anterior, posterior, and intermediate pituitary lobes.

The anterior pituitary produces at least six glycoprotein hormones, each manufactured by a different cell type. Two of these substances, *prolactin* and *growth hormone*, are under both

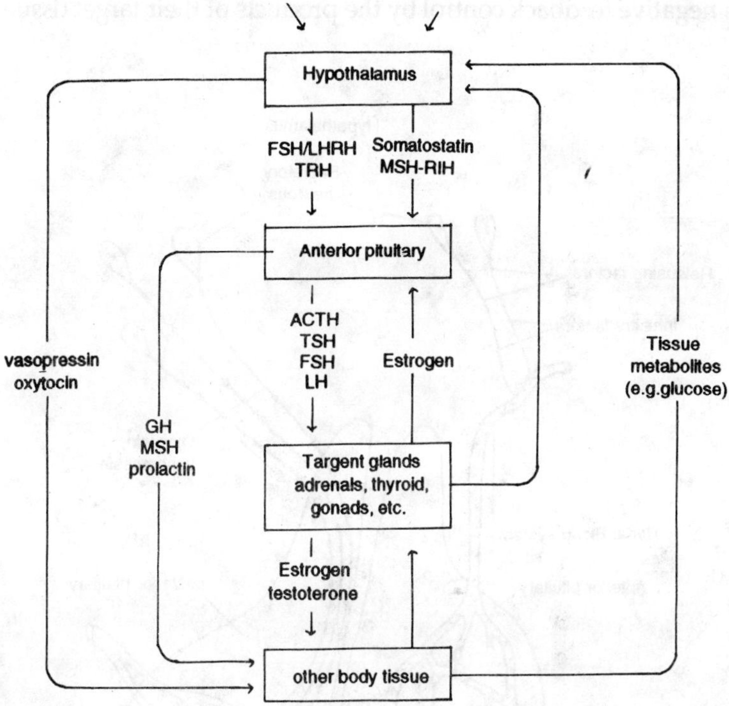

Fig. 12.7. The hierarchical organization of the endocrine system in vertebrates. Note the central roles played by the hypothalamus and anterior pituitary gland and the importance of feedback control loops. The two arrows at the top of the figure indicate the presence of input to the hypothalamus from higher centers of the central nervous system. RH, releasing hormones; IH, inhibitory hormones. All other abbreviations appear in the text.

positive and negative hypothalamic control; that is, their release is governed by both releasing and inhibitory factors. The targets of these two pituitary hormones are non-endocrine cells. By contrast, the other four anterior pituitary hormones — TSH, LH, FSH and ACTH — all have other endocrine glands as their targets. As a result, their synthesis and release is under positive feedback control by the hypothalamus and under negative feedback control by the products of their target tissues.

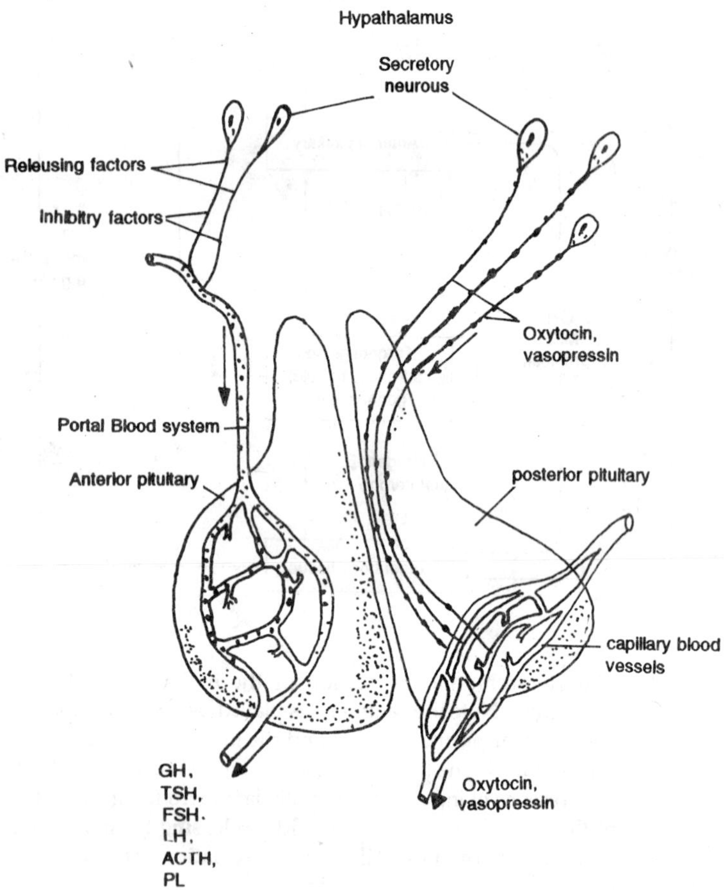

Fig. 12.8. The hypothalamic connection with the pituitary gland. On the left, the portal system, connecting the secretory neurons with the anterior pituitary, carries releasing and inhibitory

factors. On the right, the axons of secretory neurons make up the posterior pituitary. Here they secrete oxytocin and vasopressin into the bloodstream.

The intermediate lobe of the pituitary secrets melanocyte-stimulating hormone, which stimulates pigment cells to produce melanin, the pigment responsible for the dark colour of hair and freckles. The posterior lobe comprises the axonal processes of nerve cell bodies, which are located close by in the brain. These nerve cells synthesize oxytocin and vasopressin, which are transported along the axons to the posterior pituitary, where they are released into the bloodstream.

In vertebrates a series of endocrine glands situated in different places outside the brain secrete a variety of hormones, generally in direct response to hormones released from the pituitary gland. The major glands in the peripheral tissues are the gonads, the adrenal (situated next to the kidney), the thyroid in the neck, cells of the gut lining, and the pancreas, which contains specialized hormone-producing cells grouped together within its mass in small regions called Islets of Langerhans. The bulk of the pancreatic cells produces enzymes that are directly involved in digestion and released into the gut through ducts.

A variety of mechanisms trigger the release of hormones from different endocrine tissues. For example, the adrenal steroids, gonadal steroids, and thyroid hormones are under the control of the pituitary and hypothalamus, while the presence of partially digested food in the stomach triggers the cells of the epithelial lining to secrete the hormone gastrin. This small polypeptide passes into the blood-stream and then stimulates the release of digestive enzymes from the same epithelium.

A Single Hormone in Different Animals

Earlier we saw that a single hormone, such as estrogen, can produce different responses indifferent target tissues. Likewise, a particular hormone can also elicit different responses in different species. Prolactin is a good example; first identified in mammals and characterized by its ability to stimulate the proliferation of mammary gland tissue and subsequent milk production, this hormone is active in a wide variety of non-mammalian vertebrates. Thus, pituitary prolactins extracted from birds, reptiles, and amphibians all promote the milk production response in mammals. In birds, prolactin stimulates courtship and nesting behaviour; for example, roosters treated with the hormone will take care of young chicks by leading them to food and water, sheltering them, and protecting them from predators. Prolactin causes certain salamanders to return to water for breeding, while in fishes it is involved in osmoregulation and is essential in animals such as salmon species that migrate between fresh and salt water. Finally, in invertebrates prolactin produces a wide range of effects, which fall into four general categories: reproduction and parental care, osmoregulation, pigmentation of the skin, and growth. In all, researches have described over 60 different prolactin effects.

MECHANISMS OF HORMONE ACTION

Lipid-insoluble hormones and lipid-soluble (steroid) hormones exert their effects in two very different ways. On the one hand, lipid-insoluble hormones (peptides, proteins, etc.) interact with receptors on the cell surface, and there is considerable evidence that many do not have to enter cells to mediate a response. Instead, it is thought that the peptide hormone-receptor complex initiates a cascade of activation (or deactivation) of specific preexisting intracellular enzymes. In contrast, the lipid-soluble steroid hormones cross the plasma membrane and complex with receptors inside the cell; the hormone-receptor complex inside the nucleus interacts with the genome, resulting in the production of specific mRNAs and proteins. These two different modes of action are reflected in the two kinds of reactions of the different hormone types: In general, lipid-insoluble hormones work rapidly, usually within seconds to minutes; steroid hormones act more slowly, over a range of several hours.

Lipid-Insoluble Hormones and Receptors

Some lipid-insoluble hormones are believed to interact directly with receptors in the plasma membrane's outer surface. Although research results are not conclusive, three lines of evidence support this idea. In the first case, radioactively labeled hormones bind to the outer surface of cell membranes of intact cells under conditions where no radioactivity above background levels is detectable inside the cell. Also, when some hormones are covalently linked to large inert beads that cannot be taken up by the cell, they still exert their effect. (It should be noted, however, that these results do not rule out the possibility that a small amount of hormone — enough to trigger a response - enters the cell).

Second, in most cases intact cell membranes are needed to elicit a response. Hence, the addition of hormone to a suspension of disrupted cells usually fails to initiate the characteristic reaction induced by the hormone. Because internal cytoarchitecture is disrupted along with the membranes, however, it may simply be that other cell structures are essential components of the hormone response pathway.

A third line of evidence is the fact that, in some cases, the addition of hormone causes rapid changes in membrane properties, such as depolarization, hyperpolarization, and altered ionic transport. Moreover, compounds that cause similar membrane alterations sometimes mimic the hormone effect. Peptide hormones also have demonstrable effects on the permeability of membranes to small molecules and in many cases significantly alter the influx and efflux of sugars and amino acids. These changes could be due to direct or indirect effects on active or passive transport systems. The binding of hormone and receptor triggers a cascade of events.

Steroid Hormone Receptors

Unlike lipid-insoluble hormones, which bind to a receptor at the cell surface, a steroid crosses the plasma membrane and binds to specific intracellular receptor protein. The hormone-receptor complex is present in the nucleus, where it binds to the chromatin or, in some cases, directly to DNA.

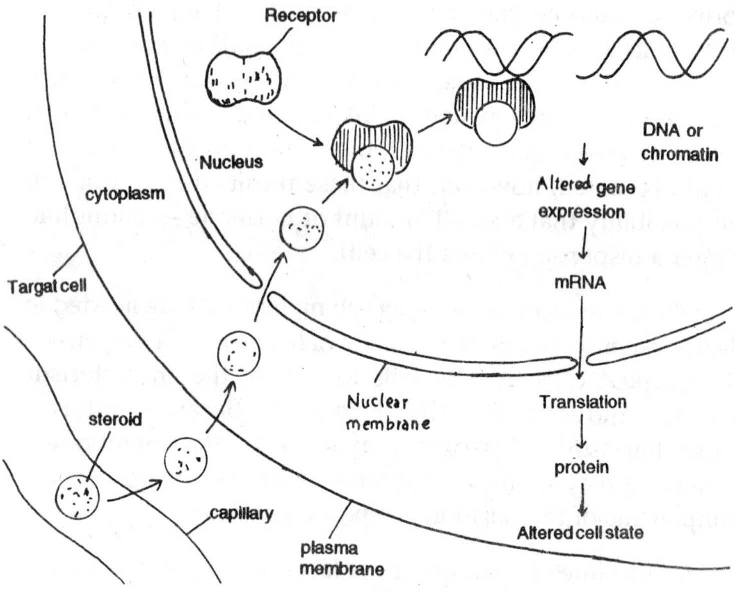

Fig. 12.9. The steps in the action of a steroid hormone.

Most studies of steroid receptors use radioactive hormones that bind to the receptor and serve as tracers for the hormone-receptor complex. These studies have suggested that a specific receptor resides only in target cells, a conclusion based on the observation that when a radioactive steroid is injected into an animal, the radioactive label is found only in the appropriate target tissue. For example, following the injection of the radioactive steroid estradiol into female rats, the radioactivity is concentrated in tissue known to be estrogen sensitive, such as that in the uterus, mammary gland, vagina and hypothalamus.

Steroid-Receptor Complexes

The steroid-receptor complex in the nucleus exerts its effect on gene transcription by binding either to chromatin or directly to specific DNA sequences. This finding comes from experiments in which a radioactively labeled steroid is located in the nucleus after the receptor complex has been formed. In experiments performed by O'Malley and coworkers using chick oviduct, a complex containing the labeled steroid progesterone was bound to chromatin. To determine the chromatin constituent responsible for binding the complex, the chromatin from oviduct cells was fractionated into its three major components: DNA, histones, and acidic proteins. It was then shown that the complex binds preferentially to a subset of acidic proteins.

Experiments in other systems suggest that the steroid-receptor complex binds directly to specific DNA sequences. For example, in *Drosophila* and other arthropods, the steroid hormone ecdysone triggers a sequence of morphological and biochemical changes that result in molting and metamorphosis. Ecdysone also causes a switch in the puffing pattern of the polytene chromosomes from the intermolt to the molting transcription pattern. To show that the ecdysone-receptor complex binds directly to the DNA of polytene chromosomes intact, *Drosophila* cells were irradiated with ultraviolet light, a procedure that cross-links the hormone-receptor complex *in situ* and permanently fixes it to its binding site. Immunofluorescent microscopy (using antibodies specific for ecdysone) was then used to visualize the fixed hormone, which was clearly located on the polytene chromosome at the sites of the hormone-induced puffs. This result shows that the hormone-receptor complex must bind to specific regions of DNA in which transcription is initiated in response to the hormone action.

Vitellogenesis and Steroid Hormones

Vitellogenin is a high-molecular-weight protein that is synthesized in the liver of many vertebrates. It is secreted into the bloodstream, taken up by developing oocytes via endocytosis,

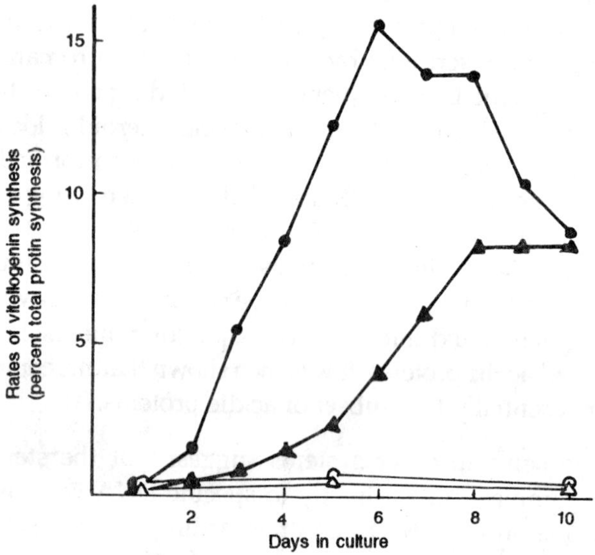

Fig. 12.10. Rates of vitellogenin synthesis in livers of male frogs following the administration of estradiol. The triangles are primary responses with (closed) and without (open) estradiol, and the circles are secondary responses with (closed) and without (open) the hormone. The experiment was carried out with *Xenopus* liver cells in culture, and the secondary response was monitored 35 days after the initial exposure to estradiol. The lag time and total synthetic rate are markedly higher in the secondary response.

and then converted into the storage yolk proteins *phosvitin* and *vitellin*. In the frog, each vitellogenin molecule consists of a dimer of two 200,000-Da protein chains plus lipid and covalently bound carbohydrate; many of the serine amino acids of the phosvitin portion of the vitellogenin molecule are phosphorylated and neutralized by calcium ions. Vitellogenin thus supplies the oocyte with a large of protein, lipid, carbohydrate, phosphate, and calcium for its future energy needs during embryonic development. The steroid estrogen controls the synthesis of vitellogenin, and normally only females produce the protein. However, if male frogs are injected with a single dose of estrogen, within a few hours their livers synthesize large amounts of vitellogenin for a few days. If a male frog is given a second injection of estrogen within one month of the first, the rate of synthesis is higher and the lag time shorter. The mechanism of this secondary response is uncertain, but the response apparently depends in part on an increase in the level of the estrogen receptor in the liver cell following the first estrogen stimulation. In unstimulated liver cells, the levels of estrogen receptor are very low. The receptor can be identified experimentally by the binding of [^{3}H] estradiol to fractionated liver cells followed by the separation of bound and unbound [^{3}H] estradiol on a Sephadex column; the receptor appears to be a protein of about 40,000 Da. In all, there are about 100 receptor sites in the cytoplasm and 100 in the nucleus of each cell; after estrogen is administered, the number of receptor sites increases about ten-fold, with most appearing in the nucleus. Researchers no consider the cytoplasmic receptors identified in cell fractionation experiments to be contaminants from the nuclear fraction and receptors to reside only in the nucleus in normal cells.

Interestingly, early tadpoles of either sex do not synthesize vitellogenin when given estrogen, whereas newly metamorphosed frogs do. Thus, we can conclude that the liver undergoes a transition during metamorphosis, from a state in which vitellogenin synthesis cannot be induced to a second state where it can by the exogenous administration of estrogen. It is known whether this change in the competence of liver cells is due to the appearance of receptor molecules, a change in the status of the vitellogenin gene, or other elements in the regulatory system.

The Induction Effects of Steroid Hormones

These are numerous examples of the introduction of specific proteins by steroids. Many of these have been shown to be the result of concomitant accumulations of mRNAs, an effect that appears to be due in part to increased transcription. This was shown by measuring vitellogenin mRNA levels in normal and estrogen-treated rooster livers by hybridizing mRNA to a cloned vitellogenin cDNA probe; the results of the experiment indicated that mRNA levels in the liver changed from undetectable (0.5 copy per cell) to 5000 copies per cell following hormone treatment.

Although most steroid hormones that have been studied act by binding first to a receptor and then to chromatin or DNA in the nucleus, there are other mechanisms. For example, treating amphibian oocytes with progesterone causes meiosis to resume, the germinal vesicle to break down, and the oocyte to mature so that it can be fertilized. Studies show that the hormone exerts it effect at the plasma membrane, because the microinjection of progesterone into oocytes does not cause meiosis and because many of the changes induced at maturation, such as alterations in the permeability of membranes and the patters of protein synthesis, can occur equally well in enucleated oocytes. Thus, in this case the major effects of a hormone do not require the involvement of the nucleus.

Hormone Action During Development

In this section we shall focus on the developmental consequences of hormone action. These include the effects of steroids on the differentiation of the oviduct and on the production of proteins by oviduct cells, the control of milk production in mammals, and amphibian metamorphosis. As we shall see, in each example the expression of specific genes is coupled to major morphogenetic events.

Effects of Hormones On the Hen Oviduct

The hen oviduct has been a useful experimental system for investigating the effect of steroid hormones on the differentiation of tissues and on the induction of large amounts of specific proteins. The embryonic rudiments of the oviduct are initially formed one on either side in male and female embryos. In the male, both of these degenerate; in the female, the right duct (together with the ovary) degenerates and the left remains as the functional oviduct. Immature chicks produce little or no estrogen or progesterone until about 100 days of age, and the immature oviduct does not produce any of the egg white proteins.

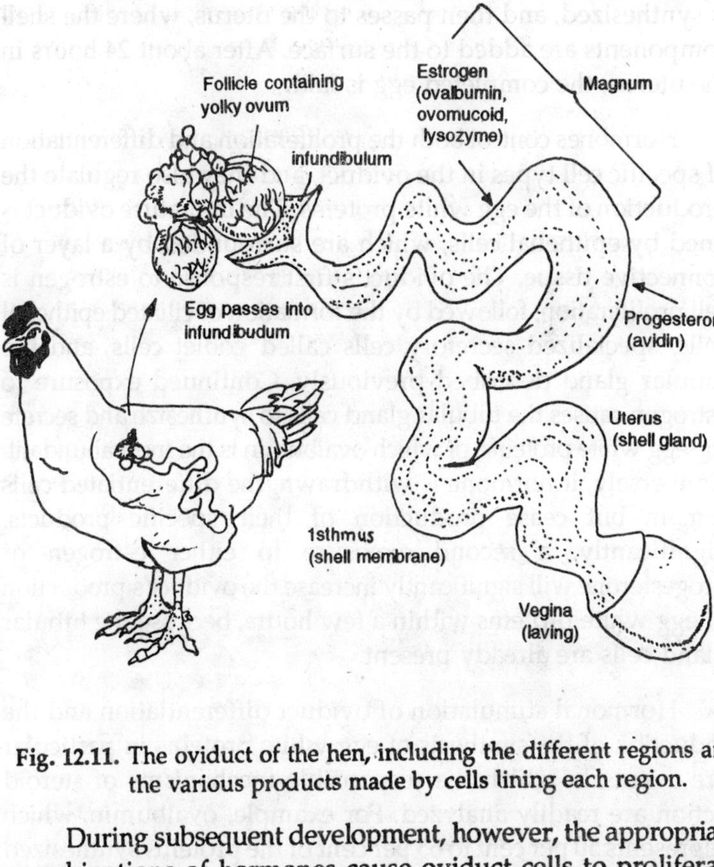

Fig. 12.11. The oviduct of the hen, including the different regions and the various products made by cells lining each region.

During subsequent development, however, the appropriate combinations of hormones cause oviduct cells to proliferate an to form a large, coiled tube with several discrete regions of cells that secrete egg white proteins in abundance in an orderly sequence. After the yolky ovum is liberated from the ovary, it enters the furthest region of the oviduct, the *infundibulum*, where fertilization occurs. The egg is then transported by peristaltic contractions to the next region, the *magnum*. Tubular gland cells in the upper region of the magnum synthesize and secrete ovalbumin, conalbumin, ovomucoid, and lysozyme, which coat the ovum. The protein avidin is produced in the lower part of the magnum. Following a 3-hour traverse in the magnum, the egg enters the *isthmus*, where the shell membrane is synthesized, and then passes to the uterus, where the shell components are added to the surface. After about 24 hours in the uterus, the completed egg is laid.

Hormones control both the proliferation and differentiation of specific cell types in the oviduct, and they also regulate the production of the egg white proteins. The immature oviduct is lined by epithelial cells, which are surrounded by a layer of connective tissue. The oviduct's first response to estrogen is cell proliferation, followed by the formation of ciliated epithelial cells, specialized secretory cells called goblet cells, and the tubular gland described previously. Continued exposure to estrogen causes the tubular gland cells to synthesize and secrete the egg white proteins, of which ovalbumin is the most abundant. Conversely, if hormone is withdrawn, the differentiated cells remain but cease production of their specific products. Significantly, a second exposure to either estrogen or progesterone will significantly increase the oviduct's production of egg white proteins within a few hours, because the tubular gland cells are already present.

Hormonal stimulation of oviduct differentiation and the induction of the synthesis of egg white proteins in particular are events in which the responsible mechanisms of steroid action are readily analyzed. For example, ovalbumin, which represents 50 per cent to 65 per cent of the proteins synthesized by the magnum, is produced by a well-defined class of cells that make up as much as 90 per cent of the magnum cell population. In the laying hen, there is a large amount roughly 30 g — of oviduct tissue, which provides an abundant source of material for isolating the various components that regulate protein synthesis.

Another advantage in studying oviduct development is

the fact that the production of egg white proteins is strictly dependent on the presence of hormones, mainly estrogen and progesterone. Thus, although these hormones are undetectable in immature chicks, exposing immature oviducts to hormones experimentally induces the appropriate developmental responses. Furthermore, it has recently been possible to culture oviduct tissue fragments *in vitro* and to obtain appropriate responses of these explants to hormonal stimulation, a technique that allows for a more controlled and detailed analysis of the effects of hormone action. Finally, experimenters have cloned DNA copies of egg white protein mRNAs and, for some proteins, regions of genomic DNA that contain the structural gene as well as flanking sequences with possible regulatory functions. By using these cloned DNA probes in RNA-DNA hybridization assays, the kinetics and nature of RNA precursor transcription and processing and of mRNA stability can be determined.

Encoding of Genes

During the development of the hen's egg, large amounts of egg white proteins are produced; however, this progious expression of proteins does not mean that multiple copies of genes are operating. Only one copy each of the ovalbumin, ovomucoid, and conalbumin genes in needed to meet the demand for these proteins' mRNAs. In fact, a transcription rate of 16 ovalbumin mRNAs per minute per cell, only one-third the theoretical maximum rate, is sufficient. The half-life of the mRNA is about 24 hours, and studies have shown that at steady state (about 3 days after estrogen stimulation) a tubular gland cell accumulates about 100,000 ovalbumin mRNAs. Each mRNA is then translated to produce about 9.2 ovalbumin molecules per minute. The combined production of ovalbumin by 7.5×10^8 oviduct gland cells over 26 hours, the amount of time needed to produce one egg, is thus 2 g, or 3×10^{19} molecules — all from a single-copy gene.

Hormones not only provided the initial stimulus for the production of mRNAs coding for egg white proteins but also are involved in the regulation of the amount of protein manufactured by cells of the oviduct. This fact can be demonstrated by examining the production of egg white protein mRNA in oviduct cells grown in culture.

In the immature oviduct, where no egg white proteins are synthesized, the mRNAs for these proteins are undetectable. As we have seen, during the first exposure to hormone, oviduct cells proliferate and differentiate, and the production of egg white proteins rises gradually over a period of days. Assays that measure specific mRNA levels show that during this time the levels of mRNA and protein synthesis rise together. For example, ovalbumin levels are undetectable (fewer than five copies per cell) in the immature oviduct and increase to about 100,000 copies per cell in the mature oviduct following hormone stimulation. If hormones are withdrawn, the tubular gland cells and their hormone receptors persist, but ovalbumin synthesis ceases to be detectable and the amount of mRNA declines. A second hormone stimulation provokes a sharper increase in the synthesis of the egg white proteins. The levels of mRNA and the rates of protein synthesis begin to rise after a short lag of 2 to 3 hours and continue to increase more rapidly than during the primary stimulation. The speed of this secondary response is probably the result of an increased concentration of hormone receptors.

Hormones also affect mRNA levels by altering their stability. It the absence of steroids, conalbumin mRNA exhibits a 3-hour half-life; however, in the presence of hormone the half-life increases to 8 hours. Measurements of ovalbumin mRNA stability suggest that it, too, it degraded more rapidly when its transcription rate falls. Thus, steroids appear to regulate the production levels of egg white proteins by increasing both the transcription rate and stability of their mRNAs.

Secondary Responses to Steroid Activation

Results from two separate experimental systems have demonstrated that steroid hormone action is more complex than the simple direct initiation of gene expression. The primary response includes a transient activation of specific genes, whose protein products in turn regulate or participate in the activation of numerous other genes. The action of these regulatory proteins is the secondary response. In the first set of experiments, the action of the insect steroid ecdysone was examined. (After the addition of ecdysone, six early puffs on polytene chromosomes could be seen within 6 to 10 minutes. The appearance of these puffs is not restricted when protein synthesis in inhibited, a finding that suggests that the activation by transcription of the six loci is, at least in part, the primary response. However, subsequent puffing at more than 100 other loci (the secondary response) and the regression of the six early puffs during the next several hours do require ongoing protein synthesis. In the second set of experiments, the role of protein synthesis was examined in the secondary response of progesterone stimulation of hen oviduct cells in culture. Here the characteristic. 2- to 3-hour lag between the administration of progesterone and the onset of synthesis of ovalbumin mRNA only occurs if protein synthesis is allowed. If inhibitors of protein synthesis (such as puromycin and cycloheximide) are given together with progesterone, the induction of ovalbumin mRNA is blocked. When the inhibitor is removed, mRNA accumulation begins, with the characteristic lag.

Hormones of Lactation

The second major example of the importance of hormones in co-ordinating the development of tissues and the production of a specific product is the formation of mammary glands and their production of milk. Mammary glands develop in both female and male mammals, but in males they remain rudimentary. Females produce milk only when required, and its production must therefore be closely monitored. In this section we shall examine how hormones control the differentiation of mammary glands in females and how milk production is regulated.

Sex Hormones Mammary Gland

In a mammalian embryo, the paired mammary glands

begin as ectodermal ingrowths into the mesoderm; in most species, simple ductal systems designed to transport milk radiate from the nipple. In mice, androgens from the fetal male gonads suppress mammary development beyond the simple duct stage. Experiment have been conducted in which the amounts and timing of androgen exposure during the fetal development of the mouse have been changed. Results show that target cells in the male acquire their ability to respond to androgen late on day 13 and lose that ability during day 15 of development *in vitro*; the degree to which the development of mammary glands in suppressed is directly related to the amount of androgens acting on the cells.

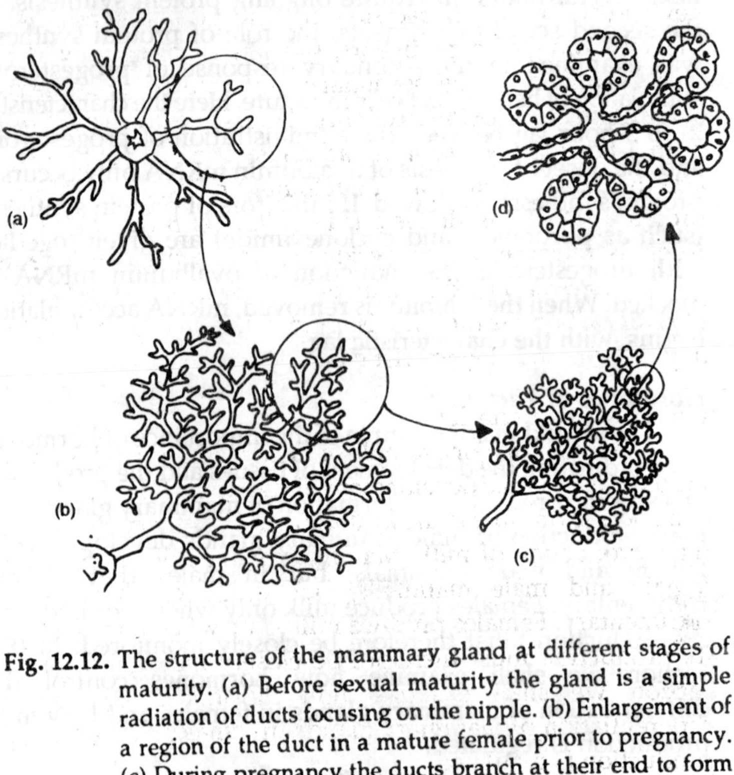

Fig. 12.12. The structure of the mammary gland at different stages of maturity. (a) Before sexual maturity the gland is a simple radiation of ducts focusing on the nipple. (b) Enlargement of a region of the duct in a mature female prior to pregnancy. (c) During pregnancy the ducts branch at their end to form many alveoli. (d) Enlargement showing alveoli.

The target tissue for androgens is the mesenchyme. Male mice carrying an X-linked mutation called testicular feminizing mutation (Tfm) are insensitive to androgens, and if mesenchyme from Tfm embryos is experimentally recombined with normal mammary epithelium and cultured in the presence of the androgen testosterone, a mammary gland forms. In contrast, Tfm epithelium plus normal mesenchyme yields a typical male response, in which the formation of mammary ducts is suppressed.

Hormonal Stimuli Functioning of the Mammary Gland

Experimental results indicate that a sequence of hormonal triggers regulates the formation and functioning of mammary glands and that the ability of a cell to respond to a particular hormone depends upon that cell having previously interacted successfully with another hormone. For instance, the mammary duct system in females first formed in the embryo proliferates and branches extensively at the time of sexual maturation, or puberty. It is known that growth hormone, prolactin, ACTH, TSH, insulin, estrogen, and progesterone are required for continued development of the mammary glands, because the growing gland ceases to develop in animals from which the pituitary has been removed. Increases in progesterone and prolactin during pregnancy may also stimulate the development of the alveoli, the secretory cells at the tips of the branched ducts.

The precise sequence of hormones that are required for proliferation and differentiation of gland cells and the production of milk has been studied in cultured explants of mammary glands. Such experiments indicate that the duct cells are maintained if they are exposed to insulin, land they will divide. Estrogens increase the proliferation of duct cells; for instance, progesterone stimulates cell division at the sites of ductal budding. Following these cell divisions, the epithelia become competent to respond to glucocorticoids such as hydrocortisone. In the continued presence of prolactin and insulin, hydrocortisone causes a dramatic increase in the RER; the nucleus and the RER then shift to the basal portion of the cell, and the Golgi apparatus moves to the apical portion. The surface area of the apex is then greatly increased by the formation of fields of microvilli. In culture, continuing exposure to prolactin and insulin causes cells in mammary tissue to begin the synthesis of milk fats, proteins, and lactose.

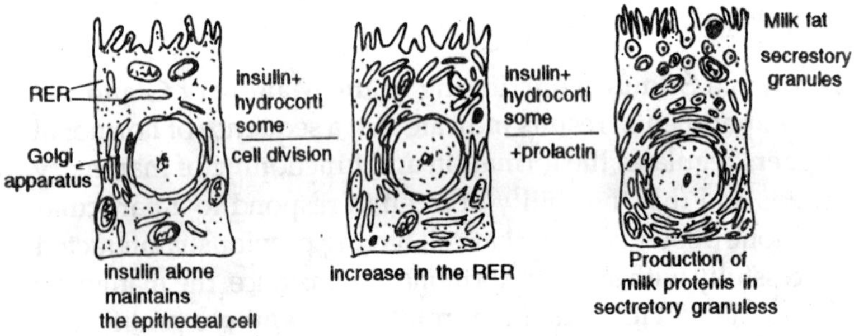

Fig. 12.13. Changes in the structure and function of duct cells in the mammary gland. Insulin and hydrocortisone stimulate the increase of the rough endoplasmic reticulum (RER), and prolactin in combination with these hormones causes the secretion or milk proteins.

The Timing of Milk Production

During pregnancy and prior to delivery, high levels of estrogen and progesterone appear to inhibit the synthesis and secretion of milk. However, the loss of the placenta after delivery results in an abrupt decrease in the circulating levels of these hormones, and the production and secretion of milk begin. Once milk production begins, the maintenance of milk production and release depends on the continued presence of prolactin, which is release in response to mechanical stimulation of the nipples by suckling. This reflex discharge of prolactin by the anterior pituitary is abolished by denervation of the nipple or by lesions in the spinal cord; in such cases, an administration of prolactin overcomes the defect.

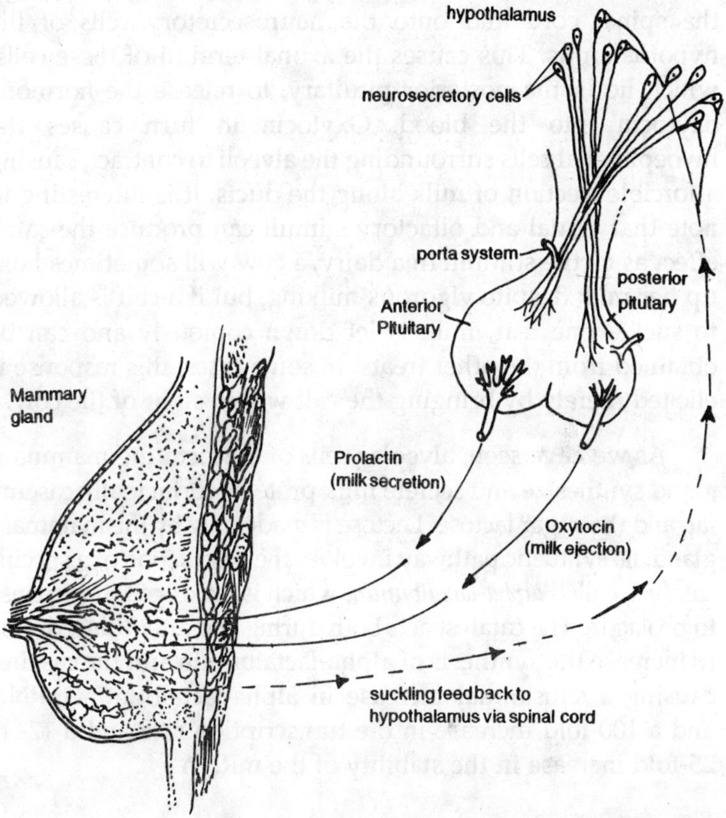

Fig. 12.14. The feedback loops controlling the release of oxytocin and prolactin through the hypothalamus to the pituitary gland.

The feedback loops postulated for the milk production mechanism are diagrammed. First, suckling triggers neural impulses that are carried to the hypothalamus, where the impulses inhibit the production of *prolactin-inhibiting factor* and stimulate the production of *prolactin-releasing factor*. The anterior pituitary then secretes prolactin, which stimulates milk production by the alveolar cells as described earlier, and milk is secreted into the lumina of the alveoli. Suckling stimulates the ejection of this milk to the outside. Impulses resulting from tactile stimulation of the nipple are then carried through the spinal cord and onto the neurosecretory cells of the hypothalamus. This causes the axonal termini of these cells, which lie in the posterior pituitary, to release the hormone oxytocin into the blood. Oxytocin in turn causes the myoepithelial cells surrounding the alveoli to contract, causing a forcible ejection of milk along the ducts. It is interesting to note that visual and olfactory stimuli can produce the same effect as tactile stimuli. In a dairy, a cow will sometimes hold up her milk despite vigorous milking, but if a calf is allowed to suckle one teat, milk is let down copiously and can be obtained from the other treats. In some cases, this response is elicited merely by bringing the calf within sight of the cow.

As we have seen, alveolar cells of the lactating mammary gland synthesize and secrete milk proteins (principally casein), fat, and the sugar lactose. Lactose is made only by the mammary gland; its synthetic pathway involves the production of a specific catalyst, called *alpha-lactalbumin*, which is produced in response to prolactin. The catalyst acts both during and after transcription to increase the synthesis of alpha-lactalbumin and the caseins, causing a substantial increase in alpha-lactalbumin mRNA and a 100-fold increase in the transcription rate and a 17- to 25-fold increase in the stability of the mRNA.

AMPHIBIAN METAMORPHOSIS

During metamorphosis in amphibians (frogs and salamanders), an orderly sequence of extensive morphological, biochemical, and behavioural changes transform the aquatic, herbivorous larva into an air -breathing, carnivorous, terrestrial young adult. Metamorphic changes occur in virtually all tissues and organs. This two-part developmental programme may be advantageous because it allows the young larva to feed heavily and grow rapidly in favourable aquatic environments, and the subsequent metamorphosis into a terrestrial form allows the adult to colonize a second, land-based habitat.

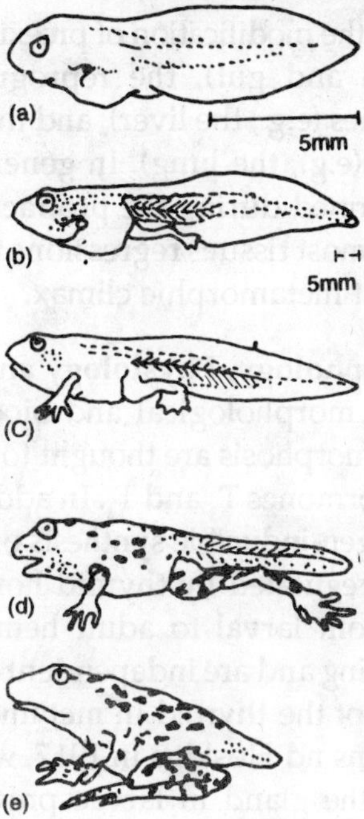

Fig. 12.15. Three basic stages in the metamorphosis of a frog: (a) premetamorphic tadpole; (b) prometamorphic tadpole; (c)-(e) metamorphic climax and the transformation into a frog.

Metamorphosis Occurs in Stages

Three stages of the metamorphic transformation can be identified. In *premetamorphosis* larval growth predominates. During *prometamorphosis,* growth continues and a few developmental changes, such as the emergence of the hind-limbs, begin. Finally, the initial development of the forelimbs marks the onset of the *metamorphic climax,* a period of extensive and dramatic changes in morphology and physiology. These changes include tail regression if frogs and the restructuring of the feeding, digestive, respiratory, excretory, locomotory, and neural systems in frogs and salamanders.

There are four general categories of change during metamorphosis: the loss of larval tissues and structures (e.g., the tail and gills), the modification of preexisting larval structures (e.g., the mouth and gut), the reprogramming of cellular metabolic activities (e.g., the liver), and the appearance of new adult structures (e.g., the lung). In general, most of the new structures are formed during the protracted prometamorphic period, whereas most tissues regressions take place during the shorted period of metamorphic climax.

Changes in Morphology, Physiology and Biochemistry

Most of the morphological and biochemical changes of amphibian metamorphosis are thought to be directly mediated by the thyroid hormones T_3 and T_4. In addition, some changes, such as the estrogen-inducible synthesis of vitellogenin, appear to be indirectly regulated by thyroid hormones; others, such as the switch from larval to adult hemoglobin, are due to chronological aging and are independent of hormonal activity. The importance of the thyroid in metamorphosis was clearly shown by Hoskins nd Hoskins in 1917, when they found that the removal of the gland in larvae prevented metamorphic change. The role of thyroid hormone can also be demonstrated by experiments in which thyroidectomized larvae are fed minced thyroid tissue or thyroid hormone; such larvae subsequently undergo metamorphosis. Similarly, T_4 pellets implanted into the tails of premetamorphic tadpoles cause localized tissues to break down and regress. The direct effect of thyroid hormone on tail regression can easily be shown in the laboratory using blocks of tail tissue cultured *in vitro*; if thyroid hormone is added to the culture medium, the characteristic histolysis and tissue reduction occur after a lag of 3 to 4 days.

In similar fashion, thyroxine (T_3) acts directly on eye tissue. For example, if thyroxine is applied to one tadpole eye, that eye will switch from the larval pigment porphyropsin to the production of the adult visual pigment rhodopsin. The untreated eye retains its larval characteristics. In a contrasting example, the switch from larval to adult hemoglobin occurs on schedule in larvae treated with the antithyroid drug thiouracil, which prevents normal metamorphosis. Clearly, some non-thyroid-related agent or a chronological indicator that is normally co-ordinated with the thyroid-induced alterations is responsible for the change to adult hemoglobin.

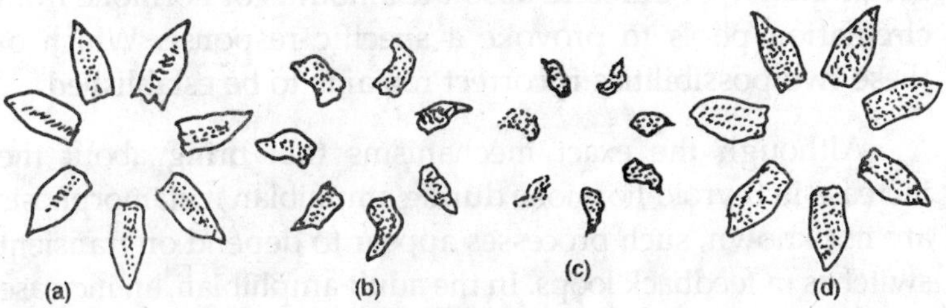

Fig. 12.16. Regression of tadpole tails in organ culture under different conditions: (a) Newly amputated tails. (b) After three days exposure to T_3. (c) After four days exposure to T_3. (d) Tails after eight days in culture without exposure to T_3.

The Metamorphic Changes and Thyroid Hormone

The levels of thyroxine circulating in the bloodstream increase throughout metamorphosis and rise sharply during climax. If a single large dose of hormone is administered to a young tadpole or to a thyroidectomized larva, the outcome is a rapid accumulation of metamorphic changes — apparently the result of most or all of the usually sequential changes starting at the same time. However, when thyroidectomized tadpoles are exposed to gradually increasing levels of thyroxine, more normal metamorphic patterns are seen. In one experiment, the administration of a small dose of thyroxin maintained a tadpole in a prolonged premetamorphic condition; when the hormone concentration was increased, prometamorphic changes were initiated. Finally, a rapid increase in the hormone level initiated the metamorphic climax.

In addition to triggering metamorphic climax, a high level of thyroid hormone must be maintained if tail resorption and other changes are to proceed to completion. After metamorphosis is complete, however, very low thyroxine levels are adequate to ensure the continued growth of the young adult. These results and those of other studies suggest that various tissues have different thresholds of responsiveness to thyroid hormones. An alternative hypothesis is that the hormones interact with different cellular processes, so that each requires the accumulation of different absolute amounts of hormone from circulation pools to provoke a specific response. Which of these two possibilities is correct remains to be established.

Although the exact mechanisms that bring about the increase in thyroid hormone during amphibian metamorphosis are not known, such processes appear to depend on transient switches in feedback loops. In the adult amphibian, an increase in circulating levels of T_3 and T_4 exerts negative feedback on the production of TSH by the pituitary and TRH (*TSH-releasing hormone*) by the hypothalamus; in the tadpole, thyroid hormone also acts as negative feedback on the pituitary.

As the beginning of metamorphosis, three events are thought to promote the increased production of thyroid hormone. First, TRH, which is already present in hypothalamic cells, may become more available to the pituitary. This probably occurs when the blood system develops more fully to connect the hypothalamus with the anterior pituitary. Indeed, the development to this connection from a simple two-layered epithelium is itself under the control of thyroxine. In does not differentiate in thyroidectomized animals, but the immersion of such larvae in gradually increasing concentrations of thyroxine causes the complicated circulatory structure to form. Second, increasing day length and warmer temperatures seem to increase the release of TRH. Finally, because the hypothalamus and pituitary switch their responsiveness to increased levels of thyroid hormone in the blood, for a time thyroxine stimulates rather than inhibits the production of TRH and TSH.

Each Tissue Responds to Increasing Hormone Levels Differently

There are many cell types within the larva that undergo specific changes during metamorphosis, but cells in some regions of the body continue to develop and are unaffected by hormones. For example, gill epithelial cells degenerate, while adjacent limb epidermis continues to flourish and grow. Likewise, the synthesis of many proteins in the liver is diminished or stopped, while many others, including the urea cycle enzymes, begin to be synthesized. Several proteins, including vitellogenin, become inducible in response to estrogen. Smooth muscle cells in the walls of blood vessels in the tail undergo autolysis in response to thyroid hormone, while smooth muscle cells in the developing limbs proliferate and participate in the formation of blood vessels. In the tail cartilage, cells degenerate, while chondrocytes in the limb buds proliferate and form the cartilage model for future bones.

As the various events unfold, the specificity of each tissue's response does not depend on its position in the animal; thus a limb transplanted to the tail of a metamorphosing tadpole will continue to grow out even as the tail regresses. Similarly, an eyecup transplanted to the tail persists and remains healthy, while the tail tissue around it is totally resorbed. On the other hand, a second tail, transplanted to an area where a hindlimb will form, regresses normally as the hindlimb grows out.

Thyroid and Other Lipid-Insoluble Hormones

The molecular interactions of thyroid hormone with cellular receptors differ from those of either the steroid or lipid insoluble hormones. The binding sites for thyroid hormone are found in the cell nucleus regardless of whether hormone is present. Thyroid hormone readily enters the cell through membrane internalization, crosses the nuclear membrane, and binds to the chromatin, where the receptor resides. The localized binding of thyroid hormone and receptors on chromatin suggests an effect on RNA transcription, and indeed thyroid hormone has been shown to stimulate the transcription of growth hormone mRNA in pituitary cell lines and of another mRNA type in hepatocytes. The hormone also triggers dramatic changes in the rates of RNA synthesis and in the pattern of protein synthesis in many cell types. Thyroid hormone also appears to affect cytoplasmic processes. For example, studies show that it stimulates metabolite uptake in chick heart cells even when the synthesis of RNA and protein is blocked. In addition, the hormone has been shown to cause tissue-specific effects on DNA synthesis, on the abundance, size, and morphology of mitochondria, and on the organization of cellular organelles.

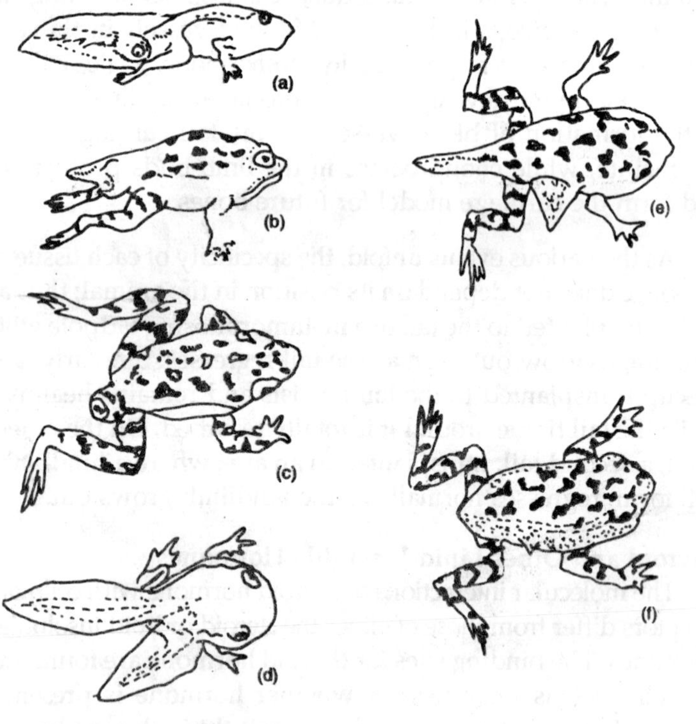

Fig. 12.17. Organ specificity during metamorphosis. (a)-(c) An eye grafted to at ail does not regress but stays healthy and attached to the frog. (d)-(f) An extra tail grafted to the flank of a tadpole regresses at the same rate as the normal tail.

Even in the degenerative response, thyroid hormones appear to trigger new patterns in gene expression. We can see the effects of such altered patterns in tail regression, which begins at the tip and progresses toward the base. Degeneration is promoted by T_4 and involves two separate processes. First, individual cells begin to degenerate as the cytoplasm condenses into dense fragments surrounded by membranes; then the chromatin condenses and the nucleus breaks down. Hydrolases produced by the cells themselves appear to carry out these processes. Later macrophages appear in large numbers, these may arise locally or invade the degenerating tissue from other areas. The macrophages then phagocytose dying cells and digest the cell's contents via the usual lysosomal pathway. The also produce several hydrolytic enzymes that break down collagen fibers and other elements of the extracellular matrix. Given the large increases in hydrolytic enzymes in tail tissue cells and the rising numbers of macrophages, it is not surprising that inhibitors of RNA and protein synthesis prevent tissue breakdown and regression.

